Cell Culture Techniques in Heart and Vessel Research

Editor: H. M. Piper

With 73 Figures and 19 Tables

Springer-Verlag Berlin Heidelberg New York
London Paris Tokyo Hong Kong

Prof. Dr. Dr. H. M. Piper
Physiologisches Institut, Abteilung für Herz- und Kreislaufphysiologie
der Universität Düsseldorf, Moorenstr. 5, D-4000 Düsseldorf 1, FRG

ISBN-13: 978-3-642-75264-3   e-ISBN-13: 978-3-642-75262-9
DOI: 10.1007/978-3-642-75262-9

Typesetting, printing and bookbinding: K. Triltsch GmbH, Würzburg
2119/3335-543210 – Printed on acid-free paper

Foreword

Classical knowledge of the heart and blood vessels was developed via anatomical, physiological, pharmacological and biochemical studies and their contribution to our understanding cannot be overestimated. However, some of the newer approaches to treatment of cardiovascular disease depend on an understanding of the cellular and molecular biology of the cells. As in all branches of science, current and future progress will depend heavily on technical resources and in vitro assay systems that reflect biological processes. The development of cultures of the chief constituent cell types of the heart and blood vessels has provided the impetus for the current wealth of information and promises to lead to new and exciting revelations both in the realm of basic mechanisms and in the world of clinical practice. Cells in culture provide a means of dissecting out the contributions of individual cell types, for assaying the effects of naturally occurring and synthetic products and for examining cell-cell and cell-molecule interrelationships. In many instances such studies have revealed fundamental properties of the cells themselves, their enzymes, receptors, transduction mechanisms, and gene regulation. In turn, these basic findings have led and will continue to lead, to concepts with potential for therapeutic actions. Thus, stemming from advances in cell culture, gene products, living cells themselves and genetically modified cells are now considered to be candidates for reducing the pain and suffering of heart and vascular diseases.

St. Louis, Missouri   Una S. Ryan
February 1990

Preface

Interest in culture models for the types of cells found in the vascular tree, i.e., cardiomyocytes, endothelial cells, smooth muscle cells, and pericytes, has grown in a multitude of special research fields. These include basic cardiovascular research (physiology, biochemistry, pharmacology, pathology), arteriosclerosis and hypertension research, hemostasiology, and work related to vascular function in all areas of organ specialization. An increasing demand for modern methods in cell biology has developed and the use of specific cell culture models is essential for most of these novel approaches.

This book provides a comprehensive and easily accessible source of culture methods for these cells in the cardiovascular tree. In addition to giving detailed methodological descriptions, the contributors discuss the benefits of the described procedures and alternative approaches. These comments may guide one through the literature. For the reader's convenience a list of those suppliers from which the authors purchased their materials has been compiled at the end of the book.

We hope that this book will serve its purpose well and further stimulate scientists in cardiovascular research and related fields to apply methods of modern cell biology.

Hans Michael Piper

Table of Contents

Smooth Muscle Cells and Pericytes

List of Contributors

R. A. Altschuld
Department of Physiological Chemistry, The Ohio State University,
Columbus, Ohio 43210-1239, USA

E. Betz
Institut für Physiologie I, Universität Tübingen, Gmelinstr. 5,
D-7400 Tübingen, FRG

T. K. Borg
Department of Pathology, School of Medicine, University of South Carolina,
Columbia, SC 29208, USA

A. Bottger
Department of Medical Cell Biology, Karolinska Institutet, Box 60400,
S-10401 Stockholm, Sweden

P. D. Bowman
Letterman Army Institute of Research, SGRD-ULT-M,
Presidio of San Francisco, CA 94129-6800, USA

M. Cantin
Clinical Research Institute of Montreal, 110 Pine Ave. West, Montreal,
Que, Canada H2W 1R7

P. D'Amore
Depts. of Pathology and Surgery, Children's Hospital and
Harvard Medical School, 300 Longwood Avenue, Boston, MA 02115, USA

P. F. Davies
Department of Pathology, Box 414, University of Chicago,
Pritzker School of Medicine, 5841 S. Maryland Ave.,
Chicago, IL 60637, USA

K. Dorovani-Zis
Department of Pathology, Division of Neuropathology, Vancouver General
Hospital, Vancouver, British Columbia, Canada V5Z1 M9

M. du Bois
Letterman Army Institute of Research, SGRD-ULT-M,
Presidio of San Francisco, CA 94129-6800, USA

P. Fallier-Becker
Institut für Physiologie I, Universität Tübingen,
Gmelinstr. 5, D-7400 Tübingen, FRG

J. Fingerle
Institut für Physiologie I, Universität Tübingen,
Gmelinstr. 5, D-7400 Tübingen, FRG

D. Gospodarowicz
Cancer Research Institute and Department of Medicine,
University of California, Medical Center, San Francisco, CA 94143, USA

A. I. Gotlieb
University of Toronto, Vascular Research Laboratory,
Department of Pathology, Toronto General Hospital, Research Centre,
200 Elizabeth Street, CCRW 1-857, Toronto, Ont, Canada M5G 2C4

R. Gronwald
Medizinische Universitätsklinik, Universität Heidelberg,
Bergheimerstr. 58, D-6900 Heidelberg, FRG

A. J. R. Habenicht
Medizinische Universitätsklinik, Universität Heidelberg,
Bergheimerstr. 58, D-6900 Heidelberg, FRG

U. Hedin
Department of Medical Cell Biology, Karolinska Institutet,
Box 60400, S-10401 Stockholm, Sweden

C. M. Hohl
Department of Physiological Chemistry, The Ohio State University,
Columbus, Ohio 43210-1239, USA

I. E. Hassinen
Department of Medical Biochemistry, University of Oulu,
Kajaanintie 52 A, SF-90220 Oulu, Finland

I. M. Herman
Department of Anatomy and Cell Biology, Tufts University,
School of Medicine, 136 Harrison Avenue, Boston, MA 02111, USA

V. W. M. van Hinsbergh
Gaubius Institute TNO, P.O. Box 612, NL-2300 AP Leiden,
The Netherlands

A. Krützfeldt
Physiologisches Institut, Universität Düsseldorf,
Moorenstr. 5, D-4000 Düsseldorf 1, FRG

S. L. Jacobson
Carleton University, Department of Biology, Ottawa, Ont,
Canada K1S 5B6

U. Janßen-Timmen
Medizinische Universitätsklinik, Universität Heidelberg,
Bergheimerstr. 58, D-6900 Heidelberg, FRG

E. G. Langeler
Gaubius Institute TNO, P.O. Box 612, NL-2300 AP Leiden,
The Netherlands

S. Mertens
Physiologisches Institut, Universität Düsseldorf,
Moorenstr. 5, D-4000 Düsseldorf 1, FRG

A. C. Nag
Department of Biological Sciences, Oakland University, Rochester,
Michigan 48063, USA

S. E. O'Connor
Department of Pathology, Box 414, University of Chicago,
Pritzker School of Medicine, 5841 S. Maryland Ave.,
Chicago, IL 60637, USA

A. N. Orekhov
Institute of Experimental Cardiology, USSR Cardiology Research Center,
Academy of Medical Sciences, 3 Cherepkovskaya Street 15 A,
Moscow 121552, USSR

A. Pinson
Laboratory for Myocardial Research, Department of Biochemistry,
The Hebrew University, Hadassah Medical School, P.O. Box 1172,
Jerusalem, Israel

H. M. Piper
Physiologisches Institut, Universität Düsseldorf,
Moorenstr. 5, D-4000 Düsseldorf 1, FRG

A. M. Rosenthal
University of Toronto, Vascular Research Laboratory,
Department of Pathology, Toronto General Hospital Research Centre,
200 Elizabeth Street, CCRW 1-857, Toronto, Ont, Canada M5G 2C4

J. Rupp
Institut für Physiologie I, Universität Tübingen,
Gmelinstr. 5, D-7400 Tübingen, FRG

U. S. Ryan
Director of Health Sciences, Monsanto Company, Mail Zone O4E,
800 N. Lindbergh Boulevard, St. Louis, Missouri 63167, USA

P. Salbach
Medizinische Universitätsklinik, Universität Heidelberg,
Bergheimerstr. 58, D-6900 Heidelberg, FRG

M. A. Scheffer
Gaubius Institute TNO, P.O. Box 612, NL-2300 AP Leiden,
The Netherlands

P. Schwartz
Zentrum Anatomie, Universität Göttingen,
Kreuzbergring 36, D-3400 Göttingen, FRG

D. M. Shasby
University of Iowa College of Medicine, Department of Internal Medicine,
Iowa City, Iowa 52242, USA

S. S. Shasby
University of Iowa College of Medicine, Department of Internal Medicine,
Iowa City, Iowa 52242, USA

R. L. Shivers
Cell Science Laboratory, Department of Zoology,
The University of Western Ontario, London, Ont, Canada N6A 5B7

V. N. Smirnov
Institute of Experimental Cardiology, USSR Cardiology Research Center,
Academy of Medical Sciences, 3 Cherepkovskaya Street 15A,
Moscow 121552, USSR

R. Spahr
Physiologisches Institut, Universität Düsseldorf,
Moorenstr. 5, D-4000 Düsseldorf 1, FRG

L. Terracio
Department of Anatomy, University of South Carolina,
Columbia, SC 29208, USA

J. Thyberg
Department of Medical Cell Biology, Karolinska Institutet, Box 60400,
S-10401 Stockholm, Sweden

P. Uussimaa
Department of Medical Biochemistry, University of Oulu,
Kajaanintie 52 A, SF-90220 Oulu, Finland

A. Volz
Boehringer Ingelheim KG, Abteilung Biochemie, D-6507 Ingelheim, FRG

H. Watanabe
Physiologisches Institut, Universität Düsseldorf,
Moorenstr. 5, D-4000 Düsseldorf 1, FRG

Heart Muscle Cells

Embryonic Chick Heart Muscle Cells*

A.C. Nag

Introduction

Many of the early studies on cultured cardiac muscle cells were carried out using chick embryonic heart tissue. Cavanaugh's [1] dissociation of chick embryo heart fragments into single cells using trypsin was a significant technical advancement over past procedures, which included culture of whole salamander hearts, fetal mouse hearts, and fragments of chick heart. This method of isolating heart cells later became one of the common choices for cellular dissociation. However, in recent years this trypsinization technique has been modified in various ways, according to need, as required by the nature of different types of tissues or organs used in the experiments. Harary and Farley [9] cultured trypsinized isolated neonatal rat heart cells in plastic petri dishes, showing spontaneous beating mammalian cardiac muscle cells in culture conditions. Subsequently, with the use of neonatal rat hearts, Kasten [15], Nag [25, 27] and others modified trypsinization methodology and carried out morphological and biochemical studies on these cultured cardiac muscle cells. Trypsinization technique was successfully adapted for the culture of embryonic chick heart cells, which have been used to demonstrate the morphology and growth rate pattern of cardiac muscle and nonmuscle cells [32].

It was observed that the preparations of heart cells obtained by trypsinization contained heterogenous cell populations as well as cardiac muscle cells. The nonmuscle cells comprised endothelial cells, smooth muscle cells, fibroblasts, macrophages, and blood cells [23]. In order to obtain a predominant population of myocytes, a screening of myocytes is needed after obtaining a heterogenous cell suspension. This screening technique is referred to as differential adhesion technique and is used often in many laboratories, including ours. This method is found to be better than other screening methods involving ficol or other gradients. The differential adhesion technique will be discussed under the appropriate heading in the text.

The cell culture system has been used to study the effects of various agents, such as hormones, growth factors, and drugs, on the growth and metabolic activities of the cells. For this purpose, formulation of culture media has been carried out very carefully so that ingredients in the culture media do not interfere with or confuse the results of the experimental agents to be tested on the cells. The requirement for such specific compositions of the culture media

* This work is supported by the National Science Foundation Grant DCB-8709594

has given rise to the formulation of serum-free synthetic media, which are essential for the investigation of the effects of various agents on the cells. The rest of the studies involving general cytological, biochemical, and physiological aspects of the cells can be investigated, using serum containing media. The appropriate culture media for different experiments are discussed separately.

Materials and Methods

Cell Dissociation Solutions

Enzymatic dissociation of heart tissue of either birds or mammals has been carried out routinely and successfully with a solution of 0.25% trypsin, 0.025% collagenase, 4% chicken serum in 96% calcium and magnesium-free Tyrode solution (CMF), or Hank's saline [3, 25, 28, 29]. The pH of the enzyme solution should be in the neutral range (7.0–7.4). The divalent cations, calcium and magnesium, are involved in the maintenance of cellular adhesion and junctional integrity. Cardiac muscle cells are adhered to one another by intercellular matrices as well as junctional complexes such as desmosomes and gap junctions. The exclusion of calcium and magnesium from the dissociation solution facilitates the loosening of cellular cohesion and thus enhances the dissociation of the myocardial tissue into single cell suspension with the help of trypsin and collagenase. The enzyme solution is sterilized by pressure filtration through a sterile GS Millipore filter (pore size: 0.22 µm) in swinny holder (Millipore Corp.). In the past, trypsin or collagenase alone were used for dissociation of heart tissue into single cell suspension. It was found that the preparations obtained with the use of a mixture of trypsin und collagenase as described above were better than the preparations by single enzymes with respect to cell yield and cell viability. It is apparent that the intercellular matrix that binds the cells in the cardiac tissue does not contain only proteins, and, consequently a single general proteolytic enzyme like trypsin does not cause satisfactory tissue dissociation.

Historically the enzymatic release of cells from tissue pieces can be traced back to its origin with the early work of Rous and Jones [33]. Trypsin, a pancreatic proteolytic enzyme, catalyzes preferentially the hydrolysis of peptide bonds between the carboxyl group of arginine or lysine and the amino group of another amino acid. For a detailed review of this enzyme, see Desnuelle's [5] review. Collagenase is a collagenolytic enzyme which is available as a crude preparation and in pure form from at least one source (Worthington Biochemical Corp.). The crude preparations are satisfactory for most purposes, and the activity of the crude preparations vary from one lot to another. The activity of purified preparation does not seem to confine to collagen only. It is difficult to remove proteolytic polysaccharidase and esterase activities from the collagenase [10, 11, 19]. The contaminating protease, esterase, and polysaccharidase activities may account for the superior quality of the tissue dissociation. The enzyme solution can be prepared in a considerable amount (100 ml), and the filtered solution can be divided into small aliquots according to the experimental needs and frozen at $-20\,°C$ or lower for future use.

Isolation of Cells

Fertile chicken eggs are needed for cell or tissue culture work. These eggs can be obtained from a local hatchery. The fertile eggs can be stored at $4°-10\,°C$ for up to 2 weeks without major lethality. Depending on the design of the experiment, 4- to 10-day-old embryos can be used conveniently for cell culture studies. Embryos older than 11 days contain bones and feathers, which interfere with the procedures. The eggs are incubated at $37.5\,°C$ to $38\,°C$ in a high humidity incubator with a device for turning the eggs at regular intervals. For most general purposes, embryos are used after 7 days of incubation. After a 7-day or any other specific incubation period, the eggs are removed from the incubator and examined over a light source for determination of viable embryos. The embryos with well-developed blood supplies are found to be healthy and active, and they are used for the experiments. The outer surface of the eggs should be cleaned before opening the shell. First the eggs are exposed for 10 min, with the blunt air sac end up, to the ultraviolet light of the laminar flow hood or any other strong ultraviolet light. The entire egg is washed with 70% alcohol or 5% Zephiran chloride. The egg is broken with an egg punch, scalpel, or forceps at the blunt or broad end containing the air sac. If a scalpel or forceps is used, this should be sterilized carefully before each use by dipping in boiling water or by flaming. A bent forceps is used to pull out the embryo by the neck, and the embryo is placed in a sterilized 100 mm glass or plastic petri dish. The heart is removed with forceps and placed in a depression glass block containing Tyrode solution (1 l: NaCl, 8 g; KCl, 0.20 g; $CaCl_2$, 0.20 g; $MgCl_2 \cdot 65\ H_2O$, 0.10 g; NaH_2PO_4; H_2O, 0.05 g; $NaHCO_3$, 1.00 g; glucose, 1.00 g $\cdot$ pH $7.0-7.4$). The whitish atria are discarded by using the sharp tip of the forceps, and the ventricles are washed of adherent blood. Using the forceps, the ventricles are then cleaned, as much as possible, of connective tissue, and washed twice with Tyrode solution, which is replaced with CMF in the subsequent steps. Ventricles are minced into approximately 1-mm pieces with fine scissors and washed briefly with CMF. The tissue mince is transferred to a 25-ml Erlenmeyer flask containing approximately 5 ml CMF. Too much tissue should not be taken in each flask or it will cause low cell yield. The flasks are then aerated with balanced air (5% CO_2 and 95% air), stoppered loosely with a silicone rubber stopper, and incubated for 10 min at $37\,°C$ on a gyratory water bath shaker. CMF is withdrawn and replaced with an equal volume of the dissociation solution. The flask is aerated with the balanced air, stoppered tightly, and incubated in a gyratory water bath shaker at $37\,°C$ (70 rpm) for 45 min. The ventricular mince is transferred to a centrifuge tube and the proteolytic digestion is stopped by filling the tube with cold nutrient ($4\,°C$) medium containing serum. The mixture of enzyme solution and nutrient medium in the tube is slowly decanted and replaced with cold fresh culture medium containing 5 µg DNAse I (Sigma Chemical Co.) per milliliter medium. The tissue is then dispersed into single-cell suspension in the above medium by pipetting. The cells are pelleted by centrifugation for 3–5 minutes at a lower speed in a table-top clinical centrifuge. The supernatant is decanted and the cells suspended by pipetting in a volume of fresh, cold nutrient medium. The cells thus are washed twice or thrice by centrifugation and filtered through a swinny filter (XX3001210, Millipore Corp.) or a wire mesh filter (mesh: 200; pore size 75–80 μm) or six layers of No. 80 Nitex. Subsequently the viability test for the isolated cells is carried out. The freshly dissociated cells are rounded (Fig. 1).

Viability of Cells

The routine test for viability of cells has been a dye exclusion test, using the dye trypan blue. With this test damaged cells become stainable, but the method is not necessarily reliable [4, 17, 24]. With the trypan blue exclusion method, 0.5 ml of cell suspension is placed in a test tube and 0.1 ml of 0.4% trypan blue is added to the cell suspension. The dye and cell suspension are thoroughly mixed with the pipette and allowed to stand for approximately 5 min. Subsequently, the cell suspension is placed on the hemocytometer for counting, which helps to obtain total cell yield as well as stained vs. unstained cells in the preparation. The other available methods for testing cellular viability are useful, but time consuming. The phase contrast microscopic examination of freshly isolated cells shows the presence or absence of smooth or irregular, broken cell surfaces, this being indicative of healthy (smooth) or unhealthy (broken) cells. The irregular cell surface is often blebby; this can happen during tissue dissociation procedures. It is known that a certain amount of extra- and intracellular damage occurs in the cells during dissociation procedures, allowing enzymes to penetrate into the cells. The proteolytic enzyme causes damage to the myofibrils, which are capable of rapid regeneration, resulting in the formation of new myofibrils [13, 22]. This reversible damage keeps the cells quiescent for some time and they begin contracting in less than 15 h. A better estimate of cell damage can be obtained by electron microscopy of freshly dissociated cells; these exhibit the extent of damage of myofibrils and other cellular organelles. The decisive test for viability is the examination of the ability of the cells to attach and grow in culture. This parameter is assessed by making high (10^6 cells/60 mm × 15 mm culture dish) and low (10^4 cells/60 mm × 15 mm culture dish) density cultures. The quantitation of the plating efficiency of these two types of cultures will indicate the extent of viability of the plated cells. It should be noted that the plating efficiency of the low density culture cannot be expected to be equal to that of the high density culture. High density facilitates better plating efficiency.

Differential Adhesion

Differential adhesion is useful in obtaining preparations rich in cardiac muscle cells, as required for critical studies of various aspects of cardiac myocytes.

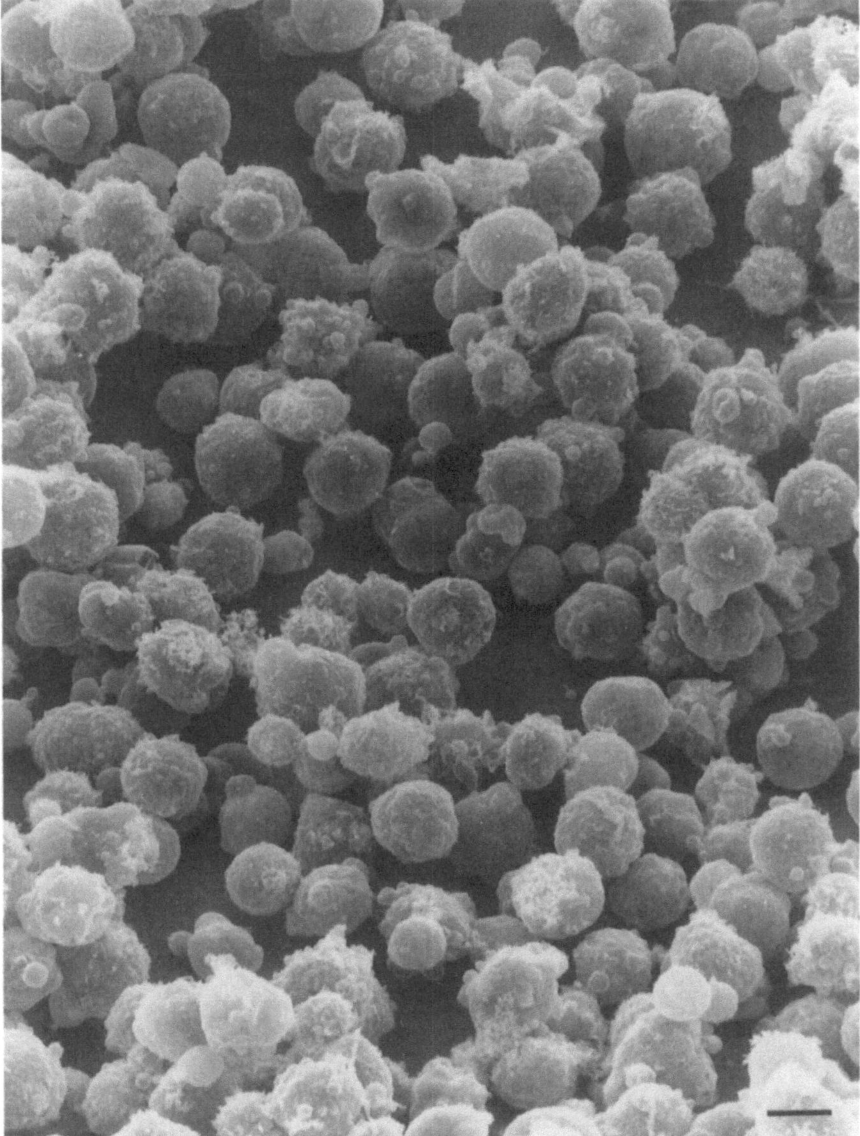

Fig. 1. Scanning electron micrograph of freshly dissociated embryonic chick heart cells. Note many cells exhibit microvilli on the cell surface. *Bar*, 9 μm

Initially this technique was used by Kasten [12, 14] for the culture of the rat neonatal cardiac myocytes and depends on the differential adhesion capability of cardiac muscle and nonmuscle cells on a substrate of, for example, plastic or glass. It was observed that when a suspension of cardiac cells settles on a substrate, the nonmuscle cells begin to attach rapidly to the surface and spread out within several minutes. On the other hand, cardiac muscle cells remain rounded for approximately 12 to 15 h before they attach to the surface and spread out. This is true for both chicken embryonic and rat embryonic/neonatal cardiac muscle cells. Kasten [14] used Erlenmeyer flasks or the Rose chamber for this technique. We used plastic cell culture dishes (60 mm × 15 mm). The results obtained with the use of different substrates are essentially the same. Approximately 95% purity with cardiac muscle cells can be achieved with the differential adhesion method. The essential steps are enumerated below.

Method

1. The isolated cell suspension washed with the culture medium, as discussed earlier, is poured into a sterile 50 ml Erlenmeyer flask or one or more Petri dishes (100 mm × 15 or 20 mm), depending on the quantity of the cell suspension.
2. The cell suspension is allowed to settle at 37 °C, preferably in a tissue culture CO_2 incubator, for 90 to 180 min. In order to provide cells with more surface area for attachment, several layers of small glass beads can be placed in the flask.
3. The cardiac muscle cell-rich suspension is withdrawn by a pipette after 90 to 180 min, leaving nonmuscle cells attached to the floor of the flask or Petri dish. The muscle cell-rich suspension is then tested for cellular viability with trypan blue, and cells are counted on the hemocytometer. The cell suspension is subsequently diluted with the medium to obtain desired cell concentration per milliliter and plated on the culture Petri dish (60 mm × 15 mm).
4. If interested in the nonmuscle cells, they can be isolated immediately by trypsinization and cultured after washing with the culture medium.

Cell Culture Media

Depending on the design of the experiments, cardiac muscle cells can be cultured in serum-containing medium or serum-free medium. The components of the serum-containing and serum-free media are described below.

Serum-Containing Medium

This culture medium consists of Eagle's basal medium with Earle's salts and without L-glutamine (88%; GIBCO), minimum nonessential amino acids (10 mM solution, GIBCO; 0.1 ml/100 ml medium), ascorbic acid (0.02 mg/ml, Sigma), insulin-transferrin-selenium mix (ITS; 0.01 ml/100 ml; Collaborative Research), fetal calf serum or calf serum (10%); (GIBCO), and penicillin-streptomycin mixture (1%).

Serum-Free Medium

This chemically defined synthetic medium is made up of Ham's F12 nutrient mixture (99%) without L-glutamine (GIBCO), fetuin (0.025%; tissue culture grade; Sigma), bovine serum albumin (1%); tissue culture grade, Sigma; Calbiochem-Behring), ascorbic acid (0.02 mg/ml; Sigma), ITS (0.01 ml//100 ml; Collaborative Research), epidermal growth factor (10 ng/ml; Collaborative Research), endothelial cell growth supplement (50 µg/ml; Collaborative Research), norepinephrine (10^{-5} M; Sigma); triiodothyronine (T_3; 10^{-5} M/ml; Sigma), and penicillin-streptomycin mixture (1%; Nag et al., 1985).

Cell Culture

Most studies on cell and molecular biology of the cardiac muscle require abundant cells, which are dependent upon the vigorous growth of the cells. One of the determining criteria for vigorous cardiac muscle cell growth is the proper initial cell density during plating. To attain optimal growth, 10^6 cells should be plated per 60 mm culture petri dish, containing 3–4 ml culture medium. It was observed that the concentrations of less than $10^4 - 10^5$ cells per dish markedly reduced the cell growth rate. In low-density culture a considerable fraction of cells fail to grow into colonies. However, this situation can be improved to a certain extent by plating cells on dishes coated with collagen (type VII, Sigma). Cells cultured in serum-free media should be plated on dishes coated with collagen, fibronectin (5 µg/cm^2 or 5 µg/ml; Collaborative Research) or laminin (Bethesda Research Laboratories) for better cell growth [28]. Cell growth usually occurs at a rapid rate until confluency is reached, at which point growth rate decreases. Cultures are maintained at 37°C in a humidified atmosphere of 5% CO_2 in air for a desired period according to the design of the experiment. Cardiac muscle cells spread out on the culture dish, assuming various shapes and sizes (Fig. 2). Although cardiac muscle cells in primary culture can be maintained for more than 4 weeks, most of the studies have been carried out with cultures incubated for periods ranging from less than 1 week to 3 weeks. To my knowledge, secondary culture with normal cardiac muscle cells has not yet been established. However, cardiac muscle cells in culture undergo mitosis and differentiation. They exhibit spontaneous contractility and form synchronized networks within a few days. For a successful culture with these cells, the CO_2 level of the incubator should be monitored routinely. Significantly high or low levels of CO_2 cause cell degeneration and death. It was observed that the pH of the culture medium falls rapidly from one day to the next when nonmuscle cells dominate the culture. Otherwise, the pH remains at neutral range for up to a week or longer, even without a medium change. Cultures should be checked routinely using a phase contrast microscope to examine the morphology and contractility of the cells. For examination, the cultures are placed on the microscope stage heated to a suitable temperature (approx. 37°C) with the aid of an air curtain incubator. However, if one checks cultures quickly, the maintenance of proper temperature on the microscope stage is not necessary.

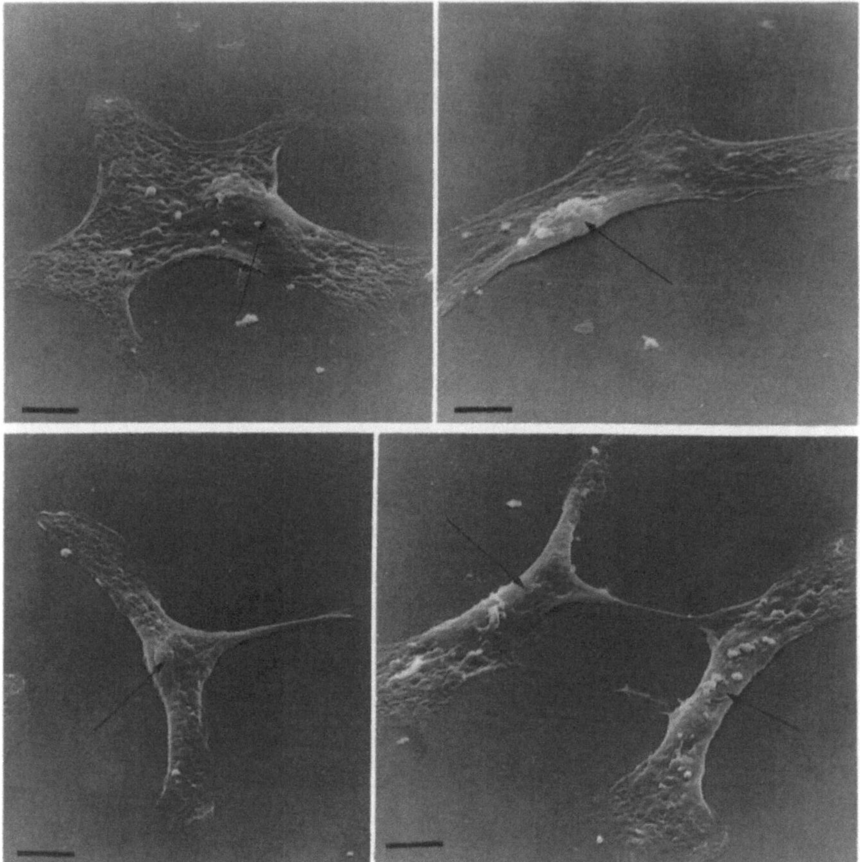

Fig. 2. Scanning electron micrograph of embryonic chick cardiac muscle cells of various shapes and sizes after 48 h of culture. *Arrows* show the thickened regions containing greater accumulation of myofibrils together with nuclei as observed with transmission electron microscopy. *Upper left bar,* 7.7 µm; *other bars,* 8.3 µm

For studying the effects of different agents such as growth factors, hormones, chemicals and drugs, the external agents are added to the media and the cells are directly exposed to these agents [8, 25, 26, 29]. In my judgment, one can determine the success of a cardiac muscle cell culture using two criteria: (1) spontaneously contracting, healthy looking cells; (2) absence or minimal presence of nonmuscle cells. These two criteria can be satisfied by following the above procedures carefully, although the total elimination of nonmuscle cells from the culture is hard to achieve. Cardiac muscle cells are routinely identified among the heterogenous heart cell populations by periodic acid schiff (PAS) staining method [27, 28, 32]. Cardiac muscle cells identified using the PAS technique are shown in Fig. 3. Muscle cells are stained for glycogen, which is abundant in cardiac myocytes, whereas nonmuscle cells are not stained because of their lack or very low concentration of glycogen. The

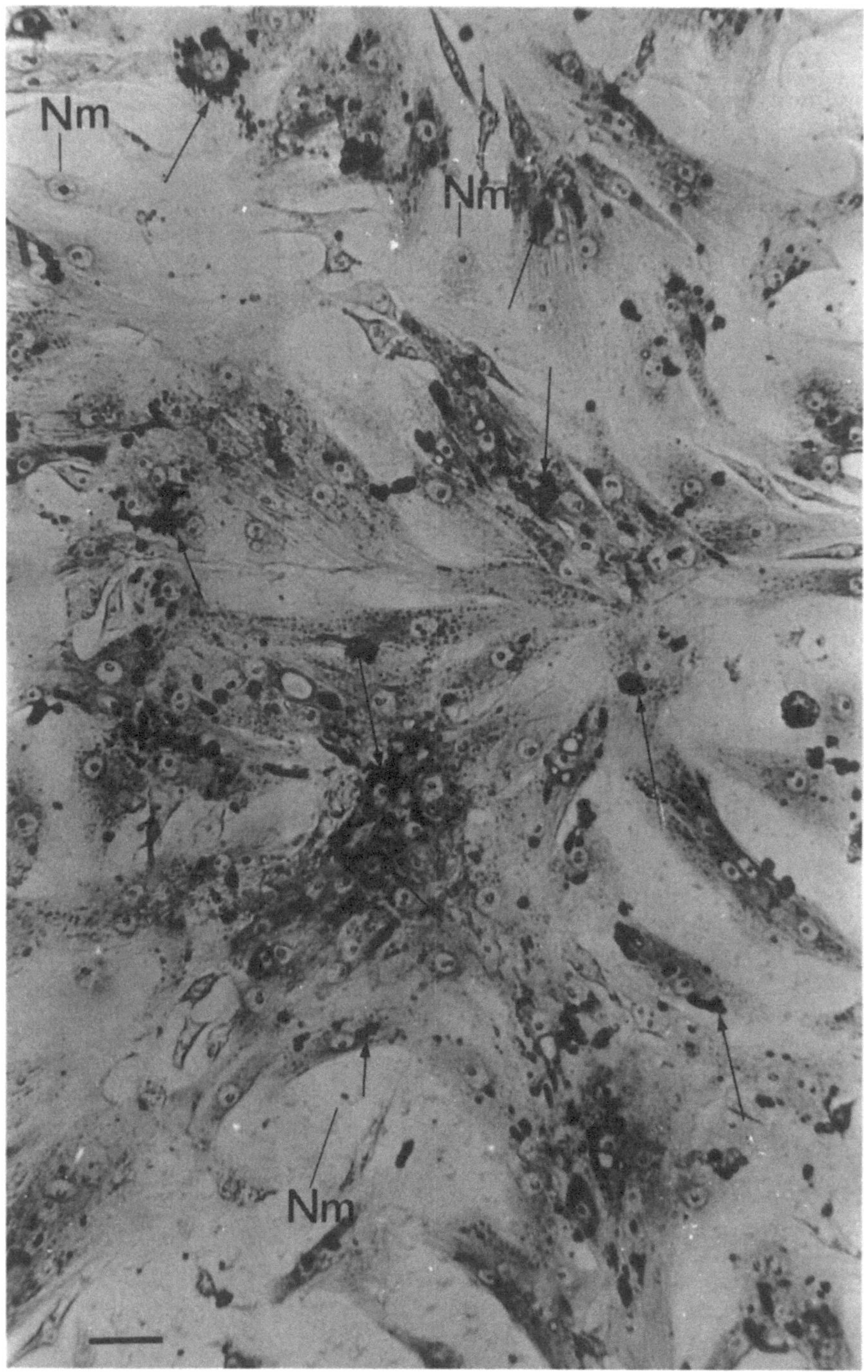

Fig. 3. Light micrograph of PAS-stained embryonic chick cardiac muscle cells in monolayer culture and unstained nonmuscle cells (*Nm*). Note black stained mass of glycogen (*arrows*). *Bar,* 40 µm

essential steps of the PAS method are:

1. The culture medium is poured off and the plates are rinsed in cold Tyrode solution three times.
2. The cells are fixed in cold (4 °C) formol:alcohol (nine parts absolute alcohol mixed with one part concentrated formulin) for 24 h at 4 °C.
3. The cells are rinsed several times in 95% alcohol and hydrated quickly (50 s in each step) to water through descending grades of alcohol (70% and 50% alcohol).
4. Fresh 1% aqueous periodic acid (Sigma) solution in distilled water is prepared and the cells are exposed to this solution for 30 min. Subsequently, the culture dishes are rinsed twice with distilled water.
5. Schiff's reagent is prepared with basic fuchsin (Polysciences), following Pearse's procedure [31] or any other histochemistry textbook. This solution should be stored in a brown glass bottle or a foil-wrapped transparent glass bottle at 4 °C. Schiff reagent should be transparent. If it becomes brown upon storage, do not use, as old Schiff reagent does not always produce the anticipated bright, purplish-red color. The cells are stained with Schiff reagent for 20 to 60 min.
6. The plates are thoroughly rinsed in water three times, not less than 2 min each time, and air dried at room temperature. The stained cells are examined under an inverted or vertical microscope, using the bright field optics.

If cardiac muscle cells are not stained well after this period, one can continue to stain for an additional 60 min. The staining quality of basic fuchsin varies from one batch to another. Freshly made periodic acid solution should be used; old solution does not work for oxidation. The rationale of the technique is in the discussion section of the text.

Discussion

Contamination of Culture

Although contamination of the embryonic chick cardiac muscle cell culture is relatively rare, it may occur due to faulty, aseptic technique. Common contaminants are bacteria or fungi, which grow well in the nutrient media that support the growth of bird and mammalian cell cultures. In order to prevent contamination one should check and take the necessary measures at each step of the procedure. Since "contamination" includes such a wide range of topics, only a few of them will be briefly highlighted here. Investigators interested in this subject in greater detail are advised to read Sect. XIV of *Tissue Culture* [16].

It was observed that the vacuum system (flask, tubing, and pump) needed to sterilize the medium and to remove culture medium from culture dishes is often contaminated with a mixed flora of bacteria and fungi. This can be prevented by washing and sterilizing the components of the vacuum system daily. The detection and elimination of bacterial and fungal contamination is time consuming, therefore it is advisable to take certain precautions:

1. Work under a laminar flow hood to reduce airborne contamination of cell cultures. The hood should be checked periodically, according to the manufacturer's instructions, for leaks and to see that filters are in good working order.
2. Maintain sterility of all apparatus and dissecting instruments. The dissecting instruments should be dipped into boiling water frequently during the manipulation and processing of heart tissue prior to the dissociation step.
3. The egg shell often contains parasites and microorganisms. The egg shell surface should be sterilized as discussed earlier.
4. Do not talk or breathe near the tissue and cultures during preparation. If preferred, use masks.
5. Sterilize enzyme solution by Millipore filtration and divide into small aliquots by dispensing aseptically into appropriately sized bottles according to the requirements of the experiments. Incubate two bottles of enzyme solution at 37 °C for at least 7 days. If both bottles remain sterile, it is assumed that the other bottles are sterile, too. Store the enzyme solution at 20 °C for future use.
6. Media prepared in the laboratory should be sterilized by Millipore filtration, tested for contamination as above, and stored at 4 °C.
7. The CO_2 incubator should be checked periodically for the growth of fungi and microorganisms. Clean the incubator with a detergent and rinse with water and Zephiran chloride. Then wipe the interior of the incubator thoroughly with a clean 70% alcohol-soaked cloth or paper towel. The incubator shelves should be cleaned and autoclaved. The distilled water for the incubator water jacket should be changed periodically. Some CO_2 incubators include an electronically regulated high temperature decontamination system, which can be used periodically to sterilize the incubator. Even with the use of the electronic decontamination system, fungi sometimes grow. In this case, clean as indicated above and use the decontamination system.

Monolayer and Aggregate Cultures

The isolation of single cells from chick heart by trypsinization of the heart tissue into single-cell suspension and the ability of such cell suspensions to grow and differentiate in monolayer [1, 2], have provided a new system for cytological, physiological, and biochemical studies of cardiac cells. Although monolayer cultures are useful in various cell studies, they are not always adequate for examining the nature of cellular interrelationships and interactions in the formation, differentiation, and function of tissues. The procedures for cellular aggregation in vitro of isolated cells in suspension [20, 21] have been developed to overcome this difficulty, in that they make it possible to study in detail the cell contacts, sorting out, morphogenetic organization of cells in multicellular systems, and cell cooperation in forming organized and differentiating tissues [7, 22, 24, 34–36]. In aggregation culture, approximately 3 million heart cells per 25 ml Erlenmeyer flask are taken, and the flasks are placed on a gyratory water bath shaker at 70 rpm at 37°C. The cells are cultured on the gyratory water shaker for 3–7 days according to the design of

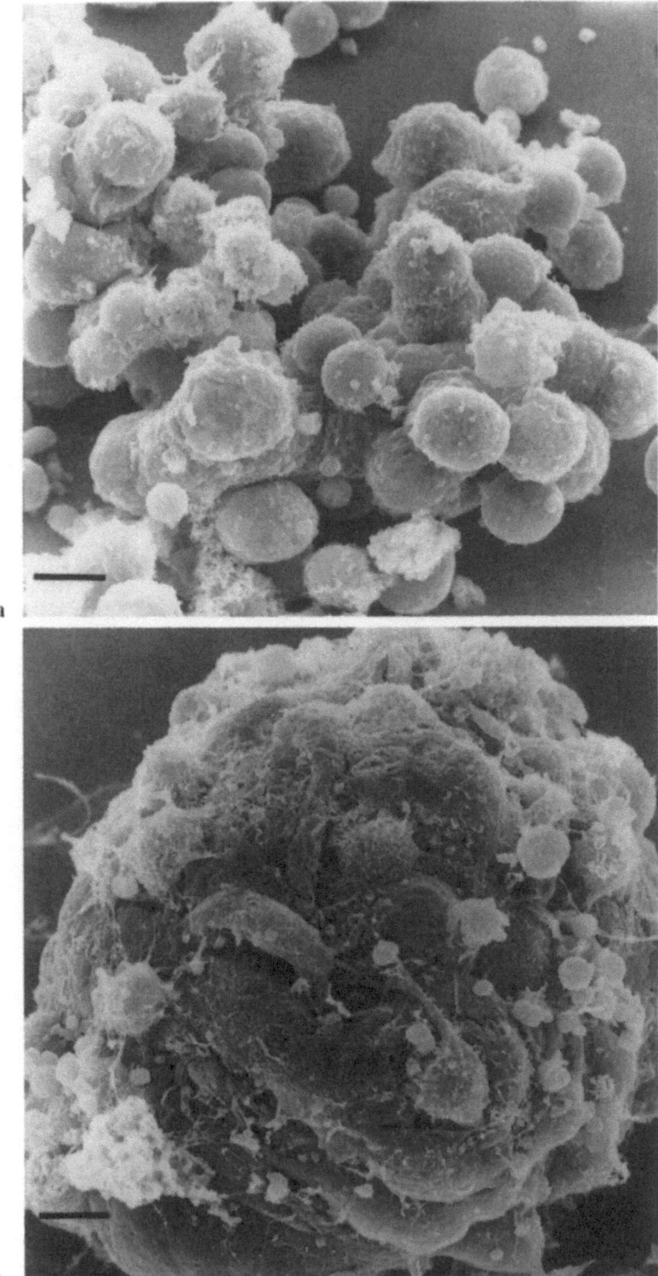

Fig. 4. a Scanning electron micrograph of an embryonic chick heart cell aggregate after 6 h of culture, showing incomplete aggregation of cells. *Bar*, 9 μm. **b** An aggregate of heart cells after 24 h of culture, showing completion of cell aggregation. *Bar*, 8.6 μm

the experiments. The movement encourages reaggregation of the suspended cells. Care must be taken with the speed of rotation and the character of shaking in order to prevent frothing of the medium or mechanical damage to the cells. Aggregation requires 24 h for completion, at which time the aggregates become spontaneously beating spherical structures (Fig. 4). The aggregate culture system has been found to be useful for certain morphological, physiological, and biochemical studies.

Morphology

Cardiac muscle cells can be distinguished from nonmuscle cells by their thicker cell bodies, which at the phase contrast microscope level are significantly denser than those of nonmuscle cells. This thickness is caused by the presence of myofibrils. The nonmuscle cell body is extremely flattened and thin. Although contractility helps in the identification of cardiac muscle cells, all cardiac muscle cells in a given dish do not beat. Electron microscopy of freshly dissociated cardiac muscle cells revealed the presence of sarcomeric segments of myofibrils and abundant free myofilaments in the cell sarcoplasm. Enzymatic dissociation of cells is responsible to a large degree for the disassembly and segmentation of the myofibrils. The freshly dissociated cells often are observed to contain many autophagic vacuoles, many of which contain disintegrated myofibrils and cellular organelles. However, the cellular damage during the dissociation step of the procedure is reversible, and the cells rapidly regenerate the myofibrils and possibly other organelles. As the culture continues, the cell bodies gradually become packed with myofibrils, mitochondria, and other organelles. When the adjacent cardiac myocytes touch one another during growth, the intercellular junctions such as desmosomes and gap junctions are found to be present between them. The intercalated discs can also be distinguished among these growing cardiac myocytes (Fig. 5). Cardiac muscle cells have a rich store of glycogen particles which are often diffused out of the cells during processing for electron microscopy, leaving empty pockets of glycogen in the sarcoplasm of the cells. The glycogen is also visible with the PAS staining technique at the light microscope level discussed earlier. The cytochemical rationale for the PAS technique includes the use of an oxidant, periodic acid (HIO_4), which breaks the C–C bonds in many structures containing 1:2-glycol groups (CHOH–CHOH) and converts them into dialdehydes (CHO.CHO). It should be noted that periodic acid also reacts with the equivalent amino or alkylamino derivatives of 1:2-glycol or its oxidation product (CHOH.CO) and converts them into dialdehydes, which in their turn are localized by combination with Schiff's reagent, giving rise to a red/purplish-red color reaction. Periodic acid is found to be superior to other reagents with regard to its property for oxidizing C–C bonds; it does not oxidize the resulting aldehydes further under the conditions usually employed.

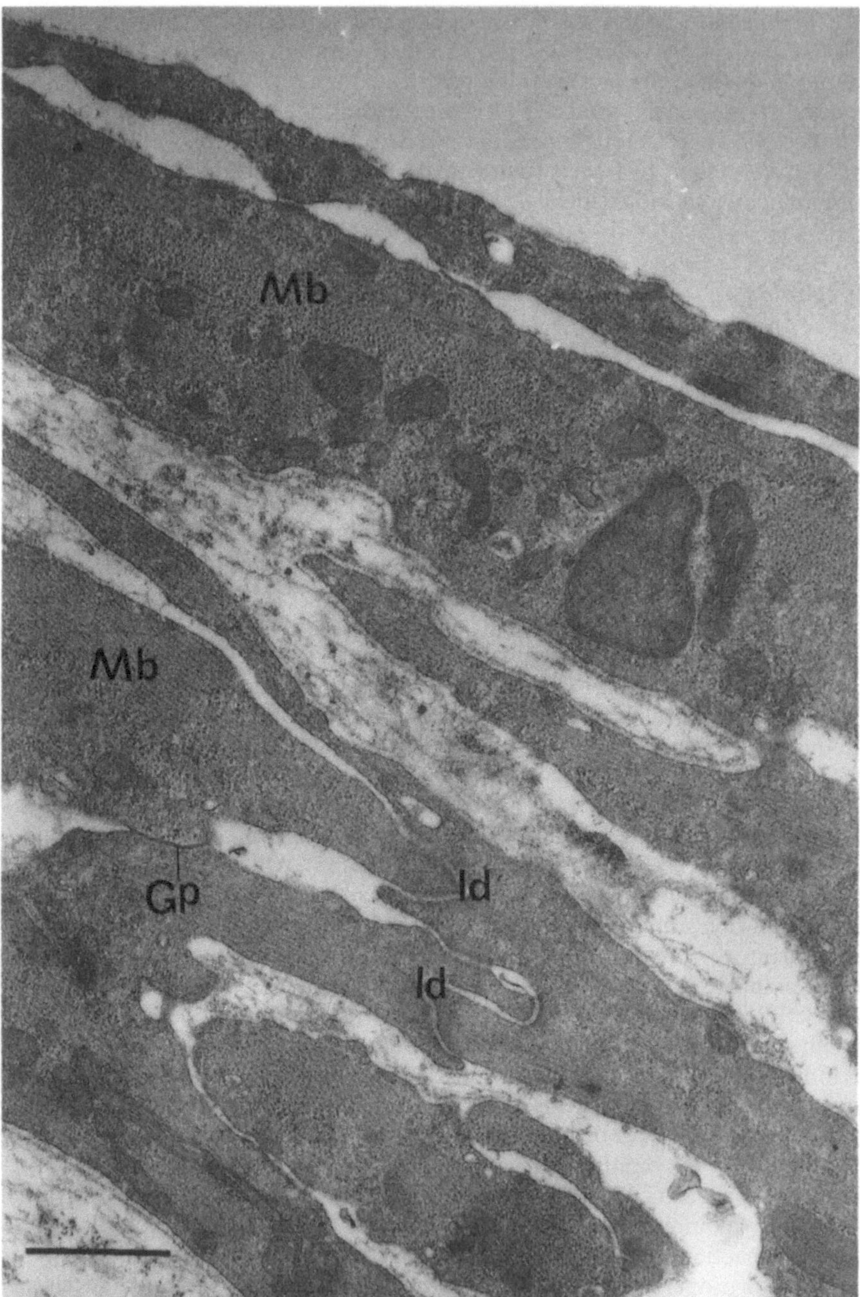

Fig. 5. Transmission electron micrograph of a portion of a monolayer of cardiac muscle cells showing myofibrils (*Mb*), intercalated discs (*Id*), and gap junction (*Gp*) after 5 days of culture. *Bar*, 1.5 μm

Cell Growth

The growth rate of embryonic chick cardiac muscle cells in culture shows an initial peak followed by a decline. The peak in growth rate is usually found to vary from days 1–3, depending on the experimental conditions [3, 28, 32]. The accurate determination of cell growth includes determination of the percentage of PAS-stained cells labeled with radioactive thymidine ($[^3H]TdR$) measured by autoradiography. When dealing with the growth pattern of heart cells, which include both muscle and nonmuscle cells, the incorporation of $[^3H]TdR$ with DNA of heart cells can be analyzed by the scintillation counting method [6, 24, 28, 29]. In addition, cell growth can be measured with a hemocytometer or automatic electronic cell counters (Coulter Counter) or by the rapid and reliable method of determination of cell protein concentration. The latter may be employed in investigations of the growth of heterogenous populations of heart cells.

Results obtained using this method unfortunately do not just reflect the growth of cardiac muscle cells. Even the differential adhesion technique does not produce a preparation of 100% cardiac muscle cells. The method for the determination of cell protein concentration as described by Oyama and Eagle [30] is a modification of the method of Lowry et al. [18]. The amount of cell growth can be determined by the cultures grown in suspension and also by the monolayer cultures. With appropriate conversion factors the values may be changed to dry weight, protein nitrogen, or cell count. The method for protein concentration is also useful for monitoring the dynamics of cell populations during the logarithmic phase of growth.

Embryonic chick heart cells grow very well in a serum-free medium and show vigorous cell proliferation [28]. One can use this culture system for various kinds of morphological, physiological, and biochemical studies. As described earlier, several growth factors, hormones, and chemicals are necessary for a successful culture in a serum-free medium. The proportion of cardiac muscle cells in ECGS-containing medium was higher than those in other media with the exception of ECGS- and ITS-containing medium, which exhibited a lower proportion of cardiac myocytes than that of the serum-containing control medium. T_3 or T_4 is required in the medium to induce optimal differentiation and contractility of cardiac muscle cells in culture.

Acknowledgements. I would like to thank Mr. Prabhakar Sreepathi for his careful work in producing electron micrograph prints.

References

1. Cavanaugh MW (1955) Growth of chick heart cells in monolayer culture. J Exp Zool 128:573–581
2. Cavanaugh MW (1957) The monolayer culture of chick myocardial cells. Arch Int Pharmacol 110:43
3. Clark WA Jr, Fischman DA (1983) Analysis of population cytokinetics of chick myocardial cells in tissue culture. Dev Biol 97:1–9

4. Claycomb WC, Palazzo MC (1980) Culture of the terminally differentiated adult cardiac muscle cells: a light and scanning electron microscopy study. Dev Biol 80:1680–1686
5. Desnuelle P (1960) Trypsin. In: Boyer PD, Lardy H, Myrback K (eds) The enzymes, vol 4. Academic, New York, p 119
6. Doyle C, Zak R, Fischman DA (1974) The correlation of DNA synthesis and DNA polymerase activity in the developing chick heart. Dev Biol 37:133–145
7. Garber B, Killar E, Moscona AA (1968) Aggregation in vitro of dissociated cells. III. Effect of state of differentiation of cells on feather development in hybrid aggregates of embryonic mouse and chick skin cells. J Exp Zool 268:455–472
8. Gospodarowicz D, Hirabayasin K, Giguere L (1981) Factors controlling the proliferative rate, final cell density and life span of bovine vascular smooth muscle cells in culture. J Cell Biol 89:568–578
9. Harary I, Farley B (1960) In vitro studies of single isolated beating heart cells. Science 131:1674–1675
10. Hilfer RS (1973) Collagenase treatment of chick heart and thyroid. In: Krause PF Jr, Patterson MK Jr (eds) Tissue culture – methods and applications. Academic, New York, pp 16–20
11. Hilfer RS, Brown JM (1971) Collagenase: its effectiveness as a dispersing agent for embryonic chick thyroid and heart. Exp Cell Res 65:246–249
12. Kasten FH (1969) High resolution filming of rhythmic and arrhythmic behavior of cultured myocardial cells with a method for quantitative analysis (Abstr). In Vitro 4:150
13. Kasten FH (1971) Cytology and cytochemistry of mammalian myocardial cells in culture. Acta Histochem [Suppl] (Jena) 9:637–647
14. Kasten FH (1972) Rat myocardial cells in vitro: mitosis and differentiated properties. In Vitro 8:128–150
15. Kasten FH, Bovis R, Mark G (1965) Phase contrast observations and electron microscopy of cultured newborn rat heart cells (Abstr). J Cell Biol 27:122A–123A
16. Krause PF Jr, Patterson MK Jr (eds) (1973) Tissue culture – methods and applications. Academic, New York
17. Levinson C, Green JW (1965) Cellular injury resulting from tissue disaggregation. Exp Cell Res 39:309–317
18. Lowry DH, Rosenborough NJ, Lewis Farr A, Randall RJ (1951) Protein measurement with folin phenol reagent. J Biol Chem 193:265–272
19. Mitchell WM, Harrington WF (1968) Purification and properties of clostridiopeptide B (clostripain). J Biol Chem 243:4683
20. Moscona AA (1961) Rotation-mediated histogenetic aggregation of dissociated cells. Exp Cell Res 22:455
21. Moscona AA (1973) Cell aggregation. In: Bittar EE (ed) Cell biology in medicine. Wiley, New York
22. Nag AC (1978) Reconstruction of mammalian heart tissue from embryonic heart cell suspension with reference to the aggregation of adult heart cells. Cytobios 23:199–223
23. Nag AC (1980) Study of nonmuscle cells of the adult mammalian heart: a fine structural analysis and distribution. Cytobios 28:41–61
24. Nag AC, Cheng M (1983) DNA synthesis in mammalian heart cells: comparative studies of monolayer and aggregate cultures. Cell Mol Biol 29:45–49
25. Nag AC, Cheng M (1984) Expression of myosin isozymes in cardiac muscle cells in culture. Biochem J 221:21–26
26. Nag AC, Lee ML (1989) Effect of amiodarone on the expression of myosin isoforms and cellular growth of cardiac muscle cells in culture. Circ Res (submitted for publication)
27. Nag AC, Crandell TF, Cheng M (1981) Competence of embryonic mammalian heart cells in culture: DNA synthesis, mitosis and differentiation. Cytobios 30:189–208
28. Nag AC, Ingland M, Cheng M (1985) Factors controlling embryonic heart cell proliferation in serum-free synthetic media. In Vitro Cell Dev Biol 21:553–562
29. Nag AC, Chen KC, Cheng M (1988) Effects of carbon monoxide on cardiac muscle cells in culture. Am J Physiol 255:C291–C296

30. Oyama VI, Eagle H (1956) Measurement of cell growth in a tissue culture with a phenol reagent (Folin-Ciocalteau). Proc Soc Exp Biol Med 91:305–307
31. Pearse AGE (1968) Histochemistry: theoretical and applied, vol 1, 3rd edn. Little, Brown, Boston
32. Pollinger IS (1973) Growth and DNA synthesis in embryonic chick heart cells, in vivo and in vitro. Exp Cell Res 76:253–262
33. Rous P, Jones FS (1916) Tissue preparations. J Exp Med 23:549
34. Shimada Y, Moscona AA, Fischman DA (1974) Scanning electron microscopy of cell aggregation: cardiac and mixed retina-cardiac cell suspensions. Dev Biol 36:428–446
35. Steinberg MS (1963) Reconstruction of tissue by dissociated cells. Science 141:401–408
36. Townes PL, Holtfreter J (1955) Directed movements and selective adhesion of embryonic amphibian cells. J Exp Zool 128:53–120

Neonatal Rat Heart Muscle Cells

A. Pinson

Introduction

Tremendous progress has been made since the pioneering, but long over-looked, observation of Burrows [5] in 1912 that single beating heart cells migrated away from embryonic chick heart isolated tissue explants. Burrows suggested that his finding supported the myogenic theory for cardiac beating activity. Several decades passed before the realization that isolated systems, such as cells in culture, might serve aas useful tools for studying biological systems began to take hold. For this reason, although tryptic dissociation, reported by Rous and Jones [68] in 1916, and Carrel's method for subculturing cells [6] from 1912 have been available for many years, science had to wait until the 1950s for these techniques to be revived by Moscona [56], who isolated cells from embryonic tissues by proteolytic digestion. He also determined which essential materials, cofactors, and vitamins were required to successfully maintain cell cultures.

From then on, the field expanded rapidly. In 1955, Cavanough [8] isolated and maintained growing, functioning heart cells from chick embryo, and in 1960, Harary and Farley [28] prepared the first cultures of neonatal heart cells. Cultures of mammalian heart cells obtained by this method continued to function, as shown by the spontaneous beating of such single cells, which became synchronous as the cells grew to make contact [8, 29].

The first objective concerning cultured heart cells was to establish not only that they beat, but also that they serve as a model system, reflecting the state of in vivo adult cells. Early studies were therefore directed towards improving techniques for keeping cells in long-term cultures, the ability of cultured cells to generate action potentials [12, 24, 44], and maintaining their specificity towards the drugs and hormones regulating beating rate [22, 28, 32, 77, 78]. Specific metabolic studies followed soon after [61, 62].

It is clear from the vast and rapidly increasing number of publications concerning heart cells in culture that such systems have become a major experimental tool in cardiac biology. At the present time the variety of techniques available is overwhelming and undoubtedly accounts for the great variation in the results in the literature. This has led to caution in the interpretation of data. This chapter will discuss the methods for culturing cardiac cells, pointing out the considerable gaps in knowledge due to the fact that optimal requirements for tissue disaggregation and nutrition have not yet been established.

Culturing Heart Cells

Culture Systems and Animals

In vitro models have been derived from organ cultures [8], tissue explants [5, 28], reaggregated single adult or embryonic cells [12, 24, 29], and from dissociated cardiomyoblast cultures. The latter are the topic of this chapter. Such cultures have been derived from chick embryo, mouse embryos, newborn rats and hamsters, and human fetuses [for review, see 64].

Cell Dissociation

The exact conditions employed in cell dissociation are crucial in achieving successful cultures. Various proteolytic enzymes have been used with highly conflicting results [64]. Cell damage by the proteolytic enzymes was minimized by carrying out the procedure in several steps and collecting the dissociated cells after each step in a serum-containing medium, which effectively inhibits further proteolysis.

The optimal enzyme requirements for cell dissociation vary with the type of tissue and the age of the animals used. In addition, it is well established that crude enzyme preparations are more efficient than purified ones. Indeed, as early as 1958, it was recognized that contaminating enzymes in crude trypsin contributed to tissue dissociation [67]. Highly purified collagenase is a poor dispersing agent, so, clearly, impurities in cruder preparations render this process more efficient [43]. Other enzymes, for example, pronase and elastase, have also been employed for making rat or chick embryo heart cell suspensions [7, 34, 45, 70, 76]. A commercial preparation of proteolytic enzymes called Viokase (Viobin Corp.) [80] has been used very successfully and is possibly more consistent in its effect than crude trypsin, which may vary in composition. Most crude enzyme preparations contain other proteolytic enzymes in varying quantities. Crude trypsin, for example, also contains chymotrypsin and elastase [71]. There are also significant variations between different batches of commercially available crude enzymes in their ability to dissociate heart cells – some of them causing severe cell damage. Some workers even claim that trypsin preferentially damages myocardial cells [36, 60, 77]. Thus Masson-Pevet et al. [51] showed that whereas there were no signs of cell injury in collagenase-dissociated cells, trypsin-treated cells were severely damaged. This is evidently reflected in cell recovery time. Gross et al. [25] also found severe damage in trypsinized preparations of dispersed cells, concluding that the recovery process takes at least 3 days. Other workers share the view that trypsin penetrates the myocardial cells and digests myofibrils [14, 25, 36, 51]. Speicher and McCarl [72] have reported a procedure for purifying crude trypsin preparations, producing a mixture enriched in chymotrypsin, elastase and trypsin, which effected cell dissociation at much lower concentrations, (0.01%). This purified nontoxic mixture of enzymes caused less damage to the isolated cells [73]. However, this enzymatic preparation is not commercially available.

Proteolytic Enzyme Preparation

Despite the problems already mentioned, crude trypsin remains the most widely used enzyme for cell dissociation. The concentrations in standard use vary from 0.05% to 0.1% [64]. These differences clearly contribute to confusion in the field as the state of cell preservation varies greatly. As mentioned by many workers, the purity of the water also influences the success of cytocultures, and in some cases it may be necessary to purify water specially for this purpose [58].

One may successfully use 0.1% trypsin (Sigma grade III) in a solution of Ham's F10 (Ca^{2+}- and Mg^{2+}-free) medium at pH 7.2–7.4 at 32°–35°C or 0.1% Viokase or phosphate-buffered saline (PBS), also Ca^{2+} and Mg^{2+} free. Our experience shows that glucose during tissue degradation and the shortest possible trypsinization cycles (10–20 min) give optimal cell preservation. Indeed, cells, if properly seeded (see below), recover within 24 h, as indicated by beating function.

Media and Sera

A large variety of media are in use in various laboratories. In addition, the frequency with which the medium is changed varies from group to group. Cell division, growth, and metabolism depend on the exact composition of the medium (the growth factors etc.) and on the length of the lag periods between media changes when less substrate may be available to the cells.

It is no simple matter to reproduce in vitro conditions resembling the finely balanced, complex neurohumoral mechanism that exists in vivo. In cultured cells, isolated from neurohumoral influences, cell division, differentiation, and metabolism are controlled by the composition of the extracellular medium, contact between cells, and the relative populations of different types of cells. The frequency of medium changes is critical in this respect [18, 58]. Substrates, such as glucose, fatty acids (FA), and amino acids, are completely taken up by the cells within 12 h of a medium change [16, 63]. At this stage cells adapt metabolically either by storing excess glucose in the form of glycogen or by exporting it from the cells as lactate [1, 18, 58], so that it is rapidly exhausted from the medium, the rate of protein synthesis becoming dependent on the extent of the amino acid pool in the medium or that formed as a product of degradation [18].

Sera from different sources, sometimes supplemented with growth factors, are commonly added to media [64], using 5%–20% serum from such widely different sources as rat fetuses, bovine fetuses or newborn calves, horses, and humans, which may be supplemented with embryo extracts. Human serum tends to be unreliable since the source cannot be controlled, and batches with toxic factors may turn up. The optimal conditions for growth and maintenance of heart cell cultures have yet to be determined. Some workers supplemented culture media with conditioned media obtained by incubating either chick heart cell cultures or fragments for 4 days with fresh medium [20, 21]. On the one hand, conditioned media retarded cell growth (the population doubling time was tripled – 72 h as opposed to 23 h in fresh medium) probably because the conditioned media added are partially exhausted of substrates, whereas, on the other hand, cell attachment and spreading was enhanced, and spontaneous beating was maintained for longer time periods [20]. The contractile and spread-promoting factors were shown by Gordon and Brice [21] to be a proteoglycan and a protein-containing material, respectively. Cells in culture secrete growth-promoting factors. Are the effects of conditioned medium on heart cell growth mediated via these factors? It would certainly be worthwile studying how the components of conditioned media influence cell growth.

Defined Media

Serum serves as a universal substitute for interstitial fluids, contributing many biologically active substances such as nutrients, hormones, growth factors, carrier molecules, and many other ill-defined components [2, 17, 18, 30]. Defined media were devised in order to gain better control over factors present in the system. The goal of a simple and fully defined culture environment can only be achieved when serum is replaced by the minimum number of well-defined substances.

Early attempts to develop serum-free, chemically defined media were only partly successful, further progress occurring in the wake of advances in the field of hormones and growth factors [23, 26, 46]. These advances, along with enhanced understanding of the nutritional demands of the cell, made it possible to replace serum in culture media in most cells, including cardiomyocytes [2, 4, 11, 27]. The compositions of serum-free, hormonally supplemented media proposed for myocardial cell cultures are summarized below.

Plating, Plating Density, and Cell Populations in Culture

At present, sterile, plastic tissue culture quality Petri dishes are used for plating, which are usually treated specifically to allow attachment and growth of the cells. However, some plastic dishes may contain metals, e.g., lead, which may have a significant effect on certain parameters (Pinson, unpublished data).

Recently, a technique using Sephadex microspheres for culturing cardiac cells has been developed [75]. However, the cell populations obtained were not pure. Seeding density has been shown to affect cell growth, division, and population by Speicher et al. [74]. They also found that cell division, as measured by ^{3}H-labeled thymidine incorporation, takes longer in cultures seeded at low (4×10^6 cells/ml) than at high (1×10^6 cells/ml) densities. However, the rates of protein synthesis were similar in both types of cultures. They concluded, as did our group [79], that comparing data from cultures seeded at different densities amounts to comparison of two systems with virtually nothing in common. Biochemical differentiation criteria in terms of enzymatic activities were established by Yagev et al. [79]. Lactic dehydrogenase (LDH) activity increases linearly in both myocyte and nonmyocardial cell-enriched cultures (Fig. 1). However, the LDH activity is threefold higher in muscle than in non-muscle cells (NMC). In cultures of mixed cell populations, however, a different pattern of enzymatic activities was observed: up to the 6th day in culture, they followed the same pattern as cultures enriched with myocytes, however, en-zyme levels subsequently leveled off due to overgrowth by NMCs with lower enzyme activities, leading to an overall decrease in enzyme activity (Fig. 1). Plating density is therefore a major factor affecting relative populations of different cells, a factor which is not always taken into account by researchers although it varied by up to an order of magnitude in different studies [64]. Indeed, many erroneous conclusions were reached in early research work because neither seeding density nor the overgrowth of NMCs was considered. Cells isolated from newborn rats consist of two populations: (1) about 80% of the initial cell suspension with a long cell cycle which rapidly lose their ability to divide once they become fully differentiated cardiomyocytes [10, 37, 47, 79]; and (2) about 20% of the initial cell suspension are NMCs with a much shorter cell cycle, which maintain their ability to divide throughout, usually referred to as "fibroblasts", but also containing other types of cells, particularly endothe-lial cells [10, 37, 47, 79]. Seeding at high density (10^6 cells/ml) assures that the culture reaches confluency and contact inhibition within 24 h, thus minimizing cell division in the NMCs and preventing them from overgrowing the car-

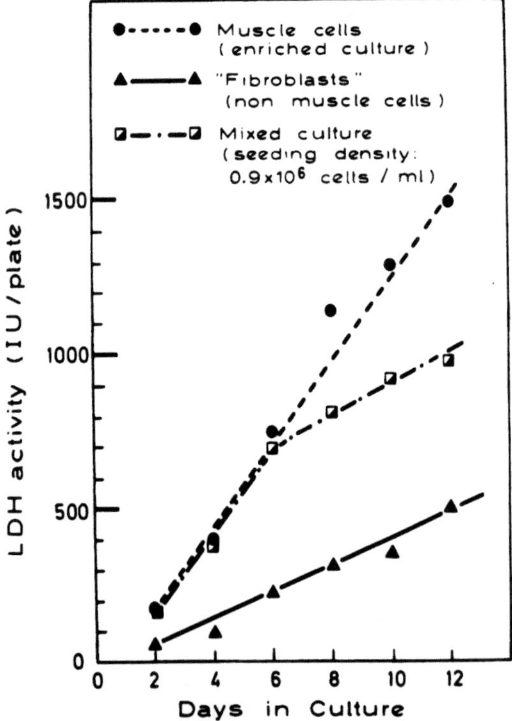

Fig. 1. Lactic dehydrogenase (LDH) activity in cardiac cells as a function of culture age. The total activity, in units per plate, was assayed according to Wroblewski and La Due in cultures of mixed type and in those enriched with either muscle or non muscle cells (fibroblasts). [From 79]

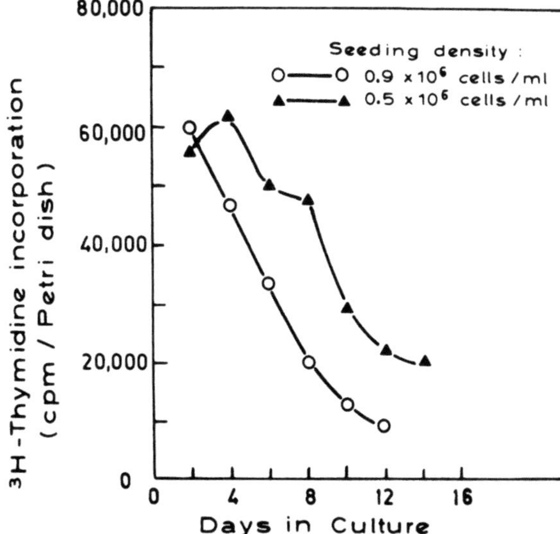

Fig. 2. ³H-labeled thymidine incorporation into trichloroacetic acid (TCA) insoluble cellular components. Cells were pulse labeled for 24 h with ³H-labeled thymidine. Proteins were precipitated after homogenization with TCA. [From 79]

diomyocytes. This was confirmed by [³H]thymidine uptake studies (Fig. 2) [47, 79]. A low seeding density, on the other hand, allows rapid "fibroblast" proliferation. The relative populations of different types of cells changed within a few days of seeding [10, 37, 47, 79]. Thus, high-density seeding is an easy way of maintaining a constant cell population in cultures for a sufficiently extended period of time. An alternative method, first proposed in the early 1970's, is the "selective adhesion" technique, which depends on the different rates of attachment of various types of cells to the substratum [3, 66].

Cell Division

Myocardial cells may be distinguished from NMCs in culture by their "dense" cytoplasm, well-developed mitochondria and Golgi system, the presence of myofibrils and intercalated discs, and their spontaneous contraction activity. The average mitotic cycle is 2.5-fold longer in myocardial cells than in NMCs [10, 37].

The general rule that differentiated cells no longer divide has limited applicability to myocardial cells, since the cardiomyoblast, which can synthesize myosin and also divide, is an intermediate state between the presumptive myoblast and the adult postmitotic cardiomyocyte. It is also difficult to define the last cell division in such a system [33]. It may well be that relatively undifferentiated myocardial cells, premyoblasts, predominate in these cultures as reported by Masse and Harary [48] since differentiated myocytes and fully developed myoblasts would be expected to be less resistant to cellular dissociation.

The fact that under optimal culture conditions the most characteristic expression of differentiation, automatic synchronous beating, occurs within 24 h, a much shorter time than reported by Gross et al. [25], supports this premise. The postnatal rat heart contains cells in different stages of development. In 1-day-old rats, roughly 55% of the cells are dividing cardiomyoblasts which contain myosin, however, the percentage decreases to 40% by day 4 [50]. This mirrors the situation in the intact animal, in which all the cardiac muscle cells are post-mitotic by day 21.

Cell Lines

As shown by Kasten and Yip [38], myocardial cells may be "banked" by multiple trypsinization-freezing-thawing cycles. Kimes and Brandt [42] claimed to have established a clonal muscle cell line from embryonic rat heart tissue. However, since then there have been no subsequent reports on this system. Another cell line, the Girardy cell line derived from human atrial appendage, has long lost any resemblance to the parent cells [19]. Since it has been shown that the percentage of postmitotic cells increases with time in culture, it is unlikely that a line of cardiomyocytes can be developed unless a means of preventing the loss of cell division, thus maintaining the intermediate cardiomyoblast, can be found. It is not possible to establish a cell line with postmitotic cells.

Selecting Cell Populations

The heart consists of a mixed population of cells with only 50% cardiomyocytes. Both in vivo and in vitro, proper functioning of the muscle cells requires the presence of NMCs. However, the interpretation of biochemical and pharmacological data is complicated by the presence of two types of cells in the Petri dish, myocardial cells and NMCs, mainly because NMCs proliferate more quickly and may eventually overgrow the culture. Several methods have been developed in order to overcome this problem.

The selective adhesion technique takes advantage of the observation that cells of the fibroblast type become more readily attached to the substratum and then spread over the surface, while the myoblasts retain their round shape for longer, and therefore remain in suspension [3, 66, 79]. Replating cells at the critical point allows preparation of a population that consists essentially of myocytes [79]. The main difficulty with this procedure lies in deciding on the length of replating time – if this phase is too brief, NMCs will preponderate, while if the preplating time is too long there will be considerable myoblast attachment to the substrate surface. Alternatively, cultures consisting primarily of cardiac endothelial cells may be produced easily by adding factors derived from tumour cells, which selectively stimulate endothelial cell growth [13].

Other techniques are based on selective elimination of the NMCs by growing the cells in a serum-free medium, by treatment with 7-β-OH cholesterol [54], or with the Ca^{2+} ionophore, A23187, which lyses cardiac NMCs [35].

Another technique is by using DNA synthesis inhibitors [9, 49]. This has been applied to the cell culture system with BUdR (5-bromodeoxyuridine), followed by UV irradiation which kills all the dividing cells [49]. However, all these chemical methods have the disadvantage that they may damage myocardial cells, for example, Ca ionophores in disrupting Ca^{2+} flux across cell membranes [59] may also have other adverse effects. In addition, the elimination of one type of cell population from the Petri dish surface while the myocytes are already in a nondividing state may lead to a nonconfluent culture, which would give rise to significant errors in expressing data [79].

Recently, McDonagh et al. [53] obtained highly enriched preparations of myocardial cells by using monoclonal antibodies raised against cell-surface adhesion factors. However, cells did not reach confluency with his method, so that its widespread practical use remains questionable.

Culturing Techniques

Equipment

The basic equipment needed for a medium-sized tissue culture unit includes: (1) an isolated UV-sterilized area, (2) a sterile hood, (3) an inverted-phase microscope, (4) an incubator at 37°C with an air/CO_2 (95% : 5%) flow, (5) sterile plastic disposable laboratory ware (Petri dishes, flasks, pipettes, and tubes from, e.g., Falcon Plastics Nunc; Costar, USA), (6) an autoclave for sterilizing glassware and filters, (7) "Trypsinators" (a waterjacketed Celstir apparatus (Wheaton Scientific)), and (8) filter holders 25–142 mm in diameter (e.g., Millipore), suitable for small and large volumes, respectively.

Reagents and Materials

The growth medium routinely used is Ham's F10 culture medium (Biological Industries; Flow Laboratories; Gibco, etc.) supplemented with 10% horse serum and 10% fetal bovine serum (Biological Industries; Flow Laboratories; Gibco), $CaCl_2 \cdot 2\,H_2O$ at 135 g/l, penicillin at 200 000 units/l and 0.2 g/l streptomycin. Another medium used is CMRL 1415ATM (Connaught Medical Research Laboratories). Ham's F10 requires incubation in a 5% CO_2 atmosphere. CMRL has a nonbicarbonate buffer and therefore can be used without CO_2 in the atmosphere [31]. F10 can only be used with 5% CO_2. Ham's F10 culture medium in the absence of Ca^{2+} and Mg^{2+} ("Solution H") is used for mincing and washing the hearts and for preparing the trypsin solution (Sigma, Israel; grade III, 0.1% weight/volume).

Isolating Cells and Preparing Cultures

Heart cells cultures are prepared by the dissociation of 1-day-old rat hearts into single cells (see Fig. 3). Sterile conditions are employed throughout. Rats are killed by decapitation and allowed to bleed. The animals are immersed in 70% ethanol and then held from the back in order to stretch the chest skin, thus revealing the position of the heart. The chest is opened by a transverse cut with sterile scissors and the hearts then aseptically removed with forceps, attempting to take the ventricles only. Excised hearts are placed in a Petri dish containing Solution H. After all the hearts are removed, they are transferred to a second Petri dish containing Solution H and minced into the smallest fragments possible, with one or two washings. Solution H is then replaced by a trypsin-containing solution, and the preparation is then transferred to a trypsinator. Some 10–15 ml of trypsin are required for 30–50 hearts. Trypsinizations are performed at 32°–35°C at a stirring rate of 150–200 rpm for 15–20 min. Fragments are then allowed to settle and the supernatant removed after each trypsinization. The first two to three trypsinizations, essentially containing cell debris, red blood cells, pericardial and endothelial cells, are rejected. Cells from subsequent trypsinizations are collected in 30-ml sterile tubes (Sterilin, UK). A few ml of growth medium are added and the cells then centrifuged at 1000 rpm. The pellet is resuspended in a small volume of growth medium. Trypsinizations are repeated until all the fragments are dissociated.

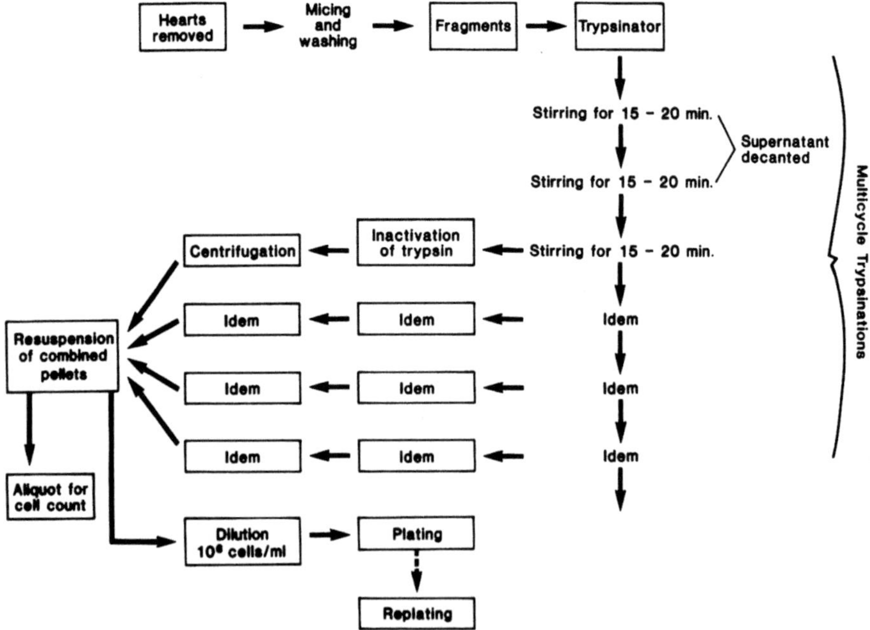

Fig. 3. Schematic representation of culture preparation methods. [From 64]

Cells in each tube are suspended by repeated aspirations in a sterile pipette and all the fractions combined in a sterile 250 ml flask, passing them through a sterile mesh in order to exclude explants. Cell suspensions are diluted with growth medium to a final density of 1×10^6 cells/ml and seeded onto Petri dishes – either 2 ml onto a 35-mm diameter Petri dish or 5 ml onto a 60-mm diameter dish or Petri dishes with different types of surfaces. After seeding, Petri dishes are shaken gently and linearly in order to achieve a more uniform distribution, and if the cultures are in Ham's F10, they are incubated at 37 °C in an atmosphere of 95% air/5% CO_2. After 24–36 h an almost confluent layer of beating cells is formed. The medium is changed every 2 days.

Separating Cell Populations

Enrichment of cells with respect to either muscle cells or NMCs may be achieved by plating the post-trypsinized cell suspension for 1 h at 37 °C. Petri dishes are then shaken and the unattached cells transferred to a new Petri dish. Under these conditions, since NMCs display more rapid attachment, the first Petri dish contains an almost pure population of NMCs, whereas the second one contains a highly enriched myoblast population (see Fig. 3). However, the replated cell densities should be corrected.

Cardiomyocyte Cultures in Serum-Free Media

The cells grown in these very useful type of cultures present qualitative differences as compared to those in serum-supplemented media [39]. The early stages involving removal of hearts, mincing, and trypsinizations are similar to those previously described. The differences lie in the growth media and requirements for coatings of Petri dishes.

The first serum-free, hormonally supplemented medium for cultured cardiomyocytes was proposed by Claycomb in 1980 [11]. Since then, several other formulations have been reported for embryonic, neonatal, and adult heart cells [40, 41, 55, 57, 65, 69]. Coating culture dishes with extracellular matrix components, such as collagen and fibronectin, is a prerequisite for efficient cell attachment and spreading in the absence of serum [11, 41, 55]. Such coating with collagen, followed by fibronectin, enhances survival and long-term performance of the cells. However, coatings of biomatrix, synthesized in vitro by endothelial cells, are inadequate since cardiomyocytes sink into such matrices, cells fail to establish intercellular contacts, and beating capacity is poor (G. Kessler-Icekson, personal communication 1989). Some investigators circumvent dish coating by preincubating dishes in serum-containing media or even plating the cells in such media during the first few hours of culture [40, 65, 69]. However, such procedures expose the cells to the short-term effects of serum, which may be detrimental.

Since they serve as the sole source of low-molecular-weight nutrients, rich nutritional basal media, supplemented with hormones and other essential

Table 1. Compositions of cardiomyocyte serum-free media

Dish coating	Basal medium	Fetuin (mg/ml)	Insulin (µg/ml)	Gluco-corticoids (μM)	Trans-ferrin (µg/ml)	BSA (%)	Selenium (ng/ml)	Ascorbic acid (mg/ml)	Reference
Fn	MEM	2.5	25	0.1	–	–	–	–	11
Fn	DMEM/F12	0.25	5	13 (5 µg/ml)	7.5	1	5	0.02	55
C	DMEM/F12	1	25	0.1	25	–	–	–	40
–	M199	–	10	–	10	–	–	–	69
S	M199	–	0.06 (10^8)	–	–	0.2	–	–	68
–	F12	0.25	5	–	5	1	5	0.02	57

BSA, bovine serum albumin; Fn, fibronectin; C, collagen; S, serum; MEM, minimum essential medium, Eagle's; DMEM, Dulbeco modified Eagle's medium; F12, Ham's F12 nutritional mixture; M199, medium 199 [From 39]

media, are generally used. Dose-response curves for optimal supplement concentrations have not been reported in the literature, and it is assumed that each research team employed the formulation giving the most satisfactory results with cultured cardiomyocytes.

Like skeletal myoblasts, cardiac myocytes require the serum glycoprotein fetuin in the culture medium [11, 15, 40, 41, 55, 57]. Although the exact role of fetuin remains obscure, it is generally held to be involved in cell attachment and spreading. Possibly, its action may be due to a contaminant [52], and, if so, it must be extremely potent since activity is retained in highly purified fetuin preparations and over a wide range of fetuin concentrations (Table 1).

Insulin is present in all serum-free media [46, 51, 52]. In addition to its regulatory effects on fatty acid and glycogen synthesis, insulin probably acts as a weak somatomedin analog [15]. Glucocorticoids contribute towards maintaining beating capacity, but restrict cell growth [52, 55]. Transferrin in such media acts as a carrier for iron and removes trace amounts of toxic metals [52, 55]. Albumin serves as a carrier of fatty acids [55, 57, 65]. The trace element selenium protects cells against oxidative damage [27]. Ascorbic acid may be beneficial although its effectiveness is limited by its susceptibility to oxidation [27].

The addition of growth factors, such as epidermal growth factor (EGF), enhances cell proliferation, primarily of NMCs [55]. Caution should therefore be exercised in using growth factors in cardiomyocyte culture media. Glutamine, for example, is omitted from such media in order to limit NMC proliferation [11, 79]. The quality of all the components, including water, is of major importance in serum-free media since they lack the detoxifying properties of serum.

Conclusions

The dual aim of this chapter was to discuss certain problems relating to cardiomyocyte culturing techniques and to describe a simple, reproducible method suitable for routine purposes. Indeed, the cells grown in Petri dishes constitute an "experimental animal." Obtaining large numbers of such cells in a simple manner is a sine qua non condition for the application of heart cell cultures to such a broad range of research fields. Although highly purified myocyte cell populations may be prepared by much more complicated methods, employing these for routine purpose is not yet possible.

The importance of standardization applies to all aspects of culture handling. Since the Petri dish is a closed system, a steady state is not maintained in the same sense as in vivo since many of the substrates become exhausted within the first 12–20 h [12, 29], while the cells continue to export metabolites into the medium. For this reason, the frequency of medium changes and a relatively short time lag between the last medium replacement and the use of the cells for experimental purposes are of critical importance for the nutritional and physiological state of the cells. From the economic point of view, it is certainly worthwhile directing research efforts towards finding the true optimal media conditions since some substrates are rapidly exhausted and serum levels of 20% are undoubtedly excessive. Cell dissociation conditions and plating densities should also be standardized. Finally, at present, the replating method combined with high seeding densities and taking cells from the areas closest to the apex is the best way for selecting homogeneous myocyte population in culture. Very little is known about the "side effects" of chemical inhibition used to limit NMC growth and should therefore be employed with caution.

Cultures grown on serum-free media are particularly suited for studying the interactions of hormones and drugs with heart muscle cells. The main drawbacks of the media described are low thresholds for toxic effects and presence of fetuin, which may introduce unknown impurities. Bearing these reservations in mind, serum-free, hormonally defined media may be used for most investigations involving cultured heart cells, and they will undoubtedly open new horizons in this field.

References

1. Anastasia JV, McCarl RL (1973) Effects of cortisol on cultured rat heart cells. Lipase activity, fatty acid oxidation, glycogen metabolism and ATP levels as related to the beating phenomenon. J Cell Biol 57:109–116
2. Barnes D, Sato G (1980) Serum-free culture: a unifying approach. Cell 22:649–695
3. Blondel B, Roijen I, Cheneval JP (1971) Heart cells in culture: a simple method for increasing the proportion of myoblasts. Experientia 27:356–358
4. Bottenstein J, Hayashi I, Hutchings S, Masui H, Mather J, McClure OB, Ohasa S, Rizzino A, Sato G, Serrero G, Wolfe R, Wu R (1979) The growth of cells in serum-free hormone supplemented media. Methods Enzymol 58:94–109
5. Burrows MT (1912) Rhythmical activity of isolated heart muscle cells in vitro. Science 3:90–92
6. Carrel A, Burrows MT (1910) Culture de tissus adultes en dehors de l'organisme. CR Soc Biol (Paris) 69:293–294
7. Cavanaugh DJ, Berndt WO, Smith TE (1963) Dissociation of heart cells by collagenase. Nature 200:261–262
8. Cavanough MW (1955) Pulsation migration and division in dissociated chick heart cells in vitro. J Exp Zool 128:573–589
9. Chacko S (1972) The effect of BUdR on the emergence of cardiac muscle cells in the developing embryo (Abstr). J Cell Biol 55:36A
10. Clark WAR, Fischmann DA (1983) Analysis of population cytokinetics of chick myocardial cells in tissue culture. Dev Biol 97:1–9
11. Claycomb WC (1980) Culture of cardiac muscle cells in serum-free medium. Exp Cell Res 131:231–236
12. Fange R, Persson H, Hesleff T (1956) Electrophysiologic and pharmacologic observations on trypsin disintegrated embryonic chick hearts cultured in vitro. Acta Physiol Scand 38:173–183
13. Fenselau A, Mello RJ (1976) Growth stimulation of cultured endothelial cells by tumor cell homogenates. Cancer Res 36:3269–3273
14. Fischmann DA, Moscona AS (1971) Reconstitution of heart tissue from suspensions of embryonic myocardial cells: ultrastructural studies on dispersed and reaggregated cells. In: Alpert NR (ed) Cardiac hypertrophy. Academic, New York, pp 125–139
15. Florini JR, Ewton DZ (1981) Insulin acts as a somatomedin analog in simulating myoblast growth in serum-free medium. In Vitro 17:763–768

16. Frelin C (1978) The growth of heart cells in culture. Evidence for a multiple activation of the pleiotypic program. Biochimie 60:627–638
17. Frelin C (1980) The regulation of protein turnover in newborn rat heart cell cultures. J Biol Chem 255:11149–11155
18. Frelin C, Padieu P (1976) Pleiotypic response of rat heart cells in culture to serum stimulation. Biochimie 58:953–959
19. Girardi AJ, Warren J, Goldman C, Jeffries B (1958) Growth and CF antigenicity of measles virus in cells deriving from human heart. Proc Soc Exp Biol Med 98:18–22
20. Gordon HP, Brice MC (1974) Intrinsic factors influencing the maintenance of contractile embryonic heart cells in vitro. I. The heart muscle conditioned medium effect. Exp Cell Res 85:303–310
21. Gordon HP, Brice MC (1974) Intrinsic factors influencing the maintenance of contractile embryonic heart cells in vitro. II. Biochemical analysis of heart muscle conditioned medium. Exp Cell Res 85:311–318
22. Goshima K (1976) Arrythmic movements of myocardial cells in culture and their improvement with antiarrhythmic drugs. J Mol Cell Cardiol 8:217–238
23. Gospodarowisz D (1979) Fibroblasts and epidermal growth factors: their uses in vivo and in vitro in studies on cell function and cell transplantation. Mol Cell Biochem 25:79–110
24. Grill WE, Rumery RE, Woodbury JW (1959) Effects of membrane current on transmembrane potentials of cultured chick embryo heart cells. Am J Physiol 197:733–735
25. Gross WO, Müller CAM, Schlotman EHM (1977) Loss of differentiation features in trypsin separated heart muscle cells. Anat Embryol (Berl) 151:341–350
26. Hale W, Wollenberger A (1970) Differentiation and behavior of isolated embryonic and neonatal heart cells in a chemically defined medium. Am J Cardiol 25:292–299
27. Ham RG, McKeehan WL (1979) Media and growth requirements. Methods Enzymol 58:44–93
28. Harary I, Farley B (1960) In vitro studies of isolated beating heart cells. Science 131:1674–1675
29. Harary I, Farley B (1960) In vitro studies on single beating rat heart cells into beating fibers. Science 132:1839–1840
30. Harary I, McCarl R, Farley B (1966) Studies in vitro on single beating rat heart cells. XI. The restoration of beating by serum lipids and fatty acids. Biochim Biophys Acta 115:15–22
31. Harary I, Hoover F, Farley B (1974) The isolation and cultivation of rat heart cells. Enzymol 32:740–745
32. Harary I, Renaud J-F, Wallace G (1976) Ca ions regulate cyclic AMP and beating in cultured heart cells. Nature 261:60–61
33. Holtzer H (1979) Myogenesis. In: Schjeide QA, de Vellis I (eds) Cell differentiation. Van Nostrand Reinhold, New York, pp 476–503
34. Houba V (1967) The use of pronase for dispersing cells. Experientia 23:572
35. Kaneko H, Goshima K (1982) Selective killing of fibroblast like cells in cultures of mouse heart cells by treatment with CA ionophore A 23187. Exp Cell Res 142:407–416
36. Kasten FH (1966) Electron microscope studies of the combined effects of trypsinization and centrifugation on rat heart cells (Abstr). J Cell Biol 31:131 A
37. Kasten FH (1972) Rat myocardial cells in vitro: mitosis and differentiated properties. In Vitro 8:128–150
38. Kasten FH, Yip DK (1974) Reamination of cultured mammalian myocardial cells during multiple cycles of trypsinization freezing-thawing. In Vitro 9:246–252
39. Kessler-Icekson G (1987) Cardiomyocytes grown in serum-free medium. In: Pinson A (ed) The heart cell in culture. CRC, Boca Raton, pp 23–28
40. Kessler-Icekson G, Wasserman L, Yoles E, Aampson SR (1983) Characterization of cardiomyocytes cultured in serum-free medium. In: Fischer G, Weiser RJ (eds) Hormonally defined media. A tool in cell biology. Springer, Berlin Heidelberg New York, p 383
41. Kessler-Icekson G, Sperling O, Rotem C, Wasserman L (1984) Cardiomyocytes cultured in serum-free medium; growth and creatine kinase activity. Exp Cell Res 155:113–120

42. Kimes BW, Brandt BL (1976) Properties of a clonal muscle cell line from rat heart. Exp Cell Res 98:367–381
43. Kono I (1969) Roles of collagenases and other proteolytic enzymes in the dispersal of animal tissues. Biochim Biophys Acta 178:397–400
44. Lehmkuhl D, Sperelakis N (1963) Transmembrane potentials of trypsin dispersed chick heart cells cultured in vitro. Am J Physiol 205:1213–1220
45. Levinson C, Green JW (1965) Cellular injury resulting from tissue disaggregation. Exp Cell Res 39:309–317
46. Lieberman I, Ove J (1959) Growth factors for mammalian cells in culture. J Biol Chem 234:2754–2758
47. Mark G, Strasser FF (1966) Pacemaker activity and mitosis in cultures of newborn rat heart ventricle cells. Exp Cell Res 44:217–233
48. Masse MJO, Harary I (1974) Role of cell division in the cytodifferentiation of rat heart cells in culture. Biochimie 56:1581–1585
49. Masse MJO, Harary I (1981) The use of 5-bromodeoxyuridine and irradiation for the estimation of the myoblast and myocyte content of primary rat heart cell cultures. J Cell Physiol 105:194–202
50. Masse MJO, Harary I (1981) The use of fluorescent antimyosin and DNA labelling in the estimation of the myoblast and myocyte population of primary rat heart cell cultures. J Cell Physiol 106:165–172
51. Masson-Pévet M, Jongsma HJ, de Bruijne J (1976) Collagenase and trypsin-dissociated heart cells: a comparative ultrastructural study. J Mol Cell Cardiol 8:747–757
52. McCarl RL, Margossian SS (1969) Restoration of beating and enzymatic response of cultured rat heart cells to cortisol acetate. Arch Biochem Biophys 130:321–325
53. McDonagh JC, Cebrta EK, Nathan RD (1987) Highly enriched preparations of cultured myocardial cells for biochemical and physiological analysis. J Mol Cell Cardiol 19:785–793
54. Mersel M, Hietter H, Luu B (1987) Differential sensitivity of heart fibroblasts and myocytes to 17-β-hydroxycholesterol. In: Pinson A (ed) The heart cell in culture, vol 3. CRC, Boca Raton, pp 125–132
55. Mohamed SNW, Holmes R, Hartzell CR (1983) A serum-free chemically defined medium for function and growth of primary neonatal rat heart cell cultures. In Vitro 19:471–478
56. Moscona A (1952) Cell suspension from organ rudiments of chick embryos. Exp Cell Res 3:535–539
57. Nag AC, Cheng M (1984) Expression in cardiac myosin isozymes in cardiac muscle cells in culture. Biochem J 221:21–26
58. Padieu P, Frelin C, Pinson A, Charbonné F, Athias P (1978) Effect of environmental factors and tissue culture methodology in producing and studying cultured cardiac cells. Recent Adv Stud Cardiac Struct Metab 12:609–620
59. Pfeiffer DR, Taylor RW, Lardy HA (1978) Ionophore A23187: cation binding and transport properties. Ann NY Acad Sci 307:402–423
60. Pine L, Taylor GC, Miller DM, Bradley G, Wetmore HR (1969) Comparison of good and bad lots of trypsin used in the production of primary monkey kidney cells: a definition of the problem and comparison of certain enzymatic characteristics. Cytobios 2:197–207
61. Pinson A, Padieu P (1974) Erucic acid oxidation by beating heart cells in culture. FEBS Lett 39:88–90
62. Pinson A, Frelin C, Padieu P (1977) Palmitate oxidation by beating heart cells in culture. Recent Adv Stud Cardiac Struct Metab 12:667–676
63. Pinson A, Degrès J, Heller M (1979) Partial and incomplete oxidation of palmitate by cultured beating cardiac cells from neonatal rats. J Biol Chem 254:8331–8335
64. Pinson A, Padieu P, Harary I (1987) Techniques for culturing heart cells. In: Pinson A (ed) The heart cell in culture, vol 1. CRC, Boca Raton, pp 7–22
65. Piper HM, Probst I, Schwartz P, Hutter FJ, Spiekermann PG (1982) Culturing of calcium stable adult cardiac myocytes. J Mol Cell Cardiol 14:397–412

66. Polinger IS (1970) Separation of cell types in embryonic heart cell culture. Exp Cell Res 63:78–82
67. Rinaldini LM (1959) An improved method for the isolation and quantitative cultivation of embryonic cells. Exp Cell Res 16:477–505
68. Rous P, Jones FS (1916) A method for obtaining suspensions of living cells from fixed tissues and for plating individual cells. J Exp Med 23:549–555
69. Simpson P, McGrath A, Savion S (1982) Myocyte hypertrophy in neonatal rat heart cultures and its regulation by serum and catecholamines. Circ Res 51:787–801
70. Smith TE, Berndt WO (1964) The establishment of beating myocardial cells in long-term culture in fluid medium. Exp Cell Res 36:179–199
71. Speicher DW, McCarl RL (1974) Pancreatic enzyme requirements for the dissociation of rat hearts for culture. In Vitro 10:30–41
72. Speicher DW, McCarl RL (1978) Isolation and characterization of the proteolytic enzyme component from commercially available crude trypsin. Anal Biochem 84:205–217
73. Speicher DW, McCarl RL (1978) Evaluation of a proteolytic enzyme mixture isolated from crude trypsins in tissue disaggregation. In Vitro 14:849–853
74. Speicher DW, Peace JN, McCarl RL (1981) Effects of plating density and in culture on growth and cell division of neonatal rat heart primary cultures. In Vitro 17:863–870
75. Uusimaa PA, Hiltunnen JK, Sormunen RT, Hassinen IEV (1988) Microcarrier culture of neonatal cardiac myocytes in metabolic studies. Cardiovasc Res 22:291–295
76. Weinstein D (1966) Comparison of pronase and trypsin for detachment of human cells during serial cultivation. Exp Cell Res 43:234–236
77. Wollenberger A (1964) Rhythmic and arrhythmic contractile activity of single myocardial cells cultured in vitro. Circ Res [Suppl 2] 15:184–201
78. Wollenberger A, Halle W (1963) Einfluß von Reserpin und Dichloroisoproterenol auf die durch Adrenalin und Digitoxin hervorgerufenen Wirkungen an Kulturen spontan schlagender isolierter Zellen des embryonalen Hühnerherzens. Monatsber Dtsch Akad Wiss Berlin 5:38
79. Yagev S, Heller M, Pinson A (1984) Changes in cytoplasmic and lysosomal enzyme activities in cultured rat heart cells: the relationship to cell differentiation and cell population in culture. In Vitro 20:893–898
80. Yasumura Y, Tashjian AH Jr, Sato G (1966) Establishment of four functional clonal strains of animal cells in culture. Science 154:1186–1189

Adult Ventricular Rat Heart Muscle Cells*

H. M. Piper, A. Volz, and P. Schwartz

Introduction

Isolated and cultured cardiomyocytes have facilitated a large variety of experimental approaches which either require cardiomyocyte material in great purity or direct experimental access to single cardiomyocytes. Techniques for the isolation of cardiomyocytes as well as those for primary cultures of such cells have been difficult to establish, due to a number of reasons including (a) In heart tissue the muscle cells are firmly connected to each other by intercalated discs [20] and the extracellular matrix network [57]; these connections are difficult to cleave without injuring the cells. (b) The cardiomyocyte is a large, polygonal, rigid cell that is more easily damaged by mechanical impact than are many other cell types. (c) The adult cardiomyocyte does not divide [40], thus the number of cells initially isolated cannot be increased in cultures [47].

The isolation of viable cardiomyocytes from the adult mammalian heart was first described by Powell et al. [45]. The rat was initially used and is still utilized most frequently and with least difficulties for the isolation of cardiomyocytes. To date, however, procedures to isolate adult cardiomyocytes have also been described for most other species of experimental animals.

At present, two types of primary culture of adult cardiomyocytes are generally used [24]; these are:

1. Serum-free short-term cultures, where the cells are rapidly attached to culture dishes which are either precoated with fetal calf serum (FCS) [41] or extracellular matrix proteins [29, 42]. Rapidly attached cells can maintain the typical elongated in vivo morphology for up to 2 weeks [56]. These cultures have been termed "the rapid attachment model" [24].
2. Serum-supplemented long-term cultures, in which the cells may attach only after a few days [7, 23, 38, 47]. After 2 weeks in serum-supplemented cultures, cardiomyocytes extensively spread. Cultures of spread cardiomyocytes can be maintained for several months. These cultures have been termed "the redifferentiated model" [24].

It is now obvious that these culture types represent only two out of many more possibilities, in which the isolated myocardial cell can either preserve or change some of its phenotypical features. Only some of the determinants of cardiomyocyte development in culture are as yet known. In this chapter,

---

* This work was supported by the Deutsche Forschungsgemeinschaft (Pi 162/2-2).

methods for the isolation and primary culture of cardiomyocytes from ventricular tissue of the adult rat heart shall be described. Recommended modifications of the isolation protocol for some other species, i.e. the rabbit, dog, guinea pig, and mouse, are also given.

Isolation of Cardiomyocytes

Problems

The basic principle of most methods used for the isolation of cardiomyocytes is the same, i.e., the heart tissue is exposed to a dissociating medium which is low in Ca^{2+} and contains proteolytic enzymes [12]. In most protocols the free Ca^{2+} concentration in the myocardial tissue is reduced to $20-100\ \mu M$, in order to dissociate the Ca^{2+}-dependent desmosome structures in the intercalated discs [37, 58]. Proteolytic enzymes are applied to cleave the connections of the individual cells with the extracellular matrix network [4, 26].

Whole hearts can be perfused in the Langendorff mode [45] or small chunks of myocardial tissue can be immersed in the dissociating medium [23, 55]. The yield of viable cells by the Langendorff method is, in general, better than by the method of immersion of tissue chunks. The isolation of cardiomyocytes has a number of inherent problems which have to be overcome by the procedure applied. These include:

1. During tissue dissociation a normal supply of oxygen and substrates to the cells cannot be maintained. This is true not only for the cells in tissue chunks immersed in the dissociating medium but also for hearts perfused with the dissociating medium, since in the latter the microvascular bed is rapidly dissolved thereby abolishing a normal microcirculation. For this reason, dissociation of tissue should proceed fairly rapidly.
2. During the isolation procedure the intracellular cation balance of the cardiomyocytes becomes disturbed. Several factors can be responsible for this. First, during cell separations sarcolemmal leaks can transiently occur. Second, the low Ca^{2+} concentration in the dissociating medium favors membrane hyperpermeability. Ischemic conditions may also reduce the activity of sarcolemmal ion pumps. For these reasons cardiomyocytes can take up Na^+ during the isolation procedure [1]. After isolation, when the Ca^{2+} concentration is again elevated to normal millimolar concentrations, a rapid Na^+/Ca^{2+} exchange across the sarcolemma can lead to Ca^{2+} overload of the cells causing the hypercontracture and deterioration of the cells [18, 28].
3. Proteolytic enzymes, used to cleave the exterior connections of cardiomyocytes in tissue, may harm the cells to be isolated. Therefore, the use of proteolytic enzymes should be as short as possible, sufficient to permit the cleavage of the cellular connections with the extracellular matrix before causing serious harm to the cell itself. A large variety of proteolytic enzymes have been successfully used for cardiomyocyte isolation [25]. Proteases are applied in mixtures the composition of which is normally poorly defined. Since the optimal combinations of defined protease activities have not yet

been identified, commercially available crude enzyme preparations must be tested to select a suitable batch.

4. During cell isolation, the myocardial cell is apparently very sensitive to impurities present in the media used for cell isolation. Organic or inorganic impurities may be present in water of presumed great purity, either because the purification process (distillation, ion exchange purification) is less effective than assumed or because the water is not properly stored and handled. Impurities in the solutions may, of course, also be due to impurities in the chemicals from which the media are prepared. These impurities can be so minimal, that they are irrelevant under other circumstances as, for example, for the performance of perfused working hearts.

For these reasons, cardiomyocytes can exhibit a number of abnormalities immediately after isolation which are often only subtle and due to structural or metabolic defects acquired during the isolation process. If these alterations are minor the cells quickly recover, if incubated under optimal supply conditions, such as in a short-term culture of the isolated cells.

Isolation Procedure for Rat Cardiomyocytes

Materials

The following materials are needed for cell isolation:

Perfusion System (Fig. 1 A)

Langendorff system, consisting of:

1. Top reservoir (100 ml), double-walled, temperature-controlled
2. A glass-coil heat exchanger with two cannulas fitted to its outlet (in the shape of an inverted Y, for the simultaneous perfusion of two hearts; replace by a single cannula for the perfusion of one heart), the distance between the top reservoir and cannulas is 100 cm
3. Connection by a double-walled, temperature-controlled glass tube between the heat exchanger and top reservoir, containing a flow reducer
4. Funnel, which can be moved below the cannulas to collect the fluid, connected with a tube leading back to the top reservoir
5. Roller pump for pumping the fluid back into the top reservoir
6. Temperature-controlled water circulator, for 37 °C temperature control of the Langendorff system
7. Pasteur pipette for gassing the top reservoir

Instruments

- 2 Scissors (coarse and fine)
- 2 Small forceps
- 2 Large Petri dishes (200 mm in diameter)
- 2 Scalpels and a watchglass (or a tissue chopper)

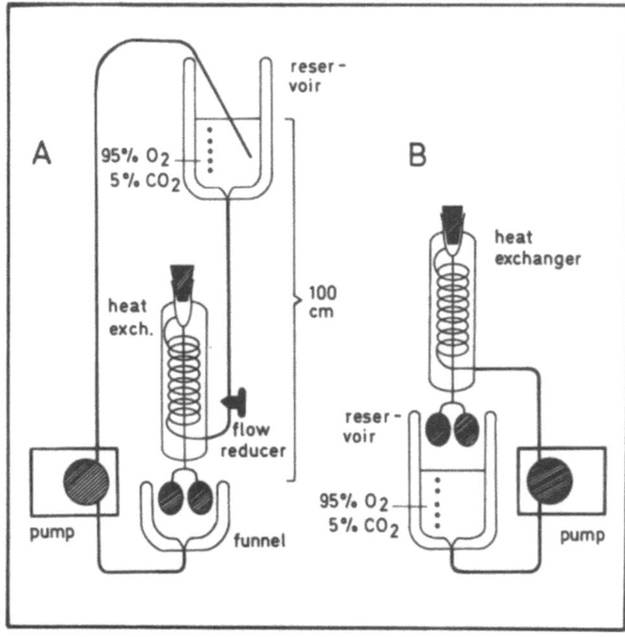

Fig. 1. Schematic representation of Langendorff perfusion systems, used for the isolation of cardiomyocytes

- Nylon mesh (mesh size 200 μm)
- 2 50-ml centrifuge tubes
- 1 50-ml glass beaker
- 1 50-ml Teflon or siliconized glass beaker
- 50-ml Erlenmeyer flask
- 4 Long centrifuge tubes (length 15 cm, diameter 1 cm)
- Plastic Pasteur pipette (large mouth or, alternatively, glass Pasteur pipette with 90° angled tip)
- Disposable 5-ml pipette with mouth about 2 mm in diameter (e.g., from Falcon or Greiner)
- Centrifuge

Media

- Buffer 1 (in mM): NaCl 110, KCl 2.6, KH_2PO_4 1.2, $MgSO_4$ 1.2, $NaHCO_3$ 25, and glucose 11, at 37 °C, continuously gassed with 95% O_2/5% CO_2 (gives pH 7.4). Prepare 1 l, warm up, and gasequilibrate before the experiment.
- Ca stock solution is 100 mM $CaCl_2$ in H_2O.
- Saline: 9 g NaCl/l, ice-cold.

- Bovine serum albumin (BSA), fraction V. For incubation of cardiomyocytes in media with low Ca^{2+}, BSA preparations freed from fatty acids
  are recommended. For preparation of the 4% BSA solution (see below) a
  simple quality may be sufficient. NB: Some BSA batches are harmful to the
  cells!
- Collagenase: Crude collagenase, from clostridium histolyticum is employed.
  An appropriate batch has to be selected. Suitable collagenases can be obtained, e.g., from Worthington, Serva, and Sigma.

Procedure

The procedure is described for the simultaneous perfusion of two rat hearts
(modifications for one heart are given in brackets).

1. Fill the Langendorff system with 120 ml buffer 1, gas upper reservoir with
   95% O_2/5% CO_2 through the tip of a Pasteur pipette. Make up collagenase
   solution immediately before starting the experiment: dissolve 30 mg in
   10 ml buffer 1, with the addition of 12.5 µl Ca stock. (The exact amount of
   collagenase has to be determined empirically for a given batch.)
2. Fill the large Petri dishes with ice-cold saline. Kill the animal by deep ether
   anesthesia. Open the chest with course scissors. Pour ice-cold saline onto
   the heart. Excise the heart with an intact aortic arch and immerse in the
   first of the Petri dishes. Free aorta from mediastinal tissue and transfer to
   second Petri dish.
3. Start the flow of the perfusion system (1 drop/s per cannula). Place beaker
   under the cannulas and remove the collecting funnel. Mount the hearts on
   the cannulas as follows: while the heart is being immersed in saline, open
   the aortic lumen with two forceps, lift the heart to a cannula, slip aorta over
   cannula (avoid penetration of aortic valve), and fix aorta with a crocodile
   clamp. Repeat the procedure with the second heart. Replace clamps by
   threads. By adjusting the flow reducer perfuse each heart with approximately 2 drops/s (10 ml/min per two hearts). Catch drops in a beaker until
   80 ml are collected. Discard the collected fluid.
4. By placing the funnel underneath the hearts and starting the pump, start
   recirculating the perfusate. Add the 10 ml of collagenase solution to the top
   reservoir. Keep the flow rate at 2 drops/s per two hearts. Continue for
   30 min. (The appropriate time depends on the concentration and activity
   of the given batch of collagenase.)
5. Near the end of the 30-min recirculation period, add 400 mg BSA plus
   30 ml recirculating medium into a 50-ml Teflon or siliconized glass beaker.
   Gas with 95% O_2/5% CO_2 through the tip of a Pasteur pipette and maintain at 37 °C.
6. Take the hearts off the cannulas, remove atria with fine scissors, and cut the
   ventricular tissue in two to four pieces. Chop these pieces with two scalpels
   into very small chunks (if available use tissue chopper and cut in
   0.7×0.7 mm pieces). Transfer minced tissue into the 50-ml beaker containing the recirculated medium. Triturate gently with the 5-ml disposable

pipette twice per min, during a 15- to 20-min incubation period (well-perfused tissue will dissolve in this time).

7. Filter the material through the nylon mesh and divide into two 50-ml centrifuge tubes. Spin tubes at 25 g for 3 min. Add 100 µl Ca stock to 50 ml of buffer 1. Aspirate supernatant from tubes, add 10 ml of this solution to each tube.

8. In a 50-ml Erlenmeyer flask, dissolve 2 g BSA in 50 ml of buffer 1 containing 500 µl Ca stock (4% BSA solution).

9. Spin tubes again, as before. Add 200 µl Ca stock to 40 ml of buffer 1. In each tube, aspirate supernatant, and resuspend the cell pellet in 10 ml of this solution.

10. Pour 4% BSA solution into the four long centrifuge tubes in equal volumes to obtain a 10-cm liquid bar. With the aid of a disposable plastic Pasteur pipette (or bent glass Pasteur pipette), gently layer, in equal distribution, the solution of resuspended cells on the top of the BSA solution.

11. Spin tubes at 15 g for 30 s or allow myocytes to settle out under gravity, i.e., leave centrifuge tubes containing the cells on the bench (intact cells form clusters which settle out first). Aspirate the supernatant and resuspend the pellet in culture medium.

Additional Step: Reduction of the Number of Damaged Cardiomyocytes by Trypsinization. The following step has to be inserted between steps 7 and 8:

7a. The number of damaged cells can be reduced by trypsinization, since damaged cardiomyocytes are more easily digested than intact ones. Add 200 mg BSA and 10 mg trypsin (trypsin 1:250) to a 50-ml Teflon or siliconized glass beaker in a 37 °C waterbath. Pour cell suspension from both tubes into beaker, fill with the resuspension medium up to 40 ml. Gently agitate (stir with a hanging bar) this solution for approximately 20 min. Continue with step 8. Depending on the activity of the collagenase and trypsin, the amount of trypsin and the incubation time must be modified. Rapid attachment of cells to FCS-treated culture dishes is not impaired when the cells are trypsinized according to this protocol.

Alternative Mode of Perfusion

For the simultaneous perfusion of more than two hearts, the above protocol can be performed on several perfusion systems simultaneously. In order to start the protocol on different systems at the same time, hearts may be isolated one after the other and perfused with buffer 1 plus 1 mM CaCl$_2$ before the isolation procedure is started.

For use in a laminar flow hood, the perfusion system can be constructed in a more compact fashion using parts of the Langendorff system described above (Fig. 1 B). The top reservoir is positioned below the heat exchanger and the cannulas so that it can be moved underneath the cannulas when recirculation is started. A tubing connects the reservoir with the inlet of the heat exchanger, via the roller pump. In this configuration the pump controls the perfusion rate. A good pump with low pulsations is recommended.

Buffer 1 can be modified using the HEPES buffer substance (4-(2-hydroxy-ethyl)-1-piperazineethanesulfonic acid), thus reducing the danger of pH changes by inadequate gassing of the media. The modified buffer 1 (1*) contains (in mM): NaCl 110, KCl 2.6, KH_2PO_4 1.2, $MgSO_4$ 1.2, HEPES 25, glucose 11; adjusted with NaOH to 7.4; 37°C; continuously gassed with 100% O_2 throughout the isolation procedure.

Isolation Procedures for Other Species

The protocol described above for the isolation of rat cardiomyocytes can also be applied to other species with some modifications. The steps to be modified are noted in the next sections. New batches of collagenase may have to be selected for other species, as we found that some of the collagenase batches suitable for the rat are not optimal for other species.

Rabbit

Rabbit hearts with a weight of 3–4 g are perfused with a single cannula using the described Langendorff system. Modify the protocol described for rat hearts as follows:

1 R. Fill the Langendorff system with 170 ml buffer 1 and gas upper reservoir with 95% O_2–5% CO_2 through the tip of a Pasteur pipette. Make up collagenase solution immediately before starting the experiment by dissolving 60 mg in 10 ml buffer 1, containing 25 µl Ca stock.
3 R–4 R. Maintain the perfusion flow at approximately 15 ml/min.

Dog

Modify the protocol described for rat hearts as follows.
– Buffer 2 comprised of (in mM): NaCl 125, KCl 2.6, KH_2PO_4 1.2, $MgSO_4$ 1.2, $CaCl_2$ 1, HEPES 10; adjust pH to 7.4 with NaOH.

2 D. Perfusable pieces of canine myocardium are obtained as follows: the heart is cut out in total from the dog's chest, placed on ice, and flushed with 1 l ice-cold saline to stop the beating. Then the left anterior descendens artery is cannulated with a small rigid tubing (nylon, Teflon) connected to a 50-ml syringe which is filled with ice-cold buffer 2. The tubing is fixed in the vessel with a snare, and the 50-ml content of the syringe is slowly injected. From the area of tissue demarcated by the washout of blood, a 3- to 4-g piece is excised which can be perfused via the cannulated vessel. The excised tissue is then transferred to a beaker containing ice-cold saline. While immersed in the fluid the cannulating tubing is cut to a 1 cm length. The tubing is then connected with a small piece of silicone tubing to the dripping cannula of the perfusion system.
Proceed otherwise as described for rabbit hearts.

Guinea Pig

Two guinea pig hearts can be perfused with volumes and flux rates as specified for two rat hearts. The amount of collagenase needed is normally less than that for the rat. In our experience only a few of the batches of collagenase suitable for the rat can also be used for the guinea pig. Modify the protocol described for rat hearts as follows:

- KB medium comprised of (in mM): KCl 70, K$_2$HPO$_4$ 30, taurine 20, glucose 20, MgSO$_4$ 5, succinic acid 5, creatine 5, ethylene glycol-bis(β-aminoethyl-ether)N,N,N',N'-tetraacetic acid (EGTA) 1, β-hydroxybutyric acid 5, Na-pyruvate 5, Na-ATP 5; adjust pH to 7.4 with KOH; and use at room temperature, i.e., 20°C (modified after Isenberg and Klöckner [21]).

4 G. By placing the funnel underneath the hearts and switching on the pump, start recirculating the perfusate. Add the 10 ml of collagenase solution containing 12.5 µl Ca stock solution to the top reservoir. Keep the flow rate at 10 ml/min/2 hearts. Continue for 30 min. Add 15 µl Ca stock after 10 and 20 min to the top reservoir.

7 G. Filter the material through the nylon mesh and divide into two 50-ml centrifuge tubes. Spin tubes at 25 g for 3 min. Discard supernatant and resuspend pellet in 50-ml KB medium at room temperature (20°C) in a 250-ml Erlenmeyer flask. Keep at room temperature for at least 45 min.

8 G. Resuspend cells every 5 min by gentle agitation of the Erlenmeyer flask. After 45 min, divide suspension into two 50-ml tubes, centrifuge at 25 g for 5 min. Add 500 µl Ca stock to 100 ml buffer 1 (37°C). After removal of the supernatant 40 ml of this buffer is added to each tube. Spin tubes at 25 g for 3 min, and discard the supernatant. Resuspend the pellet in culture medium.

Mouse

For application to mouse hearts the protocol for rats is modified with regard to several details. Buffer 1 of the rat protocol is replaced by buffer 3 (see below) in steps 1–6. The volumes and perfusion rates specified for the perfusion of one rat heart can also be used for the simultaneous perfusion of four mouse hearts, for which the following protocol is described. Perfusion cannulas are made from 20 G hypodermic needles with blunt tips. Since mouse hearts are very small it requires some practise to mount hearts on these cannulas! In our experience only few of the commercially available collagenase batches are suitable for the mouse.

- Buffer 3 comprised of (in mM): NaCl 100, KCl 10, KH$_2$PO$_4$ 1.2, morpholinopropane sulfonic acid (MOPS) 10, taurine 50, glucose 20; with pH adjusted to 7.0 with KOH; 37°C; continuously gassed with 100% O$_2$.

1 M. Fill the Langendorff system with 80 ml buffer 3 and gas upper reservoir with 100% O$_2$ through the tip of a Pasteur pipette. Make up collagenase/trypsin solution immediately before starting the experiment: dissolve 30 mg collagenase plus 10 mg trypsin (trypsin 1:250) in 10 ml buffer 3, with the addition of 12.5 µl Ca stock.

3 M – 4 M. Maintain the perfusion flow at approximately 5 ml/min/4 hearts.
7 M – 11 M. Use buffer 1* (p. 42) instead of buffer 1.

Discussion of Perfusion Protocols

In the literature, a large number of alternative protocols have been described. These have recently been reviewed by Jacobson [25]. A rational comparison of all protocols seems impossible. Nevertheless, a few comments may facilitate a categorization of the different approaches.

In most laboratories cardiomyocytes are isolated with the aid of crude collagenase, which is a mixture of proteolytic enzymes, that varies in its composition from batch to batch of commercially available preparations. Activities of enzymes other than collagenase contained in these crude preparations are important for the isolation of cardiomyocytes, since the use of purified collagenase alone is ineffective [26]. Some authors use other enzymes in combination with crude collagenase (e.g., trypsin, hyaluronidase; for a review see Jacobson [25]); some use crude protease mixtures instead of crude collagenase (pronase, [3] protease [6]). One group has reported that adult cardiomyocytes can be isolated from guinea pig heart completely without enzymes [2], but as yet little is known about the yield and the quality of the isolated cells. When cardiomyocytes are isolated from tissue chunks, the use of trypsin seems necessary [25]. In most protocols for cardiomyocyte isolation from tissue chunks trypsin is combined with crude collagenase.

In some protocols dissociating buffers with reduced Na^+ and increased K^+ concentrations are used (e.g. [1]) in order to reduce the disturbance of Na^+/K^+ homeostasis in newly isolated cells. We did not find these protocols to be advantageous compared with those described above for the isolation of cardiomyocytes from the rat, rabbit, and dog. In case Ca-tolerant cardiomyocytes are not achieved by perfusion with buffer 1, as specified above, it is certainly worth trying a perfusion medium with reduced Na^+ and increased K^+ concentration.

One of the major effects of a postincubation of isolated cells in the KB medium is probably a normalization of cellular Na^+ and K^+ contents. The usefulness of the metabolic substrates contained in this medium has not been satisfactorily established.

The concentration of Ca^{2+} in the dissociating medium is one of the important variables in the isolation protocol. It is noteworthy that the actual concentration of ionized Ca^{2+} is usually higher than the concentration of $CaCl_2$ added to the medium. This can be due to traces of Ca^{2+} present in the water (should be less than 1 μM), the glassware, the chemicals used to make up media, the enzyme material, or to an incomplete removal of Ca^{2+} from the tissue during the initial perfusion step.

In some protocols, the concentration of Ca^{2+} added to the perfusion buffer is elevated (up to 1 mM [19]) after a brief initial washout period (5 – 10 min). A relatively brief period of Ca^{2+} reduction is apparently already sufficient to loosen cell-cell contacts. Extremely low Ca^{2+} concentrations are unfavorable during the further procedure, probably because cell separation causes small membrane lesions the healing-over of which is Ca^{2+}-dependent [11]. When elevated Ca^{2+} concentrations $(0.2-1$ m$M)$ are used during tissue dissociation, the concentration of proteolytic enzymes or the exposure time can be reduced, since these enzymes are more active at higher Ca^{2+} concentrations. According to our experience, however, many batches of crude collagenase cannot be used at high Ca^{2+} concentrations, probably because harmful proteolytic enzymes become too much activated.

In some preparations, the newly isolated cells are very sticky so that they rapidly form large clusters. This often makes it difficult to handle them in suspension and it greatly impairs their attachment to culture dishes. The cells become less sticky when incubated with desoxyribonuclease [29], indicating that DNA released from broken cells is the cause.

In the above protocol, centrifugation of the cells through 4% (w/v) albumin is used to increase the percentage of intact cells. This can also be achieved by density-gradient centrifugation with Percoll (Percoll, Pharmacia) [30]. Neither of the centrifugation methods are completely selective for the intact cells. Broken cells are removed most effectively by the selective attachment of cardiomyocytes to culture dishes coated with FCS [41] or laminin [29].

Sterility

For preparation of cultures to be used on the same day, sterility of the isolation procedure is not normally needed. Even for long-term cultures sterile isolation procedures are not necessary under all circumstances. Depending on the type of microorganisms present in the laboratory, it can be sufficient to isolate cardiomyocytes with a perfusion system which is only carefully cleaned after each use (as described above). All media used during cell isolation should be sterilized by filtration. Antibiotics should be added to the media in fivefold higher concentrations than ususal during the first day. Nonsterile cell isolations are not suitable for long-term cultures when the cultures may be contaminated by fungi since fungicides are very toxic to cardiomyocytes.

Trouble Shooting

Even in laboratories with long-standing experience in the isolation of cardiomyocytes, from time to time the quality of cell preparations declines. Some general guidelines may help to shorten the search for the cause of failure.

1. Avoid the use of detergents in cleaning the glassware used for cell isolation.
2. It is helpful to keep a record of the use of all materials involved in cell preparations, particularly of all chemicals. It is also advisable to reserve chemicals only for the purpose of cell isolations. If some doubts about the purity of the chemicals arise all should be exchanged for unopened batches (they are all relatively cheap!).
3. A possible cause of failure is the quality of the water used to make up the solutions. Therefore, in case of failure, solutions should be prepared from a

source different from the usual one. In most research facilities purified water from a number of different sources can be obtained and tested. It is noteworthy, however, that in some places a multitude of these distillation or ion exchange systems is fed with water from a common primary ion-exchange purification system. If this common source releases volatile organic impurities these may be found in the water of all purification systems receiving its water.

4. It is of prime importance to keep the perfusion system clean. In general it is sufficient to flush the system after use with 1 l of water. This has to be done immediately, otherwise proteins contained in the perfusate will dry on the glass. Even if the perfusion system does not need to be sterile it is advisable to flush the glassware after use with 70% ethanol for 30 min and subsequently dry it by a stream of clean gas (e.g., filtered compressed air).

5. Some general advice might be given to beginners. When starting with cardiomyocyte isolations it is usually time saving to sort out basic difficulties by using the rat rather than any other species. The isolation procedure detailed above is relatively simple compared with others described in the literature. To use this method, therefore, makes it easier to analyse the causes for an initial failure.

Primary Cultures of Adult Cardiomyocytes

Problems

Since adult cardiomyocytes do not divide in culture [47], the number of cells initially isolated already delineates the maximum amount of cell culture material. Cultures always have to be prepared from newly isolated cell material and, therefore, the quality of a specific culture preparation depends not only on the actual culture conditions, but also on the degree of damage the cells have suffered during isolation. In addition to this basic problem, primary cultures of adult cardiomyocytes confer a number of limitations and problems, including:

1. Confluent cultures of cardiomyocytes have not been achieved, because rod-shaped cells cannot attach on the culture dish without leaving spaces in between. Spreading cardiomyocytes in long-term cultures make contacts [48], but do not cover the complete area of the culture dish since they do not divide.

2. Cardiomyocytes can be maintained in stable long-term cultures (weeks to several months) only under conditions which also cause cell spreading [44]. The spread cardiomyocyte is phenotypically different from the cell originally isolated and, for a number of investigations, may no longer represent an appropriate model for the in vivo situation.

3. Long-term cultures can be overgrown by nonmuscle cells. In order to obtain pure cultures these nonmuscle cells have to be selectively removed.

4. Most experiences of adult cardiomyocyte cultures have been obtained with cell material from the rat. In the following section, methods are described

only for this species. Whether these are equally suitable for use with cardiomyocytes from other species has not yet been investigated in detail.

Cultures of Rat Cardiomyocytes

Media

Medium 199 with Earle's salts, buffered with bicarbonate-CO_2 (or HEPES buffered, no CO_2), glutamine-free, is used with supplements: BSA 0.2%, insulin 10^{-7} M, creatine 5 mM, L-carnitine 2 mM, taurine 5 mM, penicillin 100 IU/ml, and streptomycin 100 µl/ml.

Plating of Cells

Rapid attachment of cardiomyocytes is required for cultures to be used within hours or a few days after isolation. In this section the attachment of cardiomyocytes to surfaces pretreated with FCS is described [41]. Culture dishes made of tissue culture grade polystyrene supplied by Falcon and Costar have been used successfully for the serum pretreatment in our laboratory. Some other plasticware has been shown unsuitable for this purpose [42]. Attachment substrates other than serum are discussed in Chap. 6.

For the purpose of cardiomyocyte attachment, tissue culture grade plastic materials or glass are incubated for 5–15 h with 4%–8% FCS (e.g., with 3 ml of medium 199 containing 4% FCS per 60-mm dish, overnight). FCS can be applied in culture medium or a simple buffer, like phosphate buffered saline (PBS, composition in mM: NaCl 137.0, KCl 2.6, KH_2PO_4 1.5, Na_2HPO_4 8.1, $CaCl_2$ 1.4, $MgCl_2$ 0.5, pH 7.4). Immediately before use, the medium is removed by gentle suction, and cells are plated on the still wet surfaces.

The attachment of cells on FCS is dependent on the quality of the serum batch. Fetal calf sera have to be screened for optimal attachment. With good sera and preparations containing more than 60% rod-shaped cells, 75% of all rod-shaped cardiomyocytes will attach unless the plating density is lager than 1×10^4 cells/cm^2 (2×10^5 cells per 60-mm dish). At higher densities, the number of attached cells grows less than proportionally. The maximum number of cells that can be attached is approximately 2×10^4 cells/cm^2. In very dense suspensions of cardiomyocytes cell clusters form rapidly. Clusters do not attach firmly and are lost with medium changes. Since cell clusters form also with time, when cardiomyocytes are left in suspension, the cells have to be plated immediately after isolation. High concentrations of albumin and serum in the culture medium used for cell plating impair cell attachment.

Medium Change

Culture medium should be changed between 1 and 4 h after cell plating if cultures are expected to be maintained beyond 12 h. At this time the cultures can be washed with PBS (with $CaCl_2$ and $MgSO_4$) or culture medium to remove nonattached cells and cell debris. Media must be changed very carefully in order to avoid large mechanical impact on the attached cells. When media are changed, they should be decanted from the dish rather than removed by suction. The old medium has to be immediately replaced by new medium in order to avoid drying. BSA increases the viscosity of culture media, and this gives protection to the cells during medium changes since it helps to leave a thin film of fluid behind when the medium is decanted. Since careful handling takes time, it can be advantageous to include some HEPES buffer in bicarbonate-CO_2 buffered culture media in order to stabilize the pH. After the first change, the media should normally be changed every third day.

Control of Nonmyocytes

Only small numbers of nonmyocytes are present in preparations of cardiomyocytes according to the above protocol. In the absence of serum in a glutamine-free culture medium, these cells do not proliferate. In serum-supplemented cultures, however, nonmuscle cells can overgrow the culture in 2–3 weeks. The growth of nonmuscle cells can be effectively prevented by the antimitotic agent cytosine-β-D-arabinofuranoside (Ara-C, $10 \, \mu M$) [27, 32]. A 5-day period is sufficient to remove all proliferating cells [50]. The antimitotic agent bromodeoxyuridine has been used in neonatal rat heart cell cultures [51] and may also be applicable to adult cardiomyocytes.

Reduction of nonmyocytes can also be achieved by preplating the isolated cells for 2 h on Petri dishes coated with a film of 1% gelatin. The cardiomyocytes do not attach to gelatin and are subsequently plated on FCS-treated dishes. This procedure has two disadvantages: first, during preplating cardiomyocytes form clusters and this subsequently impairs their attachment on FCS-coated dishes, and, second, the procedure does not exclude but only delays nonmyocyte growth in long-term cultures.

Medium Supplements for Cultures up to 12 Hours

If cardiomyocytes are to be used on the first day after isolation, it is advantageous to include 4% FCS in the culture medium. This is because it enables the cells to recover quickly metabolically. For cell cultures extended beyond the first 12 h even a short incubation of the cells with FCS favors an early loss of the elongated shape. Therefore, if elongated morphology is to be maintained as long as possible, the cells should not at any time be exposed to serum.

Medium Supplements for Serum-Free Cultures

Insulin: In serum-free cultures, insulin does not prolong cell survival [44] but it delays the development of cell atrophy [13].

Glutamine: In serum-free cultures, glutamine stimulates development of a spherical shape of cardiomyocytes during the first days in culture. It also stimulates nonmyocyte growth, even if this is slow in the absence of serum. Glutamine should therefore be omitted from culture media.

Creatine, Carnitine, and Taurine: Cardiomyocytes cultured in the absence of serum and creatine progressively lose creatine phosphate and thereby reduce their high energy phosphate reserve [41]. In serum-free cultures, addition of $2-5$ mM creatine preserves in vivo levels of creatine phosphate, and it increases the number of cells retaining their elongated form. In spite of the difference in metabolic function, carnitine (2 mM) and taurine (5 mM) have an effect similar to creatine (5 mM) in that they also prevent cell rounding (Fig. 2). The effects of creatine, carnitine, and taurine are additive. In the presence of all three substrates and 10^{-7} M insulin [56], the half-life of cardiomyocytes with an elongated shape is 14 days, whereas in cultures supplemented with glutamine (0.34 mM) and 10^{-7} M insulin, but neither of the three substrates, the half-life is only 2 days (Fig. 2).

Triiodothyronine: Elongated cardiomyocytes in serum-free cultures can be stimulated to contract by an electrical stimulus, but the maximal stimulus frequency, the cells are able to follow, decreases with time in culture. Newly isolated cardiomyocytes can be paced with frequencies up to 950 stimuli/min. In the absence of triiodothyronin (T_3) the maximal stimulation frequency is reduced by 30% within 6 days. When 0.8 nM T_3 is added to culture media, the maximal stimulation frequency after 6 days is the same as on day 1 [unpublished observation]. This may be due to an effect of T_3 on the expression of myosin isoforms which differ in ATPase activity [33]. Concentrations of T_3 higher than 1 nM cause cell rounding.

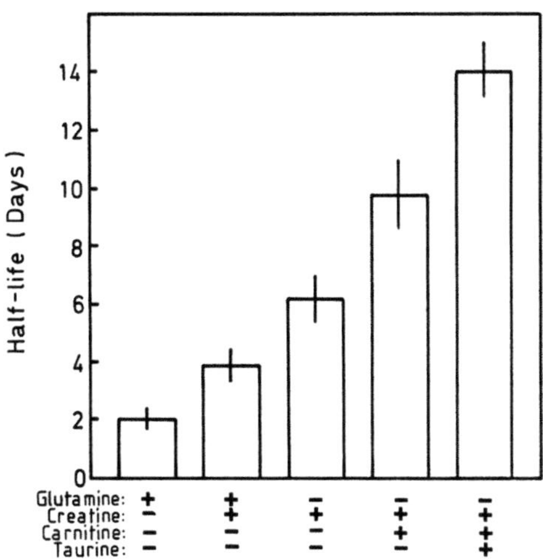

Fig. 2. Longevity of elongated cardiomyocytes in serum-free culture, expressed as half-life (time to reach 50% of initial number) of elongated cells. Cardiomyocytes from ventricular tissue of adult rats were cultured in medium 199, supplemented with 0.2% BSA, 10^{-7} M insulin, and the following substances as indicated: L-glutamine 0.34 mM, creatine 5 mM, L-carnitine 2 mM, taurine 5 mM. Means $\pm$ S.D., $n = 10$

2-Mercaptopropionylglycine: 2-Mercaptopropionylglycine (MPG) is a radical scavenger [15]. When added to culture media during the first 2 days in 1–5 mM concentration, it reduces the number of subsequently rounding cells. The presence of MPG during the first 2 days after isolation already achieves a maximal effect. Interestingly, the effect of MPG is particularly pronounced when the cell isolation is not optimal, as indicated by a large number of broken cells in the newly isolated material. Even though in such preparations attachment on FCS-coated dishes also produces a pure population of rod-shaped cardiomyocytes, the cells round rapidly in culture. This rapid rounding can be prevented by a 2-day exposure to MPG, indicating that free radical injury is involved.

Catecholamines: Catecholamines have a distinct effect on the development of adult cardiomyocytes in culture [56]. α-Adrenoreceptor agonists stimulate cell spreading even in the absence of fetal calf serum (Fig. 3). Substances tested were adrenaline and noradrenaline (both α- and β-adrenergic) and phenylephrine (α_1-adrenergic). These agents were added to 3-day-old cultures. When administered to cultures younger than 3 days, they induce early rounding of the cells. Effective concentrations of the agonists were 1–10 μM. Additions of the β-blocker propranolol (10 μM) to media containing 10 μM noradrenaline did not prevent cell spreading; in contrast, the α_1-antagonist prazosin (10 μM) inhibits cell spreading caused by 10 μM noradrenaline. These results indicate that cell spreading is related to the α_1-adrenergic potency of catecholamines. Even though the three catecholamines tested stimulate cell spreading, the morphological development of the cells is not identical: In cultures containing noradrenaline and adrenaline, cardiomyocytes develop into star-shaped cells with many long extensions (Fig. 4). Cardiomyocytes in cultures supplemented with phenylephrine spread directly, i.e., without rounding first, and become extremely flat. Phorbol esters (phorbolmyristate acetate, 10 μM) exert an effect similar to phenylephrine.

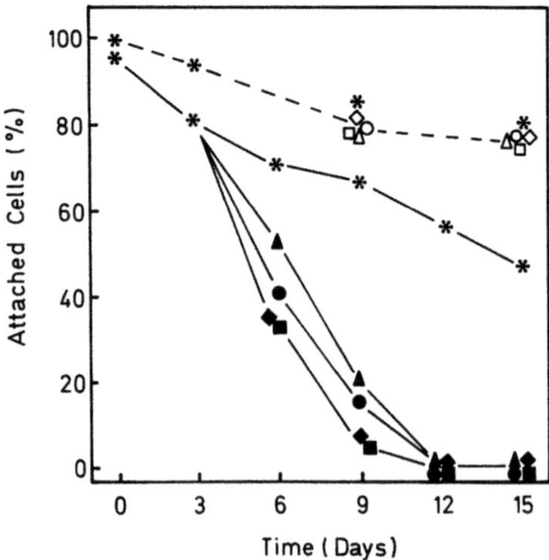

Fig. 3. Survival of cardiomyocytes in culture (*open symbols, broken lines:* of all shapes; *closed symbols, continuous lines:* of elongated shape only). Cardiomyocytes from ventricular tissue of adult rats were cultured in medium 199, supplemented with 0.2% BSA, 10^{-7} M insulin, creatine 5 mM, carnitine 2 mM, and taurine 5 mM (no further additions, *). On day 3, the following substance were added: 10 μM adrenaline (□), 10 μM noradrenaline (◇), 10 μM phenylephrine (△), 20% FCS (○). Means ± S.D., $n = 4$

In cultures where the addition of FCS, catecholamines, or phorbol esters is not made before day 3 the cellular protein content is reduced and stays reduced during the first 14 days as compared with fresh cells. In these cultures, however, the protein loss is distinctly smaller than in unstimulated control cultures [56].

After 4–6 days in cultures incubated with 10 µM concentrations of adrenaline or noradrenaline the spreading cells start beating, those incubated with phenylephrine do not. The spontaneous beating can be blocked by 10 µM propranolol, indicating that this effect is mediated through the β-receptors [56].

Cultures Supplemented with Fetal Calf Serum: When cardiomyocytes attached on FCS-pretreated plastic are incubated in medium supplemented with 20% FCS [44, 50], they retain their elongated shape only for the first 12–24 h. By the end of day 2 most cells have become spherical. On the ultrastructural level, the surface of these cells is smooth [50] unlike that of deteriorating hypercontracted cells which are covered with large blebs. By days 4 and 5, the majority of the spherical cells begin to send out finger-like extensions on the underlying surface. By day 10, in one-half of the cells a central thickening, surrounded by a polymorphically spread cell body, remains on the spherical intermediate cell form. Other cells are already uniformly spread. On day 15 and later the cells are extensively spread and, at sufficient plating density, form communicating cellular networks which beat spontaneously at a synchronous frequency. In cells maintained for 15 days or longer in such FCS-supplemented cultures, myofibrils become reorganized with a regular sarcomere structure [50].

The initial development of cardiomyocytes in FCS-supplemented cultures can be varied by changing the mode of attachment and the concentration of serum [44]. At FCS concentrations of 5% or 10%, the time course of changes is delayed as compared with 20%. When cells are attached on high concentrations of laminin, most cells do not round up before spreading, but spread directly from the attached elongated cell form [29, 44]. Direct spreading of the cells is also observed when the cells are first cultivated for at least 3 days in the absence of serum and receive FCS thereafter [56]. When cells are plated on nontreated cell culture plastic, most cells do not initially attach. They remain floating and become spherical within 24 h. But after 3–5 days these round cells also attach and start spreading. This delayed attachment only occurs in the presence of FCS. Irrespective of the mode of attachment, at the end of the second week cells cultured in the presence of 20% FCS are indistinguishable in their appearance in light and scanning microscopy [44]. Cardiomyocytes

Fig. 4. Scanning electron micrographs of single cardiomyocytes in culture. *A* Newly isolated cardiomyocyte, 4 h in basic culture medium. *B* Cardiomyocyte after 15 days, in basic culture medium. *C* Cardiomyocyte after 15 days, when 10 µM noradrenaline was present after day 3. *D* Cardiomyocyte after 15 days, when 10 µM adrenaline was present after day 3. *E* Cardiomyocyte after 15 days, when 10 µM phenylephrine was present after day 3. *F* Cardiomyocyte after 15 days, when 20% FCS was present after day 3. Composition of the basic culture medium: medium 199, supplemented with 0.2% BSA, 10^{-7} M insulin, creatine 5 mM, carnitine 2 mM, taurine 5 mM, and 0.1 mM ascorbic acid

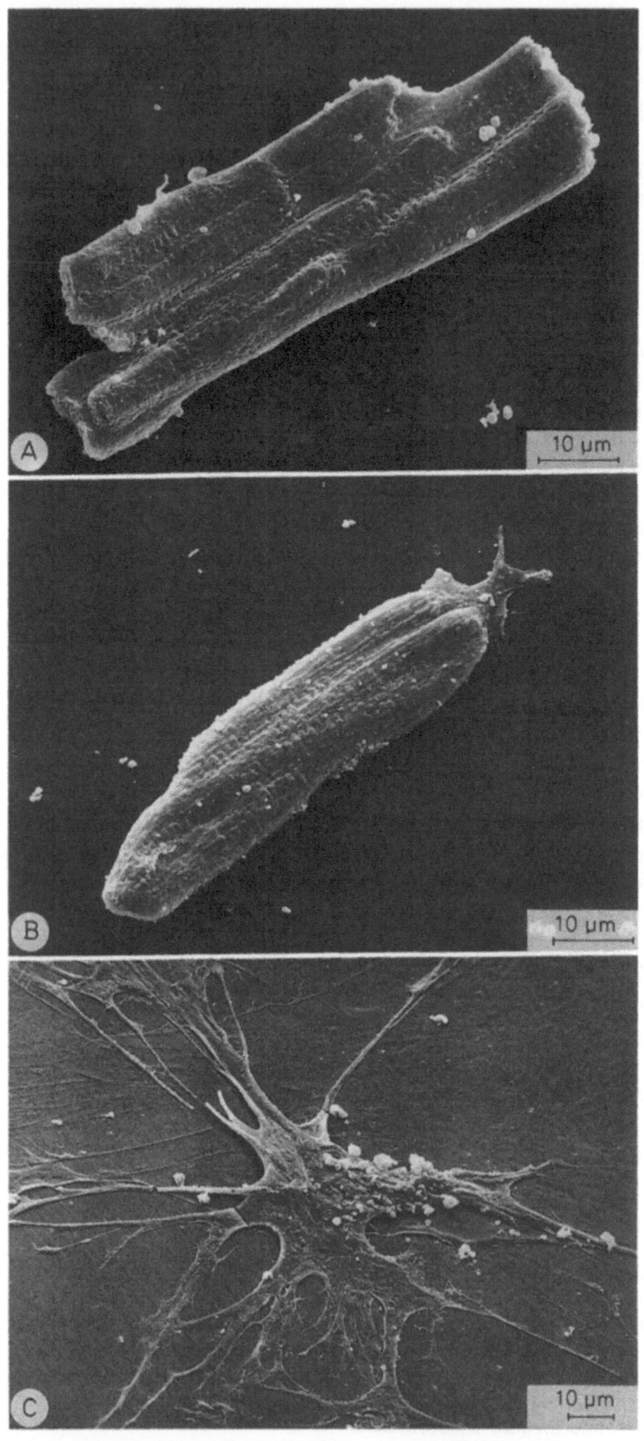

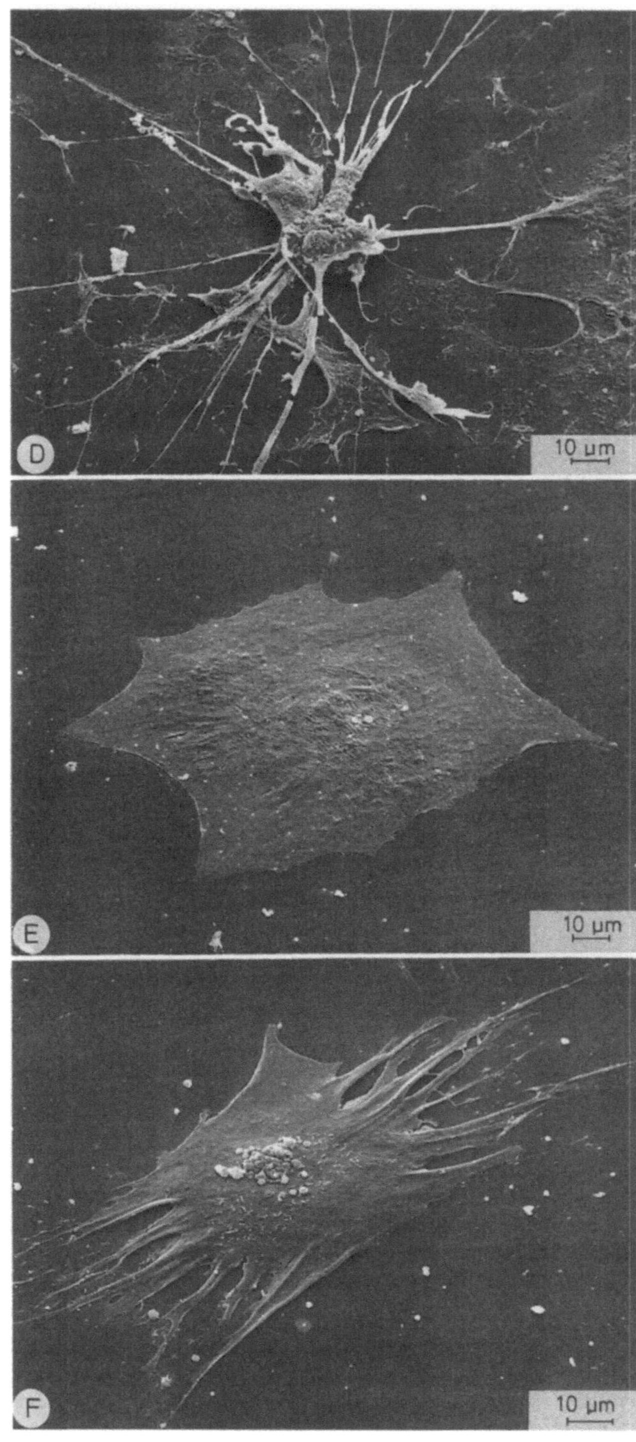
D
10 µm
E
10 µm
F
10 µm

cultured with 20% FCS from the first day double their cellular protein mass within 10 days. When 20% FCS is added to cultures not before day 3, the cellular protein contents is reduced and stays reduced as compared with fresh cells, but the protein loss is distinctly smaller than in control cultures without FCS [56].

In cultures of cardiomyocytes attached on FCS and supplemented with 20% FCS, most cells start beating 5 days after addition of serum. Where spreading cells make contacts with each other beating becomes synchronous. Spontaneous beating can be abolished by the addition of propranolol (10 µM) to the cultures [56]. This indicates that the amount of catecholamines present in FCS is sufficient to effectively stimulate β-receptors. In contrast to the effect of catecholamines, the spreading effect of FCS is apparently not related to an α-receptor stimulation by FCS since phentolamine which blocks α-receptors cannot prevent FCS-stimulated cell spreading.

Supplements Demonstrated To Be Ineffective to Prolong Cardiomyocyte Survival in Serum-Free Culture: A number of cell culture supplements have been shown to be ineffective in prolonging cell survival in serum-free cultures (Table 1). These include solute extracellular matrix proteins, which neverthe-

Table 1. Medium supplements shown to be ineffective or negatively effective for preserving the elongated shape of adult cardiomyocytes in serum-free culture

	No effect	Negative effect
Extracellular matrix proteins[a]		
Laminin	50 µg/ml	
Fibronectin	100 µg/ml	
Collagen IV	80 µg/ml	
Hormones and growth factors[a]		
Insulin	$10^{-7}\,M$	
Fibroblast GF	100 ng/ml	
Basic fibroblast GF	2 ng/ml	
Endothelial GS	150 ng/ml	
Nerve GF	50 ng/ml	
Multiplication SA	250 ng/ml	
Various[b]		
Linoleic acid	100 µM	>150 µM
Palmitic acid	100 µM	>150 µM
Glutathione	2 mM	> 5 mM
Sodium selenide	25 ng/ml	> 25 ng/ml
Fetuin		1 mg/ml
Transferrin, Fe-free		25 µg/ml
Transferrin, 30% Fe-saturated	25 µg/ml	
Transferrin, 100% Fe-saturated	25 µg/ml	

GF, growth factor; GS, growth supplement; SA, stimulating activity; palmitic and linoleic acid, fatty acids complexed in 2:1 molar ratio to bovine serum albumin
[a] This is a summary of results from [44]
[b] This is a summary of results from A. Volz (unpublished observations)

less distinctly influence cell development when used as attachment substrates. A number of hormones and growth factors tested did not prolong cell survival when given alone, but, as shown for insulin [13], they may nevertheless influence the phenotypic development of the cells in combination with other trophic factors.

Alternatives for Culturing Adult Cardiomyocytes

Cultures of the Elongated Phenotype

A short-term culture with rapid attachment of the cardiomyocytes allows recovery and purification of newly isolated cardiomyocytes. Pretreatment of tissue culture plastic dishes with dilute serum was the first method described for rapid attachment of adult cardiomyocytes which was selective for intact cells [41]. This simple and inexpensive method provided a means for preparation and maintenance of highly homogeneous populations of isolated cardiomyocytes as needed for most metabolic studies. Irrespective of the relative yield of intact cardiomyocytes in newly isolated cell material, populations of >95% elongated cells can be routinely achieved. Some factors influencing cell attachment with serum pretreatment have been reported previously [42]. The rapid attachment factor(s) contained in FCS have not yet been identified.

To date, the typical elongated in vivo cell shape for cardiomyocytes can be preserved in culture for up to 2 weeks, on FCS-coated dishes. Such a culture requires media enriched with insulin, creatine, carnitine, and taurine, but not with glutamine and FCS. Even under the most favorable conditions, however, these cultures slowly deteriorate by cell atrophy and detachment of an increasing number of cells. It is conceivable that cellular atrophy could be prevented by stretching the cells in culture, thus imitating contractile forces. It also seems possible that further beneficial additions to serum-free culture media are identified that further prolong the half-life of cardiomyocytes in elongated morphology.

The longevity of serum-free cultures crucially depends on the quality of the cell material initially isolated. Cells from isolations with a poor yield of intact cells deteriorate rapidly in serum-free cultures apparently because the cells are unable to repair major damage under these conditions. It seems questionable whether it is worth trying to "rescue" cells in such poor preparations with the use of MPG, since they may retain long-lasting deficiencies even if their morphology is improved.

Cultures of the Spread Phenotype

Cell spreading seems to be required for the establishment of cardiomyocytes in long-term culture. This may be because the isolated adult cardiomyocyte is unstable unless it spreads. Apart from catecholamines, FCS is of prime importance, relative to other factors tested, for promoting cell spreading and long-term maintenance of adult cardiomyocytes in culture. FCS can function in at least three capacities in cultures of adult cardiomyocytes: first, as a substrate for rapid attachment; second, in promoting delayed attachment of spherical cells to tissue culture plastic; and, third, in long-term maintenance of cells in culture [44].

The spread configuration of adult cardiomyocytes can be achieved either through a spherical intermediate state or as a result of spreading directly from the rod-like in vivo shape. In spread cardiomyocytes the previous cytoskeletal structures are altered but again well organized, in contrast to times when the cells assume a spherical form [16]. Therefore, spread adult cardiomyocytes have been termed "redifferentiated" in a previous review [24]. Spreading is enhanced by the presence of serum in a direct relation to its concentration.

Direct spreading is frequently observed in cardiomyocytes cultured on high laminin densities and on hepatocytes used as a living substrate [10, 49]. Cells attached on FCS also spread directly when the spreading stimulus (FCS, α-agonists) is given after the second day. On substrates to which the adult cardiomyocytes do not rapidly attach (e.g., nontreated tissue culture plastic, rat-tail collagen) the cells always become spherical before they spread. Rounding to a spherical form is also stimulated by addition of FCS or catecholamines to the culture medium before day 3. The natural extracellular matrix component laminin has been shown to be an excellent substrate for cardiomyocyte attachment [10, 17, 29, 44], but cardiomyocytes spread earlier on laminin than on FCS-coated plastic [44]. Laminin and collagen IV, which can also serve as substrates for rapid attachment, are also part of the extracellular matrix which cardiomyocytes synthesize in vitro once they have attached [54]. The great similarity of serum containing long-term cultures irrespective of the initial mode of attachment [44] is probably due to the formation of a similar extracellular matrix.

Cultured adult cardiomyocytes resemble cultured immature cardiomyocytes in their response to catecholamines. By the pattern of response of adult cardiomyocytes to subtype selective antagonists [56] and the pattern of receptor subtypes identified in binding studies [5], the α-response seems to be purely α_1 and the β-response β_1. In neonatal cardiomyocytes, α_1-receptor activation also causes spreading and β_1-receptor activation spontaneous beating [31, 52].

The hypertrophic response to α_1-activation and to phorbol esters [56] is apparently due to a common mechanism, i.e., the activation of protein kinase C [34]. The pronounced hypertrophic effect of FCS, which cannot be blocked by an α-antagonist, might be mediated by the activation of protein kinase C serum components other than catecholamines, e.g., growth factors [9].

It may be speculated that isolated cardiomyocytes have a basic tendency to hypertrophy expressed in their propensity to spread. This is because the isolation procedure stimulates the expression of the proto-oncogenes c-myc and c-fos [8] which are low in normal adult cardiac tissue and enhanced in hypertrophied myocardial cells [22, 53]. The "spreading factors," i.e., α_1-agonists, phorbol esters, and FCS seem to further enhance this process. In neonatal heart cells α_1 agonists and the phorbol ester TPA have been shown to stimulate both c-myc expression and cellular hypertrophy [53]. The spreading cardiomyocytes in culture, therefore, seem to have a number of similarities to cardiomyocytes in the hypertrophying heart. For this reason, cardiomyocyte cultures may be an interesting model to study the molecular mechanism of cardiac hypertrophy.

Spread adult cardiomyocytes possess most structural elements of the mature cell [35, 36], such as a high content of mitochondria, organized myofibrils, and cell contacts resembling the intercalated discs. They also have features which are clearly distinct from those of cardiomyocytes from the ventricular myocardium in vivo. First, the gross morphology of these cells is different from the cells in vivo. The total protein mass of spread cells cultured with FCS from the first day becomes increased compared with the newly isolated cell. Myofibrils are no longer organized in one longitudinal bundle, which makes these cells very different in their mechanical properties. Second, spread cells in cultures supplemented with serum or catecholamines are depolarized and beat spontaneously after the first week, a behavior similar to that of pacemaker cells. Third, the substrate metabolism of spread cardiomyocytes differs from that of the heart cell in vivo. In progression with time in culture the cells become increasingly glycolytic [50]. Fourth, some of the structural elements of the "redifferentiated cell" differ in their molecular nature from those of the adult cardiomyocytes. The predominant myosin isoforms (V_3) in the redifferented cells cultured with FCS are those of the embryonic heart [14, 39]. The predominant myosin isoforms of adult myocardium (V_1) can be preserved in culture by the addition of thyroid hormone [14, 39]. The example of the effect of thyroid hormone on cardiomyocyte cultures may encourage researchers to investigate even better the determinants of cardiomyocyte development in vitro. This may lead not only to better long-term in vitro models of the adult cardiomyocyte, but may also identify determinants of cardiomyocyte development in vivo under normal and pathological conditions.

Selection of a Culture Model

For many questions concerning cardiac metabolism and function, isolated cardiomyocytes represent a useful model. Cultures provide the time and a milieu in which the cells can recover from the stress of isolation. Cultures also offer the possibility for studies of extended duration. It is noteworthy, however, that cardiomyocyte cultures are not in a stable state during the first 4 weeks in culture. Their morphological and metabolic properties change. It is, therefore, advisable, whenever possible, to perform experiments always with cultures of the same age and confine experiments to a few hours. In experiments lasting for a few hours cardiomyocytes of short- and long-term cultures demonstrate the behavior of a metabolic steady state.

As a model for the myocardial cell in tissue, cultured cells with close structural resemblance to those in vivo would be the first choice, if available. For long-term maintenance of cardiomyocytes, however, the preservation of the in vivo shape is presently not possible. So far, only a few attempts have been made to study the electrophysiological and metabolic characteristics of spread cardiomyocytes in long-term cultures. Only by knowing more concerning their properties could the suitability of cardiomyocytes in long-term culture as models for the myocardial cells in tissue be evaluated.

For many experiments, short-term cultures are adequate. Under these conditions, serum-free culture on FCS-pretreated dishes seems to be the model of choice because this culture can be prepared readily and cost-effectively, it has been characterized in considerable metabolic detail, and for up to 2 weeks cardiomyocytes, under these conditions, bear close structural resemblance to the in vivo state. Even though cells are continuously lost from these cultures, those that remain apparently retain most of their typical metabolic and functional properties [24, 46], e.g., it has been shown that they maintain physiological high-energy phosphate contents [41, 56] and their insulin-receptor properties are retained [13]. These cells remain mechanically quiescent, but can be stimulated, electrically, to contract [13].

References

1. Altschuld RA, Gibb L, Ansel A, Hohl C, Kruger FA, Brierley GP (1980) Calcium tolerance of isolated rat heart cells. J Mol Cell Cardiol 12:1383–1395
2. Bailey LE, Carlos H, Amian A, Moon KET (1987) Oxidative metabolism in quinea pig ventricular myocytes protected from proteolytic enzyme activity. Cardiovasc Res 7:481–488
3. Bendukidze Z, Isenberg G, Klöckner U (1985) Ca-tolerant guinea pig ventricular myocytes as isolated by pronase in the presence of 250 μM free calcium. Basic Res Cardiol 80 [Suppl 1]:13–18
4. Berry MN, Friend DS, Scheur J (1970) Morphology and metabolism of intact muscle cells isolated from adult rat heart. Circ Res 26:679–687
5. Bode DC, Brunton LL (1989) Adrenergic, cholinergic, and other hormone receptors on cardiac myocytes. In: Piper HM, Isenberg G (eds) Isolated adult cardiomyocytes, vol 1, Structure and metabolism. CRC, Boca Raton, pp 163–202
6. Bustamante JO, Watanabe T, McDonald TF (1982) Nonspecific proteases: a new approach to the isolation of adult cardiocytes. Can J Physiol Pharmacol 60:997–1002
7. Claycomb WC, Lanson N Jr (1984) Isolation and culture of the terminally differentiated adult cardiac muscle cell. In Vitro 20:647–651
8. Claycomb WC, Lanson NA (1987) Proto-oncogene expression in proliferating and differentiating cardiac and skeletal muscle. Biochem J 247:701–706
9. Claycomb WC, Moses RL (1988) Growth factors and TPA stimulate DNA syntheses and alter the morphology of cultured terminally differentiated adult rat cardiac muscle cells. Dev Biol 127:257–265
10. Cooper G, Mercer E, Hoober JK, Gordon PR, Kent RL, Lauva IK, Marino TA (1986) Load regulation of the properties of adult feline cardiocytes. The role of substrate adhesion. Circ Res 58:692–705
11. DeMello WC (1973) Membrane sealing in frog skeletal muscle fibres. Proc Natl Acad Sci USA 70:982–984
12. Dow JW, Harding NGL, Powell T (1981) Isolated cardiac myocytes. I. Preparation of adult cardiomyocytes and their homology with the intact tissue. Cardiovasc Res 15:483–514
13. Eckel J, Van Echten G, Reinauer H (1985) Adult cardiac myocytes in primary culture: cell characteristics and insulin–receptor interaction. Am J Physiol 249:H212–H221
14. Eppenberger ME, Hauser I, Baechi T, Schaub MC, Brunner UT, Dechesne CA, Eppenberger HM (1988) Immunocytochemical analysis of the regeneration of myofibrils in long-term cultures of adult cardiomyocytes of the rat. Dev Biol 130:1–15
15. Fuchs J, Mainka L, Zimmer G (1985) 2-Mercaptopropionylglycine and related com-

pounds in treatment of mitochondrial dysfunction and postischemic myocardial damage. Drug Res 35(II):1394–1400
16. Guo JX, Jacobson SL, Brown DL (1986) Rearrangement of tubulin, actin, and myosin in cultured ventricular cardiomyocytes of the adult rat. Cell Motil Cytoskel 6:291–304
17. Haddad J, Decker ML, Hsieh LC, Lesch M, Samarel AM, Decker RS (1988) Attachment and maintenance of adult rabbit cardiac myocytes in primary cell culture. Am J Physiol 255:C19–C27
18. Haworth RA, Hunter DR, Berkoff HA (1982) Mechanism of Ca^{2+} resistance in adult heart cells isolated with trypsin plus Ca^{2+}. J Mol Cell Cardiol 14:523–530
19. Haworth RA, Goknur AB, Warner TF, Berkoff HA (1989) Some determinants of quality and yield in the isolation of adult heart cells from rat. Cell Calcium 10:57–62
20. Hoyt RH, Cohen ML, Saffitz JE (1989) Distribution and three-dimensional structures of intracellular junctions in canine myocardium. Circ Res 64:563–574
21. Isenberg G, Klöckner U (1982) Calcium tolerant ventricular myocytes prepared by preincubation in a "KB Medium". Pflügers Arch 395:6–18
22. Izumo S, Nadal-Ginard B, Mahdavi V (1988) Protooncogene induction and reprogramming of cardiac gene expression produced by pressure overload. Proc Natl Acad Sci USA 85:339–343
23. Jacobson SL (1977) Culture of spontaneously contracting myocardial cells from adult rats. Cell Struct Funct 2:1–9
24. Jacobson SL, Piper HM (1986) Cell cultures of adult cardiomyocytes as models of the myocardium. J Mol Cell Cardiol 18:439–448
25. Jacobson SL (1989) Techniques for the isolation and culture of adult cardiomyocytes. In: Piper HM, Isenberg G (eds) Isolated adult cardiomyocytes, vol. 1, Structure and metabolism. CRC, Boca Raton, pp 43–80
26. Kono T (1969) Roles of collagenases and other proteolytic enzymes in the dispersal of animal tissues. Biochem Biophys Acta 178:397–400
27. Kufe SW, Major PP, Egan EM, Beardsley GP (1980) Correlation of cytotoxicity with incorporation of ara-C into DNA. J Biol Chem 255:8997–9000
28. Lambert MR, Johnson JD, Lamka KG, Brierley GP, Altschuld RA (1986) Intracellular free Ca^{2+} and the hypercontracture of adult rat heart myocytes. Arch Biochem Biophys 245:426–435
29. Lundgren E, Borg TK, Mardh S (1984) Isolation, characterisation and adhesion of calcium-tolerant myocytes from the adult rat heart. J Mol Cell Cardiol 16:355–362
30. Maisch B (1981) Enrichment of vital adult cardiac muscle cells by continuous silica sol gradient centrifugation. Basic Res Cardiol 76:622–629
31. Meidell RS, Sen A, Henderson SA, Slahetka MF, Chien KR (1986) α_1-Adrenergic stimulation of rat myocardial cells increases protein synthesis. Am J Physiol 251:H1076–H1084
32. Mikita T, Beardsley GP (1988) Functional consequences of the arabinosylcysteine structure lesion in DNA. Biochemistry 27:4698–4705
33. Morkin E, Flink L, Goldman S (1983) Biochemical and physiological effects of thyroid hormone on cardiac performance. Progr Cardiovasc Dis 25:435–464
34. Morley SJ, Traugh JA (1989) Phorbol esters stimulate phosphorylation of eukaryotic initiation factors 3, 4B, and 4F. J Biol Chem 264:2401–2404
35. Moses RL, Claycomb WC (1982) Disorganization and reestablishment of cardiac muscle cell ultrastructure in cultured adult rat ventricular muscle cells. J Ultrastruct Res 81:358–378
36. Moses RL, Claycomb WC (1982) Ultrastructure of terminally differentiated adult rat cardiac muscle cells in culture. Am J Anat 164:113–131
37. Muir AF (1967) The effects of divalent cations on the ultrastructure of the perfused rat heart. J Anat 101:239–261
38. Nag AC, Cheng M, Fischman DA, Zak R (1983) Long-term cell culture of adult mammalian cardiomyocytes: electron microscopic and immunofluorescent analysis of myofibrillar structure. J Mol Cell Cardiol 15:301–317

39. Nag AC, Cheng M (1986) Biochemical evidence for cellular dedifferentiation in adult rat cardiac muscle cells in culture: expression of myosin isozymes. Biochem Biophys Res Commun 137:855–862
40. Overy HR, Priest RE (1966) Mitotic cell division in postnatal cardiac growth. Lab Invest 15:1100–1103
41. Piper HM, Probst I, Schwartz P, Hütter JF, Spieckermann PG (1982) Culturing of calcium stable adult cardiac myocytes. J Mol Cell Cardiol 14:397–412
42. Piper HM, Spahr R, Probst I, Spieckermann PG (1985) Substrates for the attachment of adult cardiac myocytes in culture. Basic Res Cardiol 80 [Suppl 2]:175–180
43. Piper HM, Spahr R, Schweickhardt C, Hunnenman D (1986) Importance of endogenous substrates for cultured adult cardiac myocytes. Biochim Biophys Acta 883:531–541
44. Piper HM, Jacobson SL, Schwartz P (1988) Determinants of cardiomyocyte development in long-term primary culture. J Mol Cell Cardiol 20:825–835
45. Powell T, Twist VW (1976) A rapid technique for the isolation and purification of adult cardiac muscle cells having respiratory control and a tolerance to calcium. Biochem Biophys Res Commun 72:327–333
46. Probst I, Spahr R, Piper HM (1986) Carbohydrate and fatty acid metabolism of adult cardiac myocytes maintained in short-term culture. Am J Physiol 250:H853–H860
47. Schwarzfeld TA, Jacobson SL (1981) Isolation and development in cell culture of myocardial cells of the adult rat. J Mol Cell Cardiol 13:563–575
48. Schwartz P, Piper HM, Spahr R, Hütter JF, Spieckermann PG (1985) Development of new intercellular contacts between adult cardiac myocytes in culture. Basic Res Cardiol 80 [Suppl 1]:75–78
49. Spahr R, Piper HM, Probst I, Schwartz P, Spieckermann PG (1985) Morphological dedifferentiation of adult cardiac myocytes in coculture with hepatocytes. Basic Res Cardiol 80 [Suppl 1]:83–86
50. Spahr R, Jacobson SL, Siegmund B, Schwartz P, Piper HM (1989) Substrate oxidation by adult cardiomyocytes in long-term primary culture. J Mol Cell Cardiol 21:175–185
51. Simpson P, Savion S (1982) Differentiation of rat myocytes in single cell cultures with and without proliferating nonmyocardial cells. Circ Res 50:101–116
52. Simpson P (1985) Stimulation of hypertrophy of cultured neonatal rat heart cells through an α_1-adrenergic receptor and induction of beating through an α_1- and β_1-adrenergic receptor interaction. Circ Res 56:884–894
53. Starksen NF, Simpson PC, Bishopric N, Coughlin SR, Lee WMF, Escobedo JA, Williams LT (1986) Cardiac myocyte hypertrophy is associated with c-myc protooncogene expression. Proc Natl Acad Sci USA 83:8348–8350
54. Terracio L, Lundgren E, Borg TK, Rubin K (1988) Biosynthesis and reorganization of laminin and collagen type IV in adult cardiac myocytes in vitro. In: Clark WA, Decker RS, Borg TK (eds) Biology of the isolated adult cardiac myocyte. Elsevier, New York, pp 297–300
55. Vahouny GV, Wei R, Starkeweather R, Davis C (1970) Preparation of beating heart cells from adult rats. Science 167:1616–1618
56. Volz A, Piper HM, Siegmund B, Schwartz P (1990) Longevity of adult cardiomyocetes in serum free primary culture. (Submitted for publication)
57. Weber KL, Janicki JS, Pick R, Abrahams C, Shroff SG, Bashey RI, Chen RM (1987) Collagen in the hypertrophied, pressure-overloaded myocardium. Circulation 75 [Suppl I]:40–47
58. Yokoyama HO, Jennings RB, Wartman WB (1961) Intercalated discs of dog myocardium. Exp Cell Res 23:29–44

Adult Atrial Cardiocytes in Culture *

M. Cantin

Introduction

Most of the work on culture of atrial and ventricular cardiocytes which was started 75 years ago [1–3] has been done with cells from the hearts of neonates. A study of in vitro growth of an adult animal appeared only in 1974 [4]. In several attempts, isolated cardiomyocytes from adult animals either failed to establish in culture [5–8] or produced cells lacking convincing myotypic properties [9]. Most of the published work up to now has been centered on ventricular cardiocytes, for obvious reasons. In the present review, I shall give a brief description of the techniques used to culture adult atrial cardiocytes and of the ultrastructure, DNA synthetic activity and atrial natriuretic factor (ANF) production of adult atrial cardiocytes in culture.

Techniques of Culture

Rat

Cultures can be prepared according to three methods. In the first [10], the cells are isolated subsequent to retrograde perfusion with 1 mg/ml of collagenase in Joklik medium. After the heart becomes soft, the atria are aseptically removed, minced into small pieces and subjected to sequential digestion in fresh enzyme solution in the same buffer. Dissociated cells from each digestion are recovered by brief centrifugation and plated at a density of $0.5-1.0 \times 10^5$ cells/ml in 25 cm^{-2} plastic tissue culture flasks. In order to eliminate nonmyocyte contamination, atrial cells are cultured in the absence of glutamine and in the presence of cytosine-1-β-D-arabinofuranoside. Myocytes are maintained in Eagle's minimal essential medium supplemented with preselected calf serum, vitamins and antibiotics. Cells are cultured at 37 °C in a humidified atmosphere (95% air/5% CO_2).

In the second method [11, 12] atria are excised, placed in Eagle's minimal essential medium, washed twice in Hank's solution without calcium and mag-

* This research was supported by a group grant from the Medical Research Council of Canada to the Multidisciplinary Research Group on Hypertension, by Pfizer (England), Bio-Méga Inc. (Montréal), the National Research Council of Canada and the Ministère de la Science et de la Technologie du Québec.

nesium and dissociated at 37°C in 0.1% trypsin. The cultures are enriched in cardiocytes by the selective plating technique of Kasten [8]. The dissociated cells are preplated for three 1-h periods in gelatin-coated 25-cm culture flasks to allow attachment of endothelial cells and fibroblasts. The final supernatant is plated at a density of $2 \times 10^6 - 3 \times 10^6$ cells in 6 ml of Eagle's minimal essential medium buffered with Hepes at pH 7.4 and supplemented with 10% fetal calf serum, 1% glutamine and 1% penicillin and streptomycin. The cultures are incubated at 37°C in 95% air and 5% CO_2. A third method has recently been described to prepare Ca^{2+} tolerant adult atrial and ventricular cardiocytes [13]. The procedure for isolation of cardiocytes is designed around two requirements: (1) the need to prevent tissue hypoxia during the cell isolation process and (2) avoidance of the "calcium paradox." The first problem is solved by maintaining the heart in an artificial circulating system which provides an appropriate balance of salts, vitamins, amino acids, glucose and dissolved oxygen at 37°C. Medium enters the heart through a plastic cannula placed in the aorta and exits through the vena cava. The calcium paradox is a major complication to preparations of calcium-tolerant heart cells. A calcium-free medium is required to separate the cells, but reintroduction of physiological Ca^{2+} concentrations can cause irreversible cell damage. Removal of medium Ca^{2+} from intact heart for 5 min, followed by reintroduction of 1 mM $CaCl_2$ results in marked ultrastructural damage and sarcolemmal disruption [14]. Others have observed that gradual reintroduction of Ca^{2+} to intact hearts following a period of collagenase digestion in Ca^{2+}-free medium does not cause the paradox, provided the time in Ca^{2+}-free medium is not over 20 min [15]. This has been confirmed and maintained as part of the procedure to produce cells with the least disruption of Ca^{2+} homeostasis. The yield of atrial cells is very low (10^4) so that they are not Percoll purified. But the damaged ventricular cells are separated from the viable ones by a Percoll gradient. The cell viability is assessed by morphology. The healthy atrial cells measure 75×10 μm and are trypan blue impermeant. Their ventricular counterpart has shown the right ATP/ADP ratio, the presence of receptors, intracellular Ca^{2+} stores and contractility. The cells were assessed for their utility in cell culture and electrophysiology [13].

Mouse

Cardiocyte cultures are prepared from isolated right and left atrial appendages of 50-g adult Swiss-Webster mice [16]. The animals are anesthetized with Inactin (100 ng/kg) and the hearts are quickly removed and placed into dissociation buffer (Hank's solution). Atria are cut into $1-2$ mm cubes and the cells are dispersed at 37°C with 0.25% trypsin and 0.2% EDTA in Hank's buffer. Cells from the first 60 min treatment are discarded and the procedure is repeated until all the tissue is dissociated. Freed cells are collected by centrifugation (24°C, 400 g/10 min) and plated at a density of approx. 1×10^5 cells/culture in Dulbecco's minimal essential medium, supplemented with 10% heat-inactivated fetal calf serum and 0.6 mg/ml thymidine. The cultures are incubated at 37°C in humidified air with 5% CO_2.

Guinea Pig

Guinea pigs are stunned by cervical dislocation [17]. The chest is opened and 3–5 ml of a heparin-containing solution (50 U/ml) injected into the ventricles. The heart is excised, mounted in a Langendorff apparatus and perfused at a constant flow rate (4–5 ml/min) in the "zero calcium solution" followed by a period of 30–60 min (depending on the weight of the animal) of perfusion with "enzyme solution." The right atrium is then transferred to a Petri dish (35 mm in diameter) with 2 ml of the enzyme solution, cut into 4–6 pieces and dissociated by very gently moving the pieces of tissue through the solution. The enzyme solution is then replaced by repeated washings with small volumes of culture medium. The cells are transferred into 35-mm tissue culture dishes ($\sim 10^4$ cells with 2 ml of culture medium) by means of a fire-polished wide bore Pasteur pipette. The dishes are incubated in air (37°C; 90% humidity).

Squirrel Monkey

Adult (3-year-old) female squirrel monkeys (*Saimiri sciureus*) with an average weight of 600 g are housed in individual cages and maintained on standard laboratory chow and water ad libitum [18]. The monkeys are anesthetized with Ketamine hydrochloride (30 mg i.m.) and heparinized with 500 USP units of sodium heparin. The thorax is opened, and the heart is removed and placed in 50 ml of Joklik medium [18]. The hearts are perfused with collagenase in a perfusion apparatus as described previously [19, 20]. The cells are then washed and purified as previously described [19, 20]. Atrial cells are cultured exactly as previously described [20]. They are plated at a density of 1×10^5 cells/ml in 25 cm^{-2} culture flasks in conditioned medium prepared from a rabbit corneal cell line [20]. The medium also contains 10% fetal bovine serum, vitamins, nonessential amino acids, penicillin (100 U/ml), streptomycin (100 µg/min), human transferrin (10 µg/ml), insulin (10 µg/ml) and trace minerals. The culture flasks are precoated with rat-tail collagen. To eliminate (>99%) fibroblasts and other nonmuscle cell contamination, cytosine 1-β-D-arabinofuranoside is added to each flask (14 µg/5 ml) on days 1 and 3 of cultures. After the first 7 days of culture, the cells are cultured in a medium consisting of minimal essential medium, 10% fetal bovine serum, nonessential amino acids, trace minerals, penicillin and streptomycin.

Teleost Fish

Cardiocyte cultures are prepared from the single atrium of the teleost fish *Gila atraria* [16]. Fish maintained in fresh water aquaria are anesthetized with tricaine methanesulfonate (200 mg/10 l tank water). Hearts are quickly rinsed and cells dispersed as described [16]. The dissociated cells are plated at a density of approx. 1×10^5 cells/culture. Cultures are incubated at 24°C in humidified air with 5% CO_2.

Frog

Bullfrog (*Rana catesbeiana*) atrial tissue is cleaned of as much non-cardiac cell tissue as possible and then minced into coarse fragments [21]. The fragments from one atrium are placed in 2 ml of Ca^{++}-free Ringer's digestion solution having the following composition: 111 mM NaCl, 5.4 mM KCl, 10 mM Tris (hydroxymethyl/aminomethane), 0.1% collagenase and 0.1% trypsin. The pH of the solution is adjusted to 7.0 at 25 °C by adding 12.4 N HCl. The digestion medium with tissue fragments is agitated by gentle bubbling with 95% O_2/5% CO_2, and at 30-min intervals, the supernatant is removed and replaced by 2 ml of fresh digestion solution. The supernatants from each 30-min digestion period are added to 10 ml of Ca^{2+}-free Ringer's solution without enzymes, and the heart cells are collected by gentle centrifugation. All procedures are carried out at room temperature. The cells from each 30-min harvest are resuspended in 6–10 ml of Tris-buffered Ringer's solution containing 1.8 mM $CaCl_2$, 4 mM glucose, 100 U penicillin/ml and 200 µg streptomycin/ml. Cells contained in 1–2 ml of this solution are plated into tissue culture dishes and allowed to attach.

Results

Frog

A large quantity of isolated cells can be obtained [21]. The cells are spindle shaped and have a centrally located nucleus. They range in length from 300 to 500 µm. The cell's width at the nuclear region is approx. 5 µm. The cells have a distinct, uniformly spaced sarcomere pattern. These cells give twitchlike contractions when exposed to electrical stimulation and show a transient or sustained contraction when exposed to high K-Ringer's solution.

Teleost Fish: Secretion of Atrial Natriuretic Factor

The atrial cells after 5 and 10 days of culture secrete approx. 4 ng of atrial natriuretic factor (ANF) per 1×10^5 cells whereas the amount secreted by ventricular cardiocytes in the same condition is slightly but not significantly lower [16]. Chromatographic evaluation of unextracted fish atrial and ventricular culture media by reverse phase high performance liquid chromatography (HPLC) reveals that both atrial and ventricular cells secrete immunoreactive (IR)-ANF with a retention time identical to that of authentic human ANF (Ser 99-Tyr 126). The chromatogram of media from fish atrial cardiocytes show a second peak with a longer retention time possibly representing an N-terminal extended form of ANF.

Guinea Pig

In tissue cultures, the atrial myocytes undergo a characteristic sequence of dedifferentiation [17]. The freshly isolated cells are spindle shaped, measuring between 150 and 300 µm in length and 5 to 10 µm in diameter. Within 8 to 24 h, the cells attach to the culture dish. After 2–7 days, the myocytes are completely rounded, forming spherical "cardioballs" with a diameter of 15–30 µm. The spherical stage is maintained for approx. 4–7 days. Thereafter, more and more cells become flattened. Beating atrial myocytes have been observed in culture for up to 60 days after isolation. The "cardioballs" present a single cell model for investigation of both types of autonomic transmitters and have been used to measure single ion channel current events by means of the patch-clamp technique.

Squirrel Monkey

Freshly isolated monkey atrial cardiocytes are cylindrical in shape and retain the highly organized internal ultrastructure typical of cardiac muscle cells [18]. When placed in cultures, they round up and lose much of the organized internal morphology. They attach to the surface of the culture flask, extend pseudopodlike processes and become flattened. At the ultrastructural level, the atrial cells reorganize much of the ultrastructure common to in vivo atrial cardiocytes. They possess atrial secretory granules, sarcomerically arranged myofilaments and appropriately distributed sarcoplasmic reticulum and Golgi complex. In common with cultured adult rat atrial cardiocytes [10] they possess rare, but typical elements of the transverse tubular system. When these cardiocytes are cultured in the presence of [^{3}H]thymidine (2 µCi/ml; specific activity 19 Ci/mmol) for 24 h and processed for radioautography, atrial cardiocytes were found to synthetize DNA. The majority of atrial cardiocytes are binucleated and, in those cells that incorporate [^{3}H]thymidine, both nuclei become labeled. In some cases, however, only one nucleus in binucleated cells becomes labeled.

Rat

Ultrastructure

Immediately after enzymatic dissociation, atrial muscle cells are slender and cylindrical. Within 24 h, the cells round up and assume a spherical morphology. Cardiocytes attach to the culture flasks after 48 h in culture and spread out on the growth surface after 4–5 days. The majority of cells become multipolar. At the ultrastructural level, immediately after enzymatic dissociation, most atrial cardiocytes retain myotypic ultrastructural characteristics although they rapidly undergo dissociation-induced damage and round up. Signs of dissociation-induced damage include breakdown of sarcomere organization, formation of intracellular vesicles and residual bodies, mitochondrial degeneration and formation of protrusions or blebs from the cell surface. After 4–5 days in culture, atrial cardiocytes are fully spread out. No ultrastructural difference can be found between cells cultured for 5 days or 17 days (Figs. 1–3). During this interval either one or two largely euchromatic nuclei containing one or two nucleoli are located in the central portion of each cardiocyte. Atrial secretory granules are often found in the clefts formed by nuclear lobes. Both Golgi complex and adjacent secretory granules form a discontinuous ring around the nucleus. Both structures are also present in other regions of the cytoplasm. Mitochondria are pleomorphic, ranging from irregularly spherical to long and slender. Myofilament content and organization are variable from cell to cell. Myofilament arrangements vary from well organized to almost completely disorganized. In general the most highly organized sarcomeres are located beneath the sarcolemma. Typical A, I and Z bands are present. Other cardiocytes have only partially organized sarcomeres, often with irregular Z line, while still others contain only irregular Z densities in the midst of disorganized groups of filaments. Bundles of intermediate filaments are frequently observed near Z lines. Rough endoplasmic reticulum is plentiful. Residual bodies (lipofuscin granules) and multivesicular bodies are particularly numerous along Golgi complexes but are also present in other areas of the cytoplasm. The sarcoplasmic reticulum (SR) is present as both junctional (JSR) and nonjunctional (NJSR) components. JSR is identified on the basis of close contacts between the SR and either the peripheral plasma membrane or the delimiting plasmalemma of transverse tubules. NJSR is in the form of either isolated cisternae scattered through regions of cardiocytes which do not contain myofilaments or close to sarcomere-associated elements. The most consistently appearing sarcomere-associated NJSR elements are tubules ∼40 nm wide which are located at the level of the Z line. Both internal (IJSR) and peripheral (PJSR) junctional SR (couplings) are present in these cultured atrial cardiocytes. Both PJSR and IJSR elements are separated from the plasmalemma by a 20-nm gap. IJSR elements are particularly difficult to locate due to the scarcity of transverse tubules. Transverse tubular elements are very uncommon. T-tubule lumina range from 60 to 500 nm in width. They are usually, but not exclusively located near Z lines. The exterior of the tubule membrane stains similarly to the plasmalemma. T-tubules are found associated with internal couplings. IJSR and PJSR elements are continuous with NJSR. Contiguous atrial cardiocytes are often joined by intercalated disks. Their individual junctional elements are randomly scattered along the length of the intercalated disk. Immediately subjacent to the intercalated disk plasmalemma, except in regions occupied by gap junction or peripheral couplings, is a mat of densely stained amorphous material. This mat is often continuous with distorted Z lines. Interior to the plasmalemmal densities are bundles of 10- to 12-nm filaments which are presumed to be intermediate filaments [10].

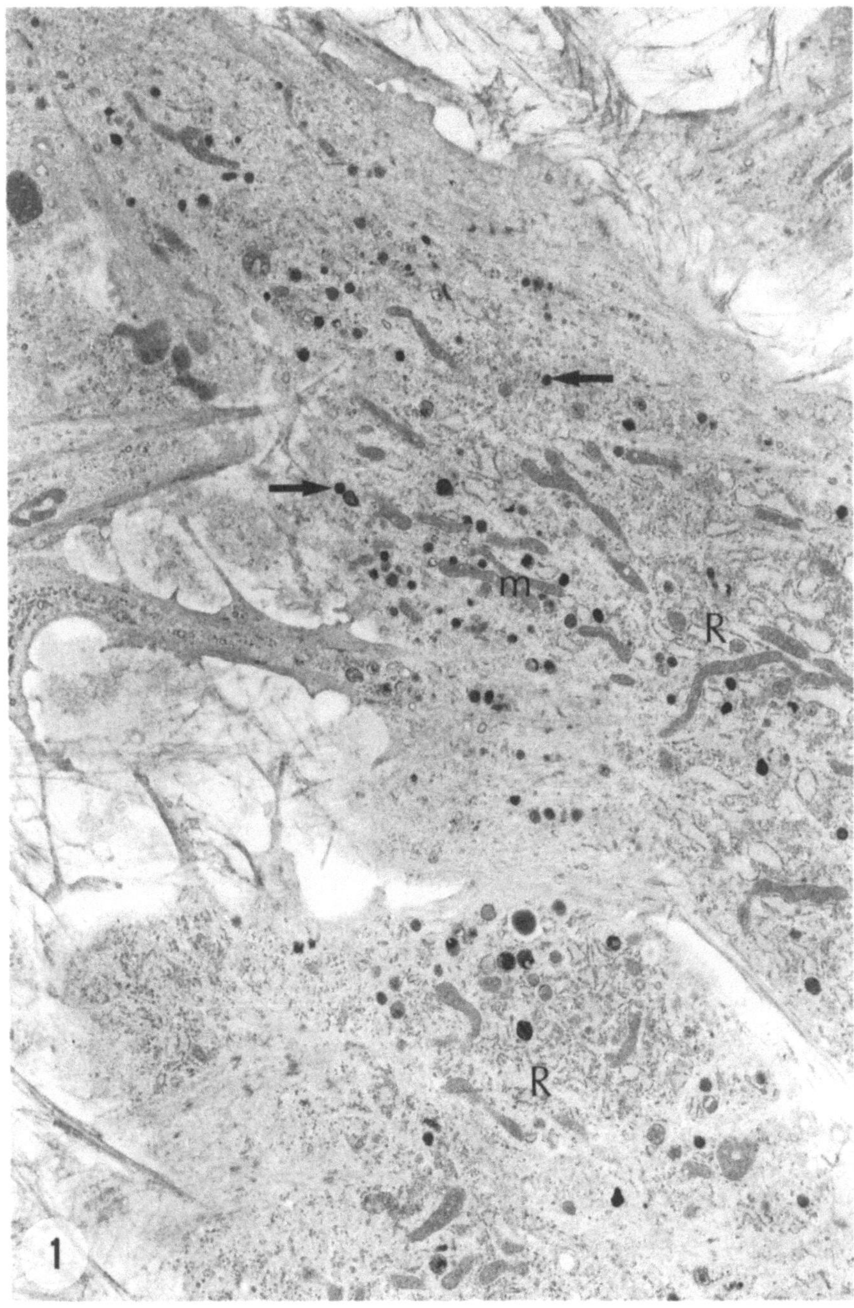

Fig. 1. Low power electron micrograph of adult rat atrial cardiocytes cultured for 12 days. Cardiocytes contain numerous secretory granules (*arrow*), large amount of rough endoplasmic reticulum (*R*), elongated mitochondria (*m*), (×5350)

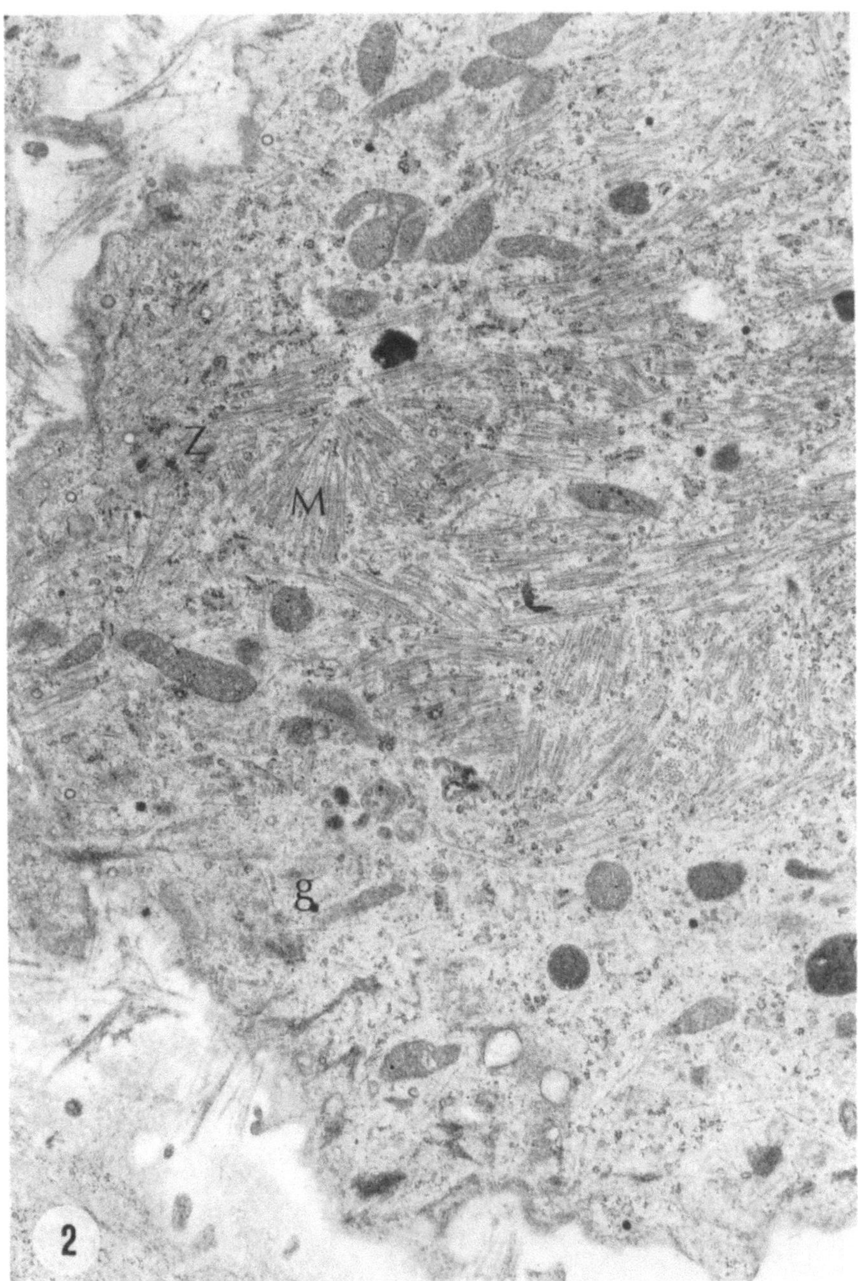

Fig. 2. Part of adult rat atrial cardiocytes cultured for 12 days. Myofilaments (M) running in several directions are present as well as abortive Z bands (Z). A few secretory granules are present (g), ($\times 12250$)

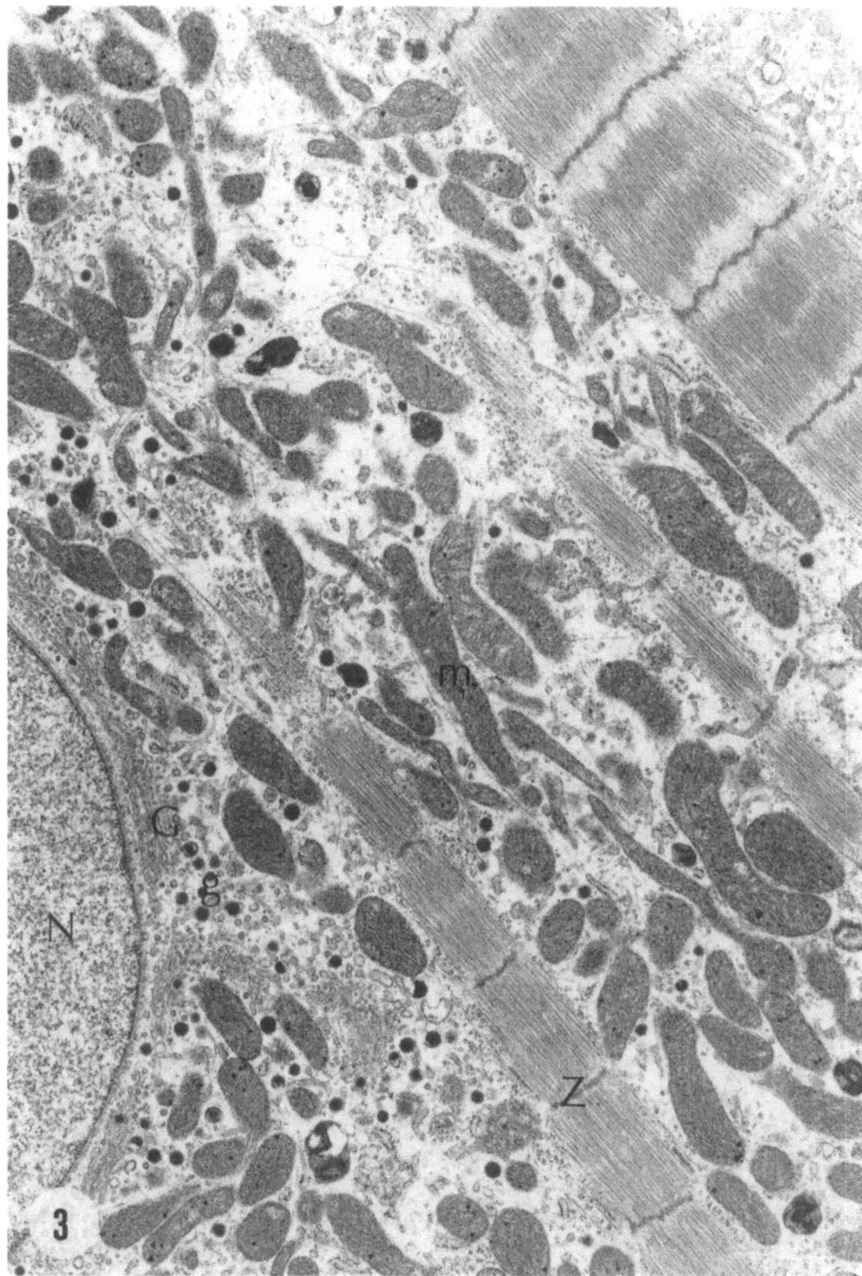

Fig. 3. Adult rat atrial cardiocytes in culture for 20 days. Note well-organized ultrastructure with nucleus (*N*), Golgi complex (*G*), secretory granules (*g*), myofilaments with Z band (*Z*) and elongated mitochondria (*m*), (×12570)

DNA Synthesis

Trypsin-dissociated atrial cardiocytes from adult (300–350 days) female rats were exposed to [3H]-thymidine (1 µCi/ml; specific activity 20.0 Ci/mmole) for sequential 24-h periods from day 2 to day 12 of culture. On day 3 and each day thereafter, cells were prepared for ultrastructural radioautography and examined with an electron microscope. Maximal incorporation occurred on day 5 when 63% of the cardiocytes were labeled. Mitotic activity was never present in more than 0.5% of the cardiocytes examined. Incorporation of [3H]-thymidine and mitoses occur only in immature cardiocytes characterized by subsarcolemmal primary filaments and Z bands with or without secretory granules. More mature cardiocytes are never labeled [11].

ANF Synthesis and Release

A comparative study has been made of the amount and form of IR-ANF present in atria versus ventricles in situ and in cultures of atrial and ventricular cardiocytes in rat of various ages from fetus to adulthood (70-day-old) [22]. The atrial content of ANF increases gradually with time and is always in the microgram range. In ventricles, where it is in the nanogram range, total ANF increases with age. Atrial cardiocytes secrete more IR-ANF than do ventricular cardiocytes at all ages investigated, the amount secreted generally decreasing with the age of the donor animal. The secretory activity of atrial cardiocytes per hour is also higher than that of their ventricular counterpart. Over this time period, both types of cells secrete more IR-ANF in the presence of serum. Cultured adult atrial cardiocytes also contain more IR-ANF than their ventricular counterpart. Typical secretory granules are decorated with gold particles when one uses the immunogold technique on Lowicryl K_4M-embedded fine sections (Fig. 4). Examination of atrial and ventricular IR-ANF by HPLC revealed that ANF is concentrated in an identical high M_r peak in both structures [22]. The amount of IR-ANF released by adult mouse atrial cardiocytes is also much higher than that released by ventricular cardiocytes [16].

ANF secretion by atrial cardiocytes in 7- to 8-day-old primary cultures from adult rats was measured by radioimmunoassay under conditions designed to separate primary effects on secretion from effects caused by contraction [23]. Abolishing contraction with 10 µm tetrodotoxin significantly reduced ANF accumulation within 2 h. Raising external Ca^{2+} concentration from 0.2 to 1.2 mM in the presence of tetrodotoxin did not increase ANF secretion. Substantial ANF secretion persisted in a nominally Ca^{2+}-free medium containing 10 mM EGTA and was not diminished by 100 µM ryanodine. In the presence of EGTA, 100 nM 12-O-tetradecanoylphorbol-13 acetate (TPA) significantly increased ANF secretion. The increment was unaffected by 100 µM ryanodine. These experiments suggest: (1) that in the absence of contraction, ANF secretion requires neither transplasmalemmal Ca^{2+} influx nor ryanodine-inhibited Ca^{2+} release by sarcoplasmic reticulum and (2) that TPA stimulates ANF secretion without requiring Ca influx or ryanodine-inhibitable sarcoplasmic reticulum Ca^{++} release.

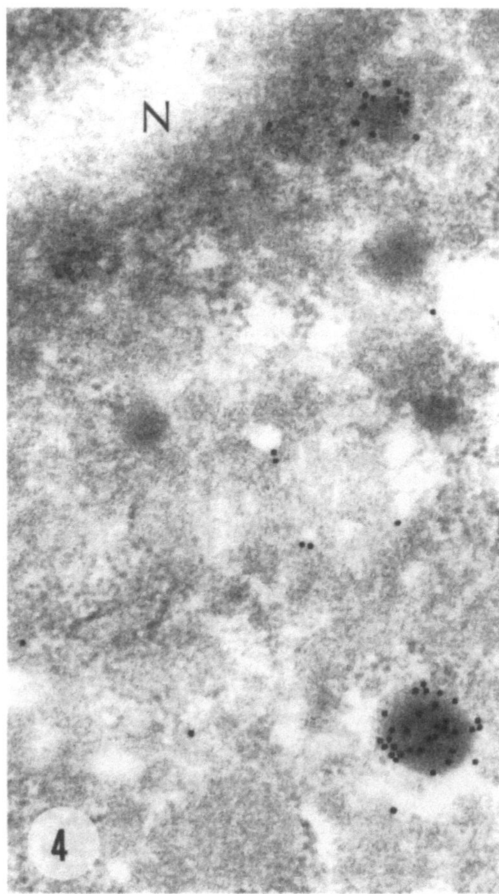

Fig. 4. Part of adult rat atrial cardiocytes cultured for 12 days, fixed with 1% glutaraldehyde and embedded in Lowicryl K_4M at $-45\,°C$. The fine section was immunostained with a primary antibody against ANF C-terminal (Ser 99-Tyr 126) and colloidal gold (15 nm)-linked anti-rabbit IgG. Nucleus (*N*). Note the presence of numerous gold particles over secretory granules indicating the presence of immunoreactive ANF in these organelles ($\times$ 68000)

The effect of monensin on microtubule-associated translocation of atrial secretory granules was studied in 5- to 7-day-old cultures of atrial cardiocytes from adult rats [24]. Secretory granules and microtubules were localized by immunofluorescence microscopy of cardiocytes double labeled with primary antibodies against ANF and α-tubulin. In control cardiocytes, fluorescence due to secretory granules is predominantly localized in the perinuclear region containing the Golgi complex. After exposure for 30 min to monensin (0.5 – 5 µm), cardiocytes transiently contain linear arrays of atrial granules associated with cytoplasmic microtubules. Thereafter, ANF fluorescence accumulates in subsarcolemmal foci at the cell periphery, while paranuclear ANF fluorescence fades. The monensin-induced redistribution of atrial secretory granules is observable in both serum-containing and serum-free media and is unaffected by reducing external Ca to low values, by inhibition of sarcoplasmic reticulum Ca^{++} release with ryanodine or by both. The redistribution is prevented by pretreatment with nocadozole which fragments microtubules and scatters Golgi complexes and the associated secretory granules throughout the cytoplasm.

Radioimmunoassay showed that monensin seemingly decreases the rate of ANF secretion into the medium from 0.15 to 0.11 fmol (h/cardiocyte). These results suggest that monensin turns on microtubules-associated translocation of secretory granules from the perinuclear area to the cell periphery by modifying the interactions between the microtubules and secretory granules. Monensin also promotes movements of secretory granules along the microtubules, but does not accelerate the release of ANF.

Conclusions

All the adult atrial cardiocytes in culture systems so far described are characterized by the transition from an initial elongate cell shape to a spherical shape. Many cells in the cultures contract spontaneously during the transition which takes several hours to days. During the gradual transition, cells lose much of their characteristic structure, but myotypy is regained during a subsequent flattening and spreading phase of cell development that follows their attachment to the culture substratum. In the cell system in which ultrastructure was studied, one can see that myotypy returns in a manner that resembles its original development during cellular differentiation in vivo. Thus, all these systems belong to the "undifferentiated model" as first described by Jacobson and Piper [25]. This state of affairs is undoubtedly due to damage induced by dissociation techniques where multiple factors can be considered: relative hypoxia, calcium intolerance and the use of enzymes which are known to penetrate within the cells during the dissociation process [26]. Great care must therefore be applied to circumvent all these potentially pathogenic agents in order to obtain intact cells following this procedure. It has already been shown that this can be done, at least for a short period of days following dissociation, for adult ventricular cardiocytes [25]. The same methodology may possibly be applied to obtain initially intact atrial cardiocytes in large numbers. Despite the initial damage, however, cells recuperate and can be used for a large number of functional studies.

DNA synthesis is now known to occur not only in adult rat atrial cardiocytes in culture [11], but also in cultured adult monkey atrial cardiocytes [18]. This is known also to occur in adult rat atrial cardiocytes in vivo following experimental production of ventricular infarction [27]. These results suggest that manipulations of cardiocytes in culture may eventually lead to the production of replicating cell lines which may find their use in the repair of myocardial damage [28].

The synthesis and release of ANF by adult rat atrial cardiocytes raise hope that this model may be used for a great number of studies on the secretory process of ANF by these cells.

References

1. Burrows MT (1916) The cultivation of tissues of the chick embryo outside the body. J Am Med Ass 55:2057–2058
2. Goss CM (1931) "Slow-motion" cinematographs of the contraction of single cardiac muscle cells. Proc Soc Exp Biol Med 29:292–293
3. Szepsenwol J (1946) A comparison of growth, differentiation, activity and action currents of heart and skeletal muscle in tissue culture. Anat Rec 95:125–146
4. Grigorev LM, Sheleunov AN (1974) Growth and secondary differentiation of explanted heart muscle of adult animals. Dokl Biol Sci 214:12–14
5. Halbert SP, Bruderer R (1971) Studies of the in vitro growth of dissociated beating rat and human heart cells. Circulation 44 [Suppl II]:174
6. Halbert SP, Bruderer R, Thompson A (1973) Growth of dissociated beating human heart cells in tissue culture. Life Sci 13:969–975
7. Harary I, Farley B (1968) In vitro studies on single beating rat heart cells. I. Growth and organization. Exp Cell Res 29:451–465
8. Kasten FH (1973) Mammalian myocardial cells. In: Kruse PT, Patterson MK Jr (eds) Tissue culture: methods and applications. Academic, New York, pp 73–81
9. Nag AG, Fishman DA, Rabinowitz M, Zak R (1974) Cell culture of adult rat cardiac muscle. J Cell Biol 63:238a
10. Moses RL, Claycomb WS (1984) Ultrastructure of cultured atrial cardiac muscle cells from adult rats. Am J Anat 171:191–206
11. Cantin M, Ballak M, Beuzeron-Mangina J, Anand-Srivastava MB, Tautu C (1981) DNA synthesis in cultured adult cardiocytes. Science 214:569–570
12. Cantin M, Tautu C, Ballak M, Yunge L, Benchimol S, Beuzeron J (1980) Ultrastructural cytochemistry of atrial muscle cells. IX. Reactivity of specific granules in cultured cardiocytes. J Mol Cell Cardiol 12:1033–1051
13. Beth M de Young, Giannattasio B, Scarpa A (1989) Isolation of calcium-tolerant atrial and ventricular myocytes from adult rat heart. Methods Enzymol 173:662–676
14. Alto LE, Dhalla NS (1979) Myocardial cation contents during induction of calcium paradox. Am J Physiol 237:H717–H719
15. Altschuld RA, Gamelin LM, Kelly RE, Lambert MB, Apel LE, Brierley GP (1987) Degradation and resynthesis of adenine nucleotides in adult rat heart myocytes. J Biol Chem 262:13527–13533
16. Baranowski RL, Westenfelder C (1989) Secretion of atrial natriuretic peptide (ANP) from fish atrial and ventricular myocytes in tissue culture. Life Sci 44:187–191
17. Bechem M, Pott L, Rennebaum H (1983) Atrial muscle cells from heart from adult guinea-pigs in culture: a new preparation for cardiac cellular electrophysiology. Eur J Cell Biol 31:366–369
18. Claycomb NC, Moses RL (1985) Culture of atrial and ventricular cardiac muscle cells from the adult squirrel monkey Saimiri Sciureus. Exp Cell Res 161:95–100
19. Claycomb NC, Palazzo MC (1980) Culture of the terminally differentiated adult cardiac muscle cell: a light and scanning electron microscopic study. Dev Biol 80:466–482
20. Claycomb NC, Lanson N Jr (1984) Isolation and culture of the terminally differentiated adult mammalian ventricular cardiac muscle cell. In Vitro 20:647–651
21. Tarr M, Trank JW (1976) Preparation of isolated single cardiac cells from adult frog atrial tissue. Experientia 32:338–340
22. Cantin M, Ding J, Thibault G, Gutkowska J, Salmi L, Garcia R, Genest J (1987) Immunoreactive atrial natriuretic factor is present in both atria and ventricles. Mol Cell Endocrinol 52:105–113
23. Iida H, Page E (1989) Determinants of atrial natriuretic peptide secretion in cultured atrial myocytes. Am J Physiol 256:C608–C613
24. Iida H, Barrow WM, Page E (1988) Monensin turns on microtubule-associated translocation of secretory granules in cultured rat atrial cardiocytes. Circ Res 62:1159–1170

25. Jacobson SL, Piper HM (1986) Cell culture of adult cardiomyocytes as models of the myocardium. J Mol Cell Cardiol 18:661–678
26. Hodges GM, Livingston DC, Franks LM (1973) The localization of trypsin in cultured mammalian cells. J Cell Sci 12:887–902
27. Rumyantsev PP (1977) Interrelation of the proliferation and differentiation processes during cardiac myogenesis and regeneration. Int Rev Cytol 51:187–273
28. Barnes DM (1988) Joint Soviet-US attack on heart muscle dogma. Science 242:193–195

Muscle Cell Cultures from Human Heart

S. L. Jacobson, R. A. Altschuld, and C. M. Hohl*

Introduction

There are few reports of work with isolated or cultured human cardiac muscle cells. With the exception of Bloom [3], early work was with cultures derived from prenatal material obtained at a variety of gestational times. At first cultures were prepared from explanted tissue; 4- to 14-week embryos [36], 8- to 12-week embryos [29], 6- to 8-week embryos [19], a 13-week embryo [11], and 7- to 10-week embryos [5]. Development of methods for isolation of single cells by treating tissue with enzymes made it possible to prepare cultures of isolated embryonic cells. Halbert et al. [14] and Thompson [35] prepared cultures of cells isolated from 13- to 20-week embryos and, similarly, Lieberman et al. [27] prepared cultures of isolated cells from 14- to 16-week embryos.

Bloom [3] attempted to isolate cells by homogenization of tissue from surgical specimens of human left atrial appendage and left ventricular myocardium. However, he was unable to obtain enough intact cells for quantitative measurements. Powell et al. [32] were able to isolate single cells from surgical specimens of right ventricle from children (4–10 years) with tetralogy of Fallot. Their technique involved simultaneous enzymatic and mechanical treatment of minced myocardium in a way similar to that used by others with embryonic tissue of experimental animals. Attempts by Powell et al. [32] to apply the same technique to the isolation of cells from papillary muscles of human *adults* undergoing mitral valve replacement were unsuccessful. Subsequently, Mitchell et al. [28] used the technique to isolate cells from surgical specimens of right ventricle from patients up to 21 years old. Although some electrophysiological measurements were made on the right ventricular cells, they were kept in a calcium ion-free Krebs Ringer solution at room temperature [32] or in a solution containing 1 mM calcium ion at 23°–28°C [28] rather than at normal physiological conditions. This may have been necessary because the cells were calcium ion intolerant, i.e., would become irreversibly hypercontracted if exposed to normal calcium ion concentrations at 37°C. Additionally, these authors did not indicate how many intact cells were ob-

* The section "Isolation of Human Ventricular Myocytes by Vascular Perfusion with Collagenase" was written by R. A. Altschuld and C. M. Hohl, Department of Physiological Chemistry, The Ohio State University, Columbus, OH 43210-1239, USA. SLJ's work was supported by a grant from the Heart and Stroke Foundation of Ontario. Work described in "Isolation of Human Ventricular Myocytes by Collagenase Vascular Perfusion" was supported in part by grants in aid from the American Heart Association, Ohio Affiliate, and USPHS-HL36240

tained with their technique. Since as little as a few tens of cells would suffice for electrophysiological measurements, it is not unlikely that cell yield was quite low.

Bustamante et al. [4] used surgical specimens of right atrium or ventricle from children (2 months to 10 years) and from adults (19 to 78 years) to prepare isolated myocytes. They report that it was much easier to disaggregate tissue from children than from adults. Apparently cells isolated by this method were calcium tolerant (2.5 mM at 35°C). They were used for intracellular recordings of electrophysiological activity. Unfortunately, neither do these authors report any data regarding cell yield.

Escande et al. [10] used a modification of the method of Bustamante et al. [4] to isolate myocytes from surgical specimens of right atrial appendage of adults (age not given). Cell yield was not stated. The cells were used for electrophysiological studies of the slow inward current. During measurements, the cells tolerated 2 mM calcium ion at *room* temperatures. No information was given about calcium tolerance at 37°C.

Glatz et al. [12] used 50-mg slices from right atrial surgical specimens of adult humans (40–70 years) to measure fatty acid oxidation. Slices of this size might contain approximately $1.5–2.0 \times 10^6$ myocytes.

Jacobson et al. [24] described the isolation and culturing of cells from surgical specimens of right atrial appendage of adult humans. The cells were observed in culture for 11 days. Subsequently (unpublished) the same method has been applied to prepare cell cultures from tissue of excised ventricles of adult heart transplant recipients. Details of this technique are presented in the Sect. "Isolating Single Cells from Small Pieces of Human Tissue."

Simpson et al. [33] reported on cultures prepared from hearts of second trimester aborted human embryos. Cells were isolated from ventricles. Cell yield was 25×10^6 cells per heart, and the culture plating efficiency was 10%.

Yang and Gao [37] report on cultures from explanted cardiac muscle derived from human embryo (10, 13, 16 weeks). They used it to study development of intercalated discs. Interestingly, they report that cells which grew out of the explants became arranged in bundles, with major axes parallel, in the manner of cells in vivo. Claycomb [6] isolated and cultured cells of human fetal hearts (14–16 weeks) obtained from a tissue bank. These cultures were used along with undissociated surgical specimens of adult human ventricle to study expression of the gene for atrial natriuretic factor.

In an abstract, Harding et al. [17] used isolated atrial myocytes to study β-receptor-mediated inotropic response of failing human hearts. Myocytes were derived from surgical specimens of right atrial appendage of 20 patients (mean age 59 years). Simultaneous enzymatic and mechanical treatment of minced tissue was used to isolate single cells, but details of the method were not given.

Most recently, Haworth et al. [18] and Hohl et al. [20] reported a method for isolation of single myocytes from excised ventricles of heart transplant recipients. This method appears, at present, to be the most efficacious way of isolating human cardiac myocytes when *large pieces* of heart tissue can be obtained. Details of such a method are presented in the Sect. "Isolation of Human Ventricular Myocytes by Collagenase Vascular Perfusion."

Methodological Principles

Sources of Human Tissue

Cardiac surgery is usually performed several times per day at hospitals in centers with populations of five hundred thousand to one million and up. Where informed consent is given by surgical patients, it is, in principle, possible to obtain tissue specimens excised in the normal course of surgery. Usually specimens are in a size range 0.1–1.0 g (see Table 1). Occasionally, where heart transplant surgery is performed, or where organ donor programs permit, large tissue specimens in the size range 5–200 g might be obtained.

Getting tissue specimens from patients undergoing surgery depends very much upon establishing a collaborative relationship with surgeon(s) and pathologist(s) who are willing to cooperate on a proposed research project. Also required are arrangements for timely notification of the laboratory that tissue will be available, as well as for receiving the tissue and transporting it to the laboratory from the hospital.

Table 1 summarizes types of surgical procedures from which tissue might be available, frequency of availability, size range of specimens, and the way that they might best be processed to obtain isolated cells. Local conditions can greatly influence the frequency with which some types of specimens become available. For example, tetralogy of Fallot (a congenital abnormality of the right ventricle) may be treated infrequently at most centers, but quite frequently at others which specialize in pediatric cardiac surgery.

Safety Considerations

Anyone working with human tissue or cells should be aware of the possibility of transmission of viral, bacterial, and fungal infections from tissue, from cells in suspension or culture, as well as from laboratory ware that has been in contact with them. High-profile possibilities include AIDS, hepatits, and tuberculosis, but there are others; and, even though most surgical patients have been well screened for infectious diseases, there is no assurance that hazardous organisms are not present.

It should be standard policy for laboratory personnel working with human tissue to wear gloves, protective cover for nose/mouth, eyes, hair, and labcoat and/or apron. Once cells are in culture, gloves and labcoat alone may provide adequate protection. Surgical soap for washing skin after the work is finished and chlorine bleach for cleaning work surfaces and spills should also be readily available. Laboratory ware should be carefully rinsed with warm water, directly into a sink drain, and subsequently submerged in boiling water for about 30 min before being washed and autoclaved for reuse.

Table 1. Tissue specimes from surgical patients

Surgical procedure or condition	Type of tissue	Relative frequency[a] of procedure	Size of specimen (g)	Probable physiological state	Recommended method of cell isolation[b]
Cardiopulmonary bypass surgery	Atrial appendage	High	0.5–1.0	Normal	TI
Cardiac biopsy	Ventricle	High	$25–50 \times 10^{-3}$	30%normal; 70% abnormal specimens	TI
Heart transplant	Recipient's ventricles	Low	5–200	Ventricles mostly disseased, some normal regions; ischemia, idiopathic cardiomyopathy, congenital abnormalities	VP
Mitral valve replacement	Papillary muscle	Medium	0.1–0.5	Mostly hypertrophied myocytes with interstitial fibrosis	TI
Correction of tetralogy of Fallot	Right ventricle	Medium – rare	0.1–0.5	Hypertrophy; some normal areas	TI
Aneurismectomy	Left ventricle	Low – medium	0.1–0.5	Areas of normal tissue at margin	TI
Hypertrophic cardiomyopathy	Interventricular septum	Low – rare	0.5–1.5	Hypertrophy, disarrary of cells	TI
Atrial tumor (mixoma)	Atrium	Low	0.1–0.5	Normal tissue at margin	TI
Unused donor heart	Atrium and ventricle	Rare	5–200	Normal	VP
Autopsy material	Atrium and ventricle	High	5–200	Probably >4 h post mortem	VP

TI, tissue immersion; VP, vascular perfusion

[a] High ≥ 1–4 times/day; medium 2–4 times/week; low 1–2 times/month; rare ≤ 1–2 times/year

[b] Refer to Sects. "Vascular Perfusion Versus Immersion Treatment, Isolating Single Cells from Small Pieces of Human Tissue, and Isolation of Human Ventricular Myocytes by Collagenase Vascular Perfusion" for details

Cell Isolation

Dow et al. [7, 8] present a comprehensive review of history and problems connected with development of techniques for isolating single cardiac myocytes from the adult heart. Regardless of the species, the isolation of heart muscle cells has been accomplished with greater ease from the hearts of prenatal and immature animals than from those of adults. Presumably this is related to the extent of development of the extracellular matrix and of intercellular junctions connecting individual cells. Eberth [9] may give the earliest account of a technique for isolating cells from adult heart. However, ways of isolating both functionally and morphologically intact cardiomyocytes from adult heart have been available for little more than a decade, and it can even be argued that this is a goal yet to be achieved.

Isolation of adult cells is, substantially, an empirical art. It is safe to say that laboratories "routinely" isolating cells from experimental animals not infrequently encounter inexplicable difficulties. Such is even more the case in preparation of cells from human adult heart, where unique problems arise: for example, (a) The age and genetic characteristics of tissue donors is essentially out of control of the investigator, and the effect of these factors is unknown. (b) Most human heart tissue is obtained from surgical patients with varying degrees of cardiovascular disease under treatment with a variety of cardioactive agents. The effect of disease and agents on cell isolation and on survival of cells in vitro is unknown. (c) The (usually) small size and unscheduled availability of human tissue samples exacerbates the difficulty of systematically testing a variety of technical conditions in cell isolation.

As might be expected, methods for isolating cells from human heart are adaptations of methods used with hearts from experimental animals. It is useful, therefore, to consider briefly some general characteristics of these methods and some problems associated with them. More comprehensive information is presented in Jacobson [22].

Methods for isolating single cardiomyocytes from adult heart have at least four features in common: (1) exposure of tissue to nominally calcium ion free-buffered saline in order to cleave desmosomal and intermediate intercellular junctions, (2) use of enzyme(s) in buffered, low-calcium-ion saline to weaken and/or remove the biochemically and morphologically complex extracellular matrix which binds cells into tissue, (3) mechanical treatment of tissue to rend intercellular gap junctions and to liberate cells from any residual connections, (4) processing of the suspension of liberated cells to remove damaged cells, cell debris, and nonmyocytes. Differential centrifugation is commonly employed to accomplish this last step.

Vascular Perfusion Versus Immersion Treatment

Cell isolation methods that have proven useful for adult human myocardium employ, at some stage, the immersion of small pieces of tissue in buffered enzyme-containing solution along with mechanical treatment such as stirring, shaking, abrading, or triturating. With small pieces of tissue, e.g., surgical specimens of atrial appendage, papillary muscle, etc., this is the primary cell-liberating treatment. For immersion treatment of adult tissue, collagenase in conjunction with a second enzyme such as trypsin must be used to obtain significant numbers of cells. Exposure of tissue to only one of these enzymes results in a *very* small yield of cells. It seems that prior or concomitant exposure to trypsin somehow primes tissue for the action of collagenase. The collagenase used is an impure preparation that contains other proteolytic enzymes. The use of purified collagenase even in conjunction with trypsin also results in a very low cell yield.

When large pieces of myocardium are available (e.g. from about 10–50 g), as from the resected heart of a transplant recipient, it may be possible to cannulate a coronary artery branch. In that case, it is advantageous to first perfuse the vasculature of the whole tissue piece with buffered saline-containing collagenase alone or, with some methods, collagenase along with another enzyme such as hyaluronidase (cf. Table 1 in [22]). Subsequent to such vascular perfusion, tissue is minced and subjected to further enzymatic and mechanical treatment by immersion. When pretreatment by vascular perfusion can be employed, it can give a considerably larger cell yield than is achieved by an immersion treatment alone.

Rapid Attachment Versus Redifferentiating Cell Culture

Two approaches to preparing cell cultures of adult cardiomyocytes have evolved. They are reviewed and discussed in detail by Jacobson and Piper [23] and by Jacobson [22]. Coating a culture substratum with an extracellular matrix protein such as laminin or collagen IV or treating it with fetal bovine serum provides a surface to which newly isolated cardiomyocytes adhere in from a few minutes to a few hours. (Such attachment substrates are the subject of another chapter, by Thomas Borg, in this volume.) On the other hand, if cells are plated on glass or on cell-culture-grade polystyrene, they remain non adherent for 1–3 days, during which time they undergo a series of well-documented degenerative morphological changes, resulting in loss of myotypic characteristics. Rapidly attached cells do not undergo such degenerative changes, *initially*; and, once cells whose attachment has been delayed become attached, they begin a regenerative process in which myotypic properties return [13]. The return of myotypy resembles, superficially, the process of differentiation during embryogenesis and therefore has come to be referred to as redifferentiation.

Some polyamino acids (e.g., polysine, polyornithine, or a commercially available polypeptide ("Cell Tak") secreted by the marine mussel *Mytilus edulis* [30] also mediate rapid attachment of cells to a substratum coated with them. However, unlike laminin or collagen IV, they are nonspecific factors which mediate, also, attachment of damaged cells, and to some extent cell debris, in addition to intact cells. Laminin, collagen IV, and fetal bovine serum-treated surfaces primarily mediate attachment only of intact cells. Once intact cells have attached, damaged cells and debris can be flushed away by changing culture medium.

Adult cardiomyocytes attached to surfaces coated by nonspecific factors show a low percentage of establishment in culture. Rapid attachment, therefore, to substrata with specific attachment factors stabilizes cell structure. Furthermore, attachment after some delay time, during which cells may synthesize their own attachment matrix [34] (as in redifferentiating cultures), appears to produce a signal which starts a regenerative process within cells.

Survival of Cells in Culture Relative to the Concentration of Serum in the Culture Medium and to the Amount of Laminin Coated on the Substratum

Cells in both rapid attachment and redifferentiating cultures become flattened and spread out by 7–14 days after plating. The speed of this process depends, in part, upon the concentration of serum in the culture medium [31]. The greater the concentration of serum, the sooner after plating the cells begin to spread. In redifferentiating culture serum must be present in culture medium *from the time of plating*. Factor(s) in serum promote the attachment of cells to the substratum, and cells which do not attach within a critical period (ca. 3–5 days) die. Rapidly attached cells can survive for some time in culture medium without serum. However, in that case, the cell population decreases rapidly, and experiments that require many cells or densely plated cells must be carried out within 24–48 h after plating. The rate of population decrease can be reduced somewhat by increasing the concentration of laminin used to coat the substrate. However, commercially available laminin is very expensive so that this approach soon becomes prohibitively expensive. It may be advantageous, for some purposes, to reduce serum concentration once cells have adhered to the substratum. However, whatever the system of culture, rapid attachment or redifferentiating, a serum concentration of 10% to 20% will be required if the cell population must be kept as high as possible for longer than a few days.

Isolating Single Cells from Small Pieces of Human Tissue

The method presented here for isolating cells from small surgical specimens of human myocardium is based upon the method of Jacobson [21] and Jacobson et al. [24]. One of the authors (SLJ) has several years of experience with it, and it usually gives good results. A potential inconvenience of the method is a requirement for a specially designed tissue dissociation vessel (TDV) in which minced tissue is exposed to enzymes and to gentle mechanical abrasion from very pliable interdigitating silicone elastomere (Dow Corning MDX-4-4210) fingers (Fig. 1). The vessel can be fabricated from detailed mechanical drawings or can be purchased at cost.[1] For purposes of producing cultures, any cell

---

[1] Drawings available from the author (S.L.J.); for information about the fabricated vessel, contact Mr. A. A. Raffler, Carleton University Science Technology Centre, Ottawa, Canada, K1S 5B6.

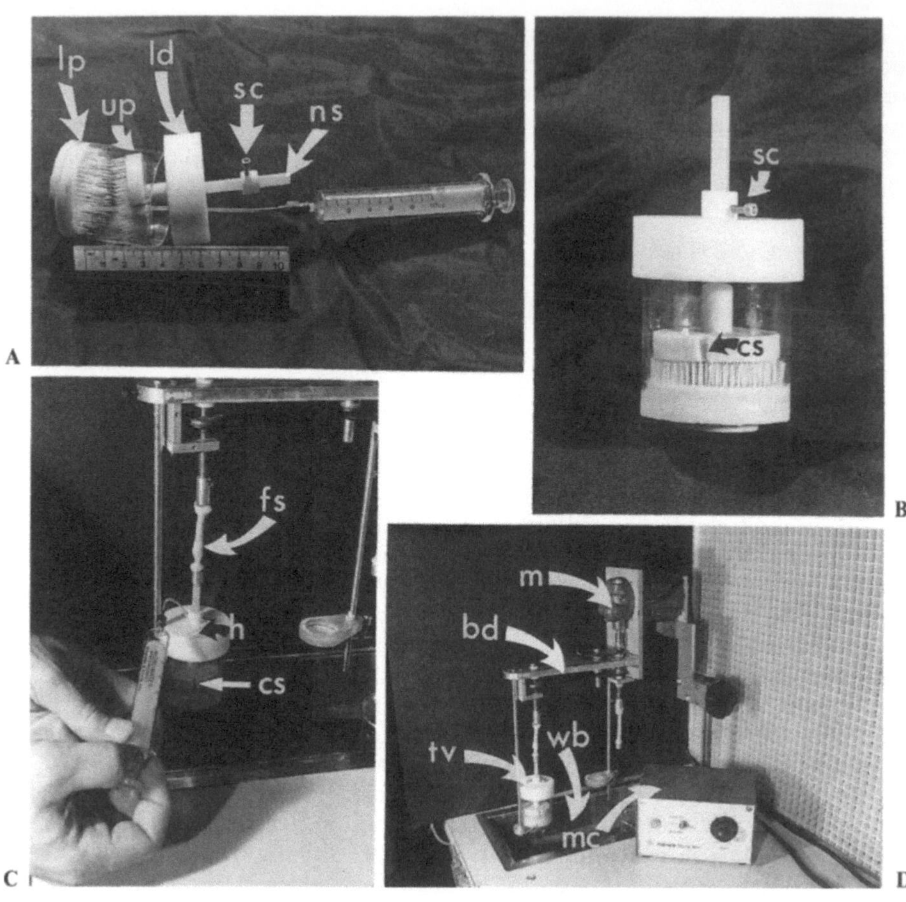

Fig. 1 A–D. The tissue dissociation vessel and motorized drive. **A** Partially disassembled vessel: *lp*, lower finger plate; *up*, upper finger plate; *ld*, Teflon lid; *sc*, setscrew collar; *ns*, nylon shaft; *syringe* and *attached tube* illustrate how vessel is filled and emptied. **B** Assembled vessel showing silicone elastomer fingers fully interdigitated. Amount of interdigitation is controlled by setting position of setscrew collar (*sc*). The cutout segment of the upper finger plate (*cs*) permits insertion of a tube to add or remove fluid from the vessel. **C** Close-up view of tissue vessel being emptied. Tube on syringe passes through hole (*h*) in Teflon lid and through cutout segment (*cs*) in upper finger plate. The shaft of the tissue vessel is driven by a flexible shaft (*fs*). **D** Overall view of apparatus to rotate tissue vessel fingers. This apparatus can accommodate two tissue dissociation vessels (TDV) and is set up in a high efficiency particulate air (HEPA) filtered work station with built-in water bath. *m*, drive motor; *bd*, belt drive to slow down rate of TDV rotation; *tv*, tissue vessel; *wb*, waterbath; *mc*, motor speed control

isolation method that results in an acceptable yield of calcium-tolerant adult cells is certainly satisfactory. Though published data make it difficult to judge which method(s) meet these criteria, it appears that the method of Bustamante et al. [4] might be a useful alternative to that described here.

Equipment Required

The equipment required for isolating cells is:

 1 Tissue dissociation vessel (TDV)
10 Luer-tip (glass) or disposable plastic syringes
10 18-gauge hypodermic needles cut to ca. 10 mm long with 80–100 mm Teflon tubes slipped over them (cut ends of needles must be smooth and free of burrs)
10 15-ml round-bottom centrifuge tubes with closures, e.g. No. 8441 Corning Glass Co.
 1 Container of ca. 15 Pasteur pipettes with 1–2 mm inside diameter (ID), firepolished tips, and rubber bulb(s), e.g. No. 1270, Bellco Glass Inc.
 2 100-mm Petri dishes (glass or plastic)
 1 100-mm bacteriological-grade plastic Petri dish
 2 150×12 mm test tubes
 1 Glass frit Buchner funnel 90 mm ID, e.g., No. 36060-600 m, Corning Glass Co.
 1 0.22 μm membrane filter and filter apparatus for sterilizing solutions
 1 150-ml Erlenmeyer flask with closure
 1 Sedgewick-Rafter cell counting chamber, e.g., No. S52, Graticules Ltd. (or a standard hemocytometer)
10 Microscope slides
 1 Source of gaseous, sterile-filtered oxygen
 1 Source of vacuum for aspiration of fluids from centrifuge tubes
 1 $35\,^\circ$C H_2O bath
 1 Centrifuge
 1 Motorized reversible variable speed (5–20 RPM) drive for TDV, e.g., Dyna-mix 43, Fisher Scientific Co.
 1 Phase contrast microscope (preferably inverted)

Note. The TDV (see Fig. 1) comes in two sizes, 25 and 50 mm in diameter. We usually use the larger size, and the method described here assumes its use. If the piece of tissue is very small, e.g., 100 mg, the smaller vessel may be advantageous. Use of the smaller vessel requires two changes: (1) that volume of solution to fill it be reduced from about 8 ml to about 2 ml and (2) that rotation speed be increased from 8 to 12 RPM. The TDV is fabricated entirely from autoclavable, nontoxic materials. The wall is glass, the finger plates are cast medical-grade silicone elastomer (Dow Corning MDX-4-4220), the lid is Teflon, and the shaft is nylon.

The TDV requires a motorized drive capable of turning at about 8 RPM. Most speed-controlled motors cannot turn that slowly. The least expensive, readily available (in North America) mechanism to turn at about 8 RPM is a motor for a barbecue rotisserie. Its speed is not adjustable, and it is not reversible. However, it could serve for initial trials with the cell isolation method described here. Other possibilities include a reversible speed-controlled electric hand drill, or a laboratory mixer with a speed control. These two latter devices require further speed reduction. We use pulleys connected by a rubber drive belt to accomplish this in our laboratory.

Materials Required

The following materials must be available for this cell-isolating procedure:

- Crude collagenase e.g. Type IA or I, Sigma Chemical Co.
- Trypsin (diphenylcarbanyl chloride, DPCC, treated), e.g. Type XI, Sigma Chemical Co.
- Bovine serum albumin fraction V, e.g. Cat. No. A9647 Sigma Chemical Co.,
- Medium 199 powder with Hank's salts and glutamine, e.g. No. 400-1200, Gibco Laboratories
  1 M CaCl$_2$ solution in culture-grade water
  N-(2 hydroxyethyl) piperazine-N'(2-ethanesulfonic acid); (HEPES)

Note: Trypsin and collagenase must be tested and chosen for their efficacy in cell isolation. It would be very unrealistic to expect material purchased without testing to be efficacious. One of the most likely causes of poor cell isolation is inferior enzyme(s). We normally order small quantities of as many lots of enzyme as a supplier(s) has available. Lots are tested, and when/if a satisfactory one is found, a supply, sufficient for approximately a year's work, is purchased. It is possible that with some lots of enzyme, better results can be achieved with somewhat higher or lower (e.g., factor of 2) concentrations. Efficacy of enzymes for cell isolation does not, however, correlate with activity measured for a specific lot.

Various solutions are needed for cell isolation studies. A summary of these is in Table 2.

Table 2. Solutions needed for cell isolation

Ingredient	BSA/199 (g/l)	Saline 1 (g/l)	Saline 2A (g/l)
NaCl	–	5.26	7.94
KCl	–	2.24	0.37
NaHCO$_3$	0.84	0.18	0.18
Glucose	–	1.00	1.00
Sucrose	–	14.36	–
HEPES	5.96	0.48	0.48
Phenol red	–	0.02	0.02
BSA	40.00	–	2.00 g
M199 power	11.00	–	–
pH 7.2–7.4 in equilibrium with	Air	Air	Oxygen

Solutions are prepared with culture-grade water. Adjust pH with 1 N NaOH or HCl. HEPES, N-(2 hydroxyethyl)piperazine-N'(2-ethanesulfonic acid); BSA, bovine serum albumin

Advance Preparation

Notice of the availability of tissue, especially from heart transplant surgery, can be quite short. It is convenient and necessary to have ready in advance as much as possible of the paraphernalia for cell isolation and culture preparation. We always keep one set of equipment sterilized and designated for use with human tissue. When notice is received that tissue will be available, the equipment is set up before going to the hospital. In addition, a supply of required sterilized solutions is kept ready in the laboratory refrigerator. All cell isolation and culture procedures are performed in a high efficiency particulate air (HEPA) filtered laminar flow work station. Solutions are sterilized by passage through 0.22-μm pore membrane filters, and equipment is sterilized in an autoclave.

Transporting the Tissue

We have experimented with a variety of sterile media for transporting tissue specimens from hospital to laboratory. These include oxygenated blood from the heart-lung machine (at 20 °C), lactated Ringer or medium 199, oxygenated and maintained at 20° or 4 °C. The outcome of cell isolation and culturing seems to be independent of transportation medium and conditions. This is perhaps due, however, to the fact that only 15–30 min are required to deliver tissue to our laboratory. If transport time must be longer, it may be useful to try to approximate the cardioplegic conditions used in Altshuld's vascular perfusion method described in "Isolation of Human Ventricular Myocytes by Collagenase Vascular Perfusion."

   If an immersion treatment is to be used to isolate cells, it will be necessary to mince the tissue once it arrives at the laboratory. Therefore, if practical, fat and connective tissue should be trimmed from the specimen at the hospital and the specimen should be cut into relatively small pieces there (e.g., 1–5 mm in size). The small size of pieces should help keep extracellular space of the sample, as nearly as possible, in equilibrium with the transport fluid.

Tissue Preparation Studies

At the laboratory tissue is transferred to a Petri dish filled with saline 1 at 20 °C. Fat and connective tissue are trimmed from the specimen if that was not done at the hospital. The specimen is thoroughly rinsed and transferred to a second dish of fresh saline 1. In the second dish the tissue is minced. We have successfully used paired razor blades, scalpels, or very sharp scissors. If available, a tissue chopper such as that designed by McIlwain (model 2340-100-2 marketed e.g., by Brinkman Canada Ltd.) can be used, which saves a considerable amount of time. Tissue is cut, by whatever method, into pieces with dimensions of about 1 mm.

Trypsin Treatment

Minced tissue is transferred to the tissue dissociation vessel (TDV) and 8 ml of saline 1, prewarmed to 35°C, containing 0.25% trypsin (w/v), are added. The vessel is closed and the elastomeric fingers are set for just-complete interdigitation by setting the position of the set-screw collar (Fig. 1). The shaft of the TDV is connected to the motorized drive, the TDV is placed in a 35°C water-bath, and the drive is adjusted to rotate at about 8 RPM.

After 7 min of treatment with trypsin, TDV rotation is stopped with the cutout segment of the rotor plate aligned with the hole in the cover. A 10-ml syringe fitted with an 18-gauge needle stub and Teflon tube is used to withdraw the enzymatic medium (zeroth harvest). The tube is introduced through the cover-hole and cutout segment. Progress in cell isolation is monitored at each harvest. A drop of fluid withdrawn from the TDV is placed on a microscope slide and is examined in phase contrast at a magnification of about 100-fold. Isolated myocytes and cell debris can readily be seen. It would be very unusual to find anything but red blood cells and some cell debris in the zeroth harvest, so this material is normally discarded.

Collagenase Treatment

Immediately after emptying the TDV, a fresh syringe with needle stub and tube is used to transfer 8 ml of prewarmed saline 1, containing 0.1% collagenase (w/v), to the TDV and treatment at 8 RPM and 35°C is resumed. After 15 min, the TDV is stopped and, using a fresh syringe, the fluid is withdrawn (first harvest). The TDV is refilled again and rotated for another 15 min.

Harvesting and refilling the TDV should be repeated at 15-min intervals about seven times, or until cell yield per harvest has decreased to the point where further work is not worth the number of additional cells obtained. Periodically (ca. 7-min intervals), the direction of rotation of the TDV is reversed to minimize the tendency of tissue to accumulate at the center or periphery of the vessel. After the third harvest, the collagenase concentration is reduced to 0.05% by adding sufficient saline 1 to accomplish a twofold dilution of the remaining enzyme solution.

Material from the first harvest is transferred to a round-bottom, 15-ml centrifuge tube and spun 2 min at 50 g and 20°C. Supernatant is aspirated and the sedimented material (often only a few cells, for the first harvest) is resuspended in 2–5 ml of saline 2A taken from a bacteriological-grade plastic Petri dish containing 20 ml of saline at 20°C. A Pasteur pipette is used to take up the saline and the cell suspension and to resuspend cells.

Cardiomyocytes are particularly subject to mechanical damage. Whenever cells in suspension are taken up or ejected through an orifice (syringe, pipette, etc.), the flow velocity must be kept as low as is practical and the orifice must be about 1 to 2 mm in diameter with smooth edges.

Resuspended material from the harvest is added to the saline 2A in the bacteriological Petri dish. The dish is kept under an inverted glass – frit Buchner funnel. A tube connected to the funnel stem delivers a steady flow of oxygen over the surface of the saline 2 A in which cells from all harvests will be pooled.

Pooled cells quickly settle to the bottom of the Petri dish, but they do not stick to the bacteriological grade plastic. At each harvest, 2–5 ml of cell-free saline 2 A is taken from the dish for resuspending harvested cells and then returned to it along with the resuspended cells.

Raising Calcium Ion to 1 mM

After the last harvest about 10 ml of saline are removed from the cell-pool dish, and sufficient 1 M CaCl$_2$ (2 µl) is added to raise the calcium ion concentration *in the 20 ml* of saline 2 A in the dish to 100 µM. The 10 ml are returned (uniformly spread) to the cell-pool dish. Pooled cells stand in 100 µM calcium ion for 5 min, and then calcium is raised again, stepwise, to 250 µM and 500 µM each time for 5 min. After 5 min in 500 µM calcium ion, by swirling and tipping the dish, pooled cells are washed to one side and are transferred with a Pasteur pipette to two 15-ml round-bottom centrifuge tubes. The pooled cells are spun for 3 min at 20 g and 20 °C. The supernatant is aspirated, and cells are resuspended in 4 ml of saline 2 A with 1 mM calcium ion.

Final Sedimentation at 1 g

BSA/199 columns are prepared in two test tubes (12 mm × 150 mm) filled to about 100 mm. The resuspended cells, in 1 mM calcium, are carefully layered on a column of BSA/199 and allowed to sediment at 1 g for 20–30 min. When 1 g sedimentation is complete, intact cells are visible in a region near the bottom of each tube. Supernatant is aspirated and cells are resuspended in ca. 5 ml of medium appropriate for use with the experimental work that is planned. Cell yield is mesured by counting an aliquot of the cell suspension. Medium is then added to bring the suspension to the required population density. Cell yields vary from very few cells to about $1-3 \times 10^5$ intact calcium-tolerant cells per gram of wet tissue. A yield of $3-8 \times 10^4$ is usual.

Isolation of Human Ventricular Myocytes by Collagenase Vascular Perfusion

Isolated adult ventricular cardiomyocytes are becoming the model of choice for examining excitation-contraction coupling in normal and failing human myocardium [16, 18, 25, 26]. These cells also offer the potential for in-depth studies of metabolic and hormonal responses of human ventricular cardiomyocytes per se, without the complications of uncertain perfusion, inadequate control of the extracellular environment, and interference from other cell types [20]. Therefore, we have developed a large-scale method for isolating viable,

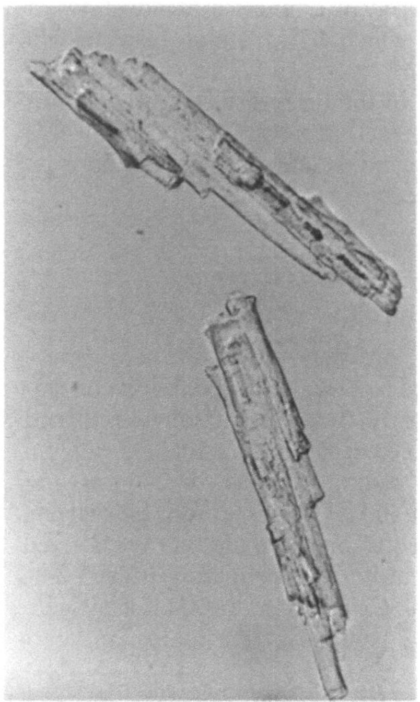

Fig. 2. Typical rod-shaped left ventricular cardiomyocytes from the explanted heart of a transplant recipient with idiopathic dilated cardiomyopathy. These cells closely resemble normal human myocytes except for having a slightly "pitted" appearance

calcium-tolerant human ventricular cardiomyocytes from the cardiectomized ventricles of transplant recipients and from consented organ donors, one whose next of kin had given informed consent, whose hearts could not be used for transplantation. These isolated human ventricular cardiomyocytes have proved suitable for both single cell experiments and for those requiring larger numbers of cells in suspension. With minor modifications, the method outlined in this section should also be applicable to the preparation of human ventricular cells for cell culture.

The solutions required in isolating human ventricular myocytes by collagenase vascular perfusion are summarized in Table 3.

Tissue Processing

Immediately after cardiectomy, the excised ventricles are placed in a plastic bag containing ice-cold saline. Each main coronary artery is located, cannulated, and flushed with $50-100$ ml of $0°-4°C$ cardioplegic solution. The heart is then placed in an ice-filled styrofoam cooler and transported to the laboratory. This cardioplegia, which is patterned after that used to preserve harvested donor hearts for transplantation, maintains myocardial ATP stores for up to 6 h [15]. However, it is preferable to use the ventricles within 30 min of cardiectomy.

Table 3. Solutions required (concentrations in unless otherwise stated) mM

Ingredient	Modified stanford cardio-plegic solution	Medium 199	Buffer A	Buffer B	Buffer C
NaCl	–	–	60	118	118
KCl	30	–	16	4.8	4.8
MgSo$_4$	–	–	3.25	1.2	1.2
NaHCo$_3$	22.4	26	25	25	20
KH$_2$Po$_4$	–	–	1.2	1.2	1.2
CaCl$_2$	–	–	–	–	1.0
Glucose	225	–	11	11	11
Mannitol	5	–	80	–	–
Pyruvate	–	–	5	5	5
Taurine	–	–	10	–	–
Adenosine	–	5 μM	5 μM	–	–
EGTA	–	–	50 μM	–	–
Insulin	–	–	150 IU/l	150 IU/l	150 IU/l
BSA[a]	–	–	1 mg/ml	1 mg/ml	20 mg/ml
Amino acids[b]	–	–	1 ml/100 ml	1 ml/100 ml	1 ml/100 ml
Collagenase[c]	–	–	1–2 mg/ml	1–2 mg/ml	–
HEPES	–	–	–	–	5
Medium 199[d]	–	11 g/l	–	–	–
Heparin	–	1.0 IU/ml	–	–	–
Penicillin-streptomycin	–	–	100 U/ml; 100 µg/ml	–	–

EGTA, ethylene glycol tetra-acetic acid; BSA, bovine serum albumin; HEPES, N-(2 hydroxyethyl)piperazine-N' (2-ethanesulfonic acid)

[a] Fraction V. Cat. No. 81-003, ICN Immuno Biologicals
[b] MEM (Cat. No. 320-1140 AG) and BME (Cat. No. 320-1052 AG) 100 × concentrated amino acids, Gibco Laboratories
[c] Types I and IV, Worthington Biochemicals
[d] Powdered culture medium with Hank's salts and glutamine, Cat. No. 400-1200, Gibco Laboratories

Cannulation

Upon arrival in the laboratory, the heart should be inspected very carefully to locate a segment of left coronary artery that has not been severed or otherwise damaged during excision. One should also look for a vessel that does not appear to be occluded by atherosclerotic plaque. Inspection of a recent coronary angiogram can be of help in this regard. A patent segment of the left coronary artery perfusing apparently viable (preferably lean) muscle tissue is then dissected out and cannulated. The vessel is perfused with the medium 199 to flush the vascular bed and define the area of perfusion. The cannula is secured with several strong sutures, the perfused area (10–20 g) dissected out, and any severed arterial vessels large enough to shunt flow are ligated.

Perfusion Apparatus

The tissue is transferred to a perfusion chamber. Any typical Langendorff apparatus that allows recirculation of the perfusion medium should suffice, provided there is a means to support the tissue segment and still allow drainage of the effluent. Systems designed for rat liver perfusion are also satisfactory. The apparatus should provide adequate temperature control so that the perfusate and tissue are held close to 37 °C. There should be parallel perfusion lines and reservoirs for switching buffers, and each perfusion reservoir should accommodate a pH electrode, an air stone or other gassing device, and the tubing necessary for recirculation of 100–150 ml perfusate. Finally, the system should be designed so that the tissue is easily accessible throughout the procedure.

Washout of Extracellular Calcium, Addition of Collagenase

The coronary vasculature is cleared of residual blood and medium 199 with non-recirculated Ca^{2+}-free buffer A, 20 ml/min, which is gassed with 95% O_2/5% CO_2 to maintain a pH of 7.3–7.4 at 37 °C. After complete washout of the calcium-containing medium 199, recirculation is begun and collagenase, 1 mg/ml types I and IV, and bovine serum albumin (BSA), 1 mg/ml, BSA fraction V, are added to the perfusate. Washout of medium 199 is taken as complete when NaOH added to an aliquot of collected effluent does not impart a pink tinge to it (due to increased pH and the presence of phenol red). Batches of commercially available collagenase are extremely variable; several lots of each type should be screened and an appropriate mixture formulated. Adult swine hearts can be used for screening purposes; these have most closely resembled human hearts in terms of the response to collagenase digestion.

Collagenase Perfusion, Filtration

Shed fat and cellular debris are removed continuously with an on-line, 142-mm in diameter Millipore sterilizing filter apparatus with an AP 20-12750 prefilter and 0.45-μm filter. Also, a bypass in the circuit to allow for changes of filters during perfusion is advantageous in the case of excessively fatty or fibrotic myocardium. During perfusion with buffer A, samples from a cut edge of ventricle should be aspirated periodically with a plastic disposable pipette and examined by low power light microscopy. At first these samples contain only a few damaged cells, but as the tissue begins to soften, clusters of dissociated and partially dissociated rod-shaped cells appear.

Addition of Calcium to Perfusion Medium

When the tissue first becomes slightly soft to the touch and aspirated samples show some separation of intercellular connections, perfusion is switched to buffer B, which is also saturated with 95% O_2/5% CO_2. Calcium is then added incrementally over several minutes until a concentration of 1 mM is attained. The rate at which calcium is restored to the perfusate is determined by the behavior of the aspirated rod-shaped cells. In the presence of vigorous spontaneous contractile activity, calcium repletion should be done very slowly and each subsequent addition made after the initial burst of activity subsides. With relatively quiescent cells, calcium repletion can be done rapidly and in two or three steps [2]. Although buffer A was formulated to inhibit sodium loading during the period of calcium-free perfusion, there is probably some increase in cytosolic sodium, as evidenced by the presence of spontaneous contractile activity [1]. During perfusion with buffer B, flow rate is gradually increased to 35–40 ml/min. If there is an unacceptable increase in perfusion pressure, 5 μM adenosine can be added to produce vasodilation.

Cell Dispersion

Perfusion is continued until there is a precipitous drop in perfusion pressure at a flow rate of 30–40 ml/min, at which point the adequately perfused portions of the tissue should have a thick mucoid consistency and a healthy red-brown color. Tissue is removed from the apparatus, and fat and poorly perfused areas are trimmed away. The softened tissue is abraded with a scalpel and the resulting slurry suspended in 25 ml perfusate supplemented with 2% BSA. It is then incubated 15–30 min in an orbital shaking waterbath at 37°C under a stream of water-saturated 95% O_2/5% CO_2. The upper layer of liberated fat is carefully decanted and discarded. Dispersed cells are filtered through cheese-cloth and centrifuged at 30 g for 1–2 min. Suspending medium is replaced and the cells are washed several times by unit-gravity sedimentation with buffer C. Again, for each centrifugation or settling step, the floating lipid layer is carefully removed and discarded. Hearts from individuals with idiopathic dilated cardiomyopathies, in particular, liberate appreciable quantities of lipid material during collagenase digestion, even when superficial fat is apparently absent from the tissue segment.

Removal of Damaged Cells

Settling and resuspension are repeated until an acceptable percentage of viable rod-shaped cells is obtained. It should be noted, however, that the yield of viable calcium-tolerant human myocytes tends to be lower and more variable than for swine or canine cells prepared by the same procedure. Age and disease-related processes seem to be a major factor. However, it has been possible to isolate approximately 2×10^7 viable calcium-tolerant cells from a normal human tissue segment with this procedure [20] and good quality cells, but in lower yield, have also been prepared from failing human hearts (Fig. 2).

Preliminary Characterization of Isolated Cells

The quality of the isolated cell preparation can be evaluated by standard cell counting techniques. A small sample of the cell suspension is treated with an equal volume of 0.3% trypan blue and 2% fresh glutaraldehyde dissolved in Krebs-Henseleit buffer and examined by bright-field light microscopy using an improved Neubauer hemocytometer. At least 400 cells are counted, and the percentage excluding trypan blue and maintaining a normally relaxed "rod-shaped" appearance are tabulated. The use of trypan blue is especially helpful since some severely failing human hearts can yield a significant number of non-viable rod-shaped cells. Metabolic integrity can be evaluated in several ways, but measurement of ATP and total adenine nucleotides is especially convenient [2]. An expected range of values for normal and failing human myocardium based on biopsy data from normal and failing human hearts has been established [15]. For normal human myocardium, ATP is in the 28–40 nmol/mg total Lowry protein range; in end-stage failure this value falls to 10–20 nmol/mg. In the absence of ischemic or preparative damage to the rod-shaped viable cells, ATP content (nmol/mg myocyte protein) divided by fractional cell viability is nearly equal to the ATP content of comparable intact myocardium [2].

Methodologic Problems

The major difficulty with the collagenase vascular perfusion method of cell isolation is the large variability among hearts with respect to the effects of collagenase. Tissue from younger individuals and those with nonischemic cardiomyopathies is digested more rapidly than the fibrotic ventricles of older individuals with ischemic heart disease. Thus, no specific time has been given for several important steps. In general, softening occurs between 30 and 60 min, with full digestion requiring an additional 30–60 min of perfusion. However, longer perfusion times have been used, and these have produced good quality myocytes. A similar problem of variable responses to collagenase digestion has been encountered with other methods.

Preparing Cell Cultures from Isolated Human Cells

Methods described here are for preparation of both rapid attachment and redifferentiating types of cultures. Matters common to both types of culture are discussed first.

General Procedures

Sterilization

Glass cover slips used as culture substrata must be sterilized. We do this by placing coverslips in a covered tray rack that is held in a dry-heat oven at

200 °C for at least 1 h. This method of dry-heat sterilization is preferred to steam sterilization for anything that comes into intimate and/or prolonged contact with cells. In our experience, sterilization in an autoclave can result in the deposition of toxic materials apparently present on the walls of the autoclave and/or in the steam. This problem is reduced, but not eliminated if materials placed in the autoclave are well wrapped.

Water Supply

A source of pure water is required for cell culture (and for cell isolation). Several filtration systems now on the market provide satisfactory water. An alternative is distillation followed by deionization, and finally, by distillation in a glass still. One can, in general, use any "pure" water, as long as cells cultured in it can be established and maintained for sufficiently long periods.

Medium Changes

Cell culture media are changed periodically. The change period depends primarily on cell population density. We change media in cultures with up to 10^5 cells per 35-mm dish, once per week or more often if the media, judging by the color of the phenol red pH indicator, appears to be less than about pH 7. When medium is changed, only about 75% of the amount in a plate is replaced. This has the advantage, as opposed to 100% exchange, of allowing some or any desirable cell product(s) that may "condition" the medium to remain. Also, where pH depends upon a pCO_2 different from atmospheric pCO_2, cells are presented with less of a pH shock than if all of the medium was removed from the plates.

Serum Supplementation of Culture Medium

Much of the time it will be necessary to supplement culture medium by adding fetal bovine serum or one of the commercially available "serum substitute" products. It should be noted that serum substitutes, although they offer better lot-to-lot reproducibility, often contain some serum as well as other hormones and growth factors. Whether serum or one of the substitutes is used, it is essential that they be tested for efficacy in cell attachment *and* maintenance of cells in one's own cultures before a quantity is purchased. It would be wrong to assume that serum or serum substitutes are of constant efficacy (from lot to lot of any one supplier) for culturing human cardiomyocytes. If cells do not do well in culture, serum is one of the most likely causes. Serum should be tested at several concentrations (e.g., 5%, 10%, 15%, 20%). If cells establish or survive less well in cultures with increased concentration of a serum, it is likely that it has toxic component(s), and the lot should be rejected.

Control of Nonmyocytes

When adult cardiomyocytes are isolated, various types of nonmyocytes are isolated as well. Myocytes do not proliferate in culture, whereas many nonmyocytes do. The result is that if cells are to be maintained in culture for more than 2–4 days, the nonmyocyte population may become the dominant cell type in culture. There are several methods for controlling the population of nonmyocytes. These are reviewed in Jacobson [22]. We normally employ the antimitotic agent cytosine β-D-arabinofuranoside (ARA-C) to control nonmyocytes in cultures of human cells. ARA-C is added to the culture medium for the first 7 days, at a concentration of 10 μM. After 7 days, medium is changed to medium without ARA-C.

Antibiotics

It is possible to prepare sterile cultures of human cardiomyocytes without using antibiotics in the culture medium. However, the excision of tissue and subsequent handling of it in the operating room cannot be depended upon to be done in a fully sterile manner. For this reason, it is advisable to use antibiotics in culture medium for at least the initial 24–48 h. We employ penicillin (50 units/ml) and streptomycin (50 µg/ml).

Equipment and Materials

The following implements are needed:

35-mm plastic cell culture dishes, e.g., Falcon 1003
22 mm × 22 mm glass microscope coverslips (pretreated as described for type of culture to be prepared)
Culture medium 199 with Hank's salts and glutamine, e.g., Gibco Laboratories 400–1200
Fetal bovine serum
Laminin (commercially available, e.g., Collaborative Research Inc.; Bethesda Research Laboratories)
Penicillin (Na salt)
Streptomycin
3% solution (w/v) HCl
1% solution (w/v) Na_2CO_3
Cytosine β-D-arabinafuranoside, e.g. Sigma Chemical Co., Cat. No. C6645
Source of high purity (culture grade) water

Preparation of Rapid Attachment and Redifferentiating Cultures

We find it convenient to culture cells on 22-mm glass coverslips in 35-mm Petri dishes. This facilitates handling cultures for morphological and electrophysiological work where cells need to be transferred from one container to another.

The description presented here presumes this method will be used; there is no reason to expect other sizes and types of materials would be unsatisfactory, but they would require testing.

Preparing the Laminin Substratum (Rapid Attachment Culture)

Except for the high cost of laminin, it would be sensible to coat the substratum as heavily as possible. We have had good results with glass coverslips and culture plastic surfaces coated by exposure to a solution of 100 µg of laminin per ml of saline 1. If desired, this concentration can be changed two-to fivefold. Increased concentration would give a higher percentage and more sustained cell attachment and vice versa. Laminin solution is applied to a culture plastic or glass surface, at 20 °C, for at least 1 h before plating cells. About 0.1 ml of solution is required per cm^2 of surface. Before coating, we routinely clean glass surfaces by soaking in 3% HCl for about 1 hour, at 20 °C. This is followed by profuse rinsing in double distilled water and drying at 100 °C. Glass surfaces are sterilized in a hot air oven before coating. Laminin coating solution is sterilized by passage through 0.22-µm pore size filter on a syringe. (See also Borg and Terracio, this volume).

Preparing a Glass Substratum (Redifferentiating Culture)

We routinely treat glass coverslips to increase the percentage of plated cells that attach in redifferentiating cultures. Treatment consists of soaking in 3% HCl (in culture-grade water) for 1–24 h, as convenient. This is followed by profuse rinsing in culture-grade water and then soaking for 1 h in a 1% sodium carbonate solution. Finally, coverslips are profusely rinsed and dried at 100 °C. They are sterilized in a hot-air oven after treatment.

Preparing Culture Medium and Plating Cells

We use culture medium 199 with Hank's salts and glutamine. It is prepared from powder as per the manufacturer's instructions. We have used, from time to time, other complete culture media and the choice of medium does not seem very important, as long as one of the commonly employed complete media is used.

   If planned experiments require within about 36 h only *most* of the cells plated, or within about 72–96 h only 5%–10% of the cells plated, supplementation of medium with fetal bovine serum may not be required. If planned experiments require longer-term survival of more cells, it will be necessary to supplement medium 199 with from 5%–20% (v/v) serum. The higher the concentration of serum, however, the more quickly cells will begin to flatten and spread in culture. Serum is required for dependable attachment of cells to culture plastic or glass that has not been coated with a cell-attachment material.

The density at which cardiomyocytes are plated depends upon experimental requirements. If all the available area of a culture surface were covered, in an orderly fashion, by a single layer of cells at the time of plating, there would be space for about 500 cells/mm². Randomly oriented cells plated at about 100/mm² would begin to show a significant amount of overlap. For rapid attachment of cells, contact with the substratum is required; and 100 cells/mm² would be a practical upper limit on plating density. Cultures plated at from 50 to 100 cells/mm² would have a very large percentage (ca 50%–80%) of their cells in contact with neighbors. For experiments requiring cells not in contact with their neighbors, a plating density of about 5–10 cells/mm² is usually satisfactory.

In our experience, plating efficiency for human cells varies widely, from about 10%–60%. Reason(s) for the variability is unknown, but it is reasonable to suspect variability in the condition of tissue specimens.

Once a plating density has been decided upon, cells are suspended in a predetermined volume of the appropriate medium. For 35-mm dishes, we normally plate 1.5–2.0 ml of cell suspension per dish. A 10-ml pipette is a convenient means of transfering cell suspension to culture plates. Cardiomyocytes, being very large, sediment rapidly at 1 g. To be sure of uniform plating density from culture plate to plate, it is necessary to (gently) agitate the cell suspension *immediately* before filling the pipette. The pipette should be held as horizontally as is practical, to minimize sedimentation of cells, and should be discharged (gently) as quickly as possible. Cultures are incubated at 37 °C in an atmosphere of 95% air/5% CO_2.

Acknowledgements. Dorothy Jacobson's invaluable help in preparing the manuscript is gratefully acknowledged. Thanks are due Michael Weber for expert technical assistance and Dr. Virginia Walley and the University of Ottawa Heart Institute for supplying human heart tissue specimens (to SLJ). The authors (RAA and CMH) thank Dr. Gerald Brierley for his continuing support, the Ohio State University Transplant Program for assistance with the hearts, and Dr. Robert Haworth for his technical advice at the inception of the project.

References

1. Altschuld RA, Gibb L, Ansel A, Hohl C, Kruger FA, Brierley GP (1980) Calcium tolerance of isolated rat heart cells. J Mol Cell Cardiol 12:1383–1395
2. Altschuld RA, Gamelin LM, Kelley RE, Lambert MR, Apel LE, Brierley GP (1987) Degradation and resynthesis of adenine nucleotides in adult rat heart myocytes. J Biol Chem 262:13527–13533
3. Bloom S (1970) Phylogenetic differences in spontaneous contractility of isolated heart muscle cells. Comp Biochem Physiol 37:127–129
4. Bustamante JO, Watanabe T, Murphy DA, McDonald TF (1982) Isolation of single atrial and ventricular cells from the human heart. Can Med Assoc J 126:791–793
5. Chang TD, Cumming GR (1972) Chronotropic responses of human heart tissue cultures. Circ Res 30:628–633
6. Claycomb WC (1988) Atrial-natriuretic-factor mRNA is developmentally regulated in heart ventricles and actively expressed in cultured ventricular cardiac muscle cells of rat and human. Biochem J 255:617–620

7. Dow JW, Harding NGL, Powell T (1981) Isolated cardiac myocytes. I. Preparation of adult myocytes and their homology with the intact tissue. Cardiovasc Res 15:483–514
8. Dow JW, Harding NGL, Powell T (1981) Isolated cardiac myocytes. II. Functional aspects of mature cells. Cardiovasc Res 15:549–579
9. Eberth CJ (1866) Die Elemente der quergestreiften Muskeln. Virchows Arch Pathol Anat Physiol 37:100–124
10. Escande D, Coulombe A, Faivre JF, Coraboeuf E (1986) Characteristics of the time-dependent slow inward current in adult human atrial single myocytes. J Mol Cell Cardiol 18:547–551
11. Garrey WE, Townsend SE (1948) Neural responses and reactions of the heart of a human embryo. Am J Physiol 152:219–224
12. Glatz JFC, Jacobs AEM, Veerkamp JH (1984) Fatty acid oxidation in human and rat heart. Comparison of cell-free and cellular systems. Biochem Biophys Acta 734:454–465
13. Guo J-X, Jacobson SL, Brown DL (1986) Rearrangement of tubulin, actin, and myosin in cultured ventricular cardiomyocytes of the adult rat. Cell Motil Cytoskeleton 6:291–304
14. Halbert SP, Bruderer R, Thompson A (1973) Growth of dissociated beating human heart cells in tissue culture. Life Sci 13:969–975
15. Hammer DF, Starling RC, Myerowitz PD, Kelley RE, Altschuld RA (1989) ATP and total adenine nucleotide depletion in failing human myocardium (Abstr). J Mol Cell Cardiol [Suppl 2] 21:S. 28
16. Harding SE, Jones SM, O'Gara PO, Poole-Wilson PA (1988) Human and rabbit ventricular cells – comparison of inotropic responses to calcium and isoprenaline. J Mol Cell Cardiol [Suppl 5] 20:S. 72
17. Harding SE, Jones SM, Poole-Wilson PA (1989) Reduced inotropic response to beta-adrenoceptor stimulation in single atrial myocytes from failing human hearts. J Mol Cell Cardiol [Suppl 2] 21:S. 51
18. Haworth RA, Goknur AB, Perkoff HA (1988) Contractile function of isolated rat and human heart cells. In: Clark WA, Decker RS, Borg TK (eds) Biology of isolated adult cardiac myocytes. Elsevier, New York, pp 386–389
19. Höfer K (1931) Gewebskulturen vom schlagenden embryonalen Menschenherzen. Med Klin 30:1107–1109
20. Hohl CM, Hu B, Wimsatt DK, Fertel RH, Starling RC, Brierley GP, Altschuld RA (1989) Isolated myocytes from a normal human heart. In: Sperelakis N (ed) Frontiers in smooth muscle research. Liss, New York
21. Jacobson SL (1977) Culture of spontaneously contracting myocardial cells from adult rats. Cell Struct Funct 2:1–9
22. Jacobson SL (1989) Techniques for isolation and culture of adult cardiomyocytes. In: Piper HM, Isenberg G (eds) Isolated adult cardiomyocytes vol 1: Structure and metabolism. CRC Press, Boca Raton, pp 44–80
23. Jacobson SL, Piper HM (1986) Cell cultures of adult cardiomyocytes as models of the myocardium. J Mol Cell Cardiol 18:661–678
24. Jacobson SL, Banfalvi M, Schwarzfeld TA (1985) Long-term primary cultures of adult human and rat cardiomyocytes. Basic Res Cardiol [Suppl 1] 80:79–82
25. Li Q, Biagi B, Hohl C, Starling R, Altschuld R (1989) Effects of isoproterenol and caffeine on calcium transients and action potentials in human ventricular cardiomyocytes. In: Sperelakis N (ed) Frontiers in smooth muscle research. Liss, New York
26. Li Q, Biagi B, Starling R, Hohl C, Altschuld R, Stokes B (1989) Characteristics of calcium transients and electrophysiology in human ventricular myocytes. Biophys J 55:488 a
27. Lieberman M, Adam WJ, Bullock PN (1980) The cultured heart cell: problems and prospects. Methods Cell Biol 21 A:187–203
28. Mitchell MR, Powell T, Sturridge MF, Terrar DA, Twist VW (1982) Action potentials and second inward current recorded from individual human ventricular muscle cells. J Physiol (Lond) 332:51–52P

29. Morosow BD (1929) Explantationsversuche an getrockneten und wiederbelebten Herzen der Menschen- und Hühnerembryonen. Arch Exp Zellforsch 8:154–160
30. Picciano PT, Benedict CV (1986) Muscle adhesive protein: a new cell attachment factor. In Vitro Cell Dev Biol 22:24A
31. Piper HM, Jacobson SL, Schwartz P (1988) Determinants of cardiomyocyte development in long-term primary culture. J Mol Cell Cardiol 20:825–835
32. Powell T, Sturridge MF, Suvarna SK, Terrar DA, Twist VW (1981) Intact individual heart cells isolated from human ventricular tissue. Br Med J 283:1013–1015
33. Simpson PC, Pulliam YD, Tsao TC (1986) Cultured fetal human heart myocytes to study viral myocarditis and cardiomyopathy (Abstr). Circulation [Suppl 2] 74:162
34. Terracio L, Lundgren E, Borg TK, Rubin K (1988) Biosynthesis and reorganization of laminin and collagen type IV in adult cardiac myocytes in vitro. In: Clark WA, Decker RS, Borg TK (eds) Biology of isolated adult cardiac myocytes. Elsevier, New York, pp 297–300
35. Thomson A (1977) The effect of diphtheria toxin on pulsating rabbit, guinea pig, human and rat heart cell cultures. J Mol Cell Cardiol 9:945–956
36. Timofejewsky A (1928) Über Tuberkuloseinfektion von Gewebskulturen. Arch Exp Zellforsch 6:230–234
37. Yang P, Gao J (1987) The primary myocardial tissue culture of human fetus: scanning electron microscopic observations on development and formation of intercalated disc in cardiac muscle cells. Biol Abstr 84:AB224

Attachment Substrates for Heart Muscle Cells*

T. K. Borg and L. Terracio

Introduction

The ability to respond to external stimulation is a fundamental property of all cells. The characteristics of the surrounding extracellular matrix (ECM) are critical to the regulation of the phenotype of the cell, which in turn is intimately associated with the particular function(s) of the cell. Changes in the chemical and physical composition of the ECM can cause altered cellular function as evidenced during the development of an organism or in progression of various disease conditions. Changes in the composition of the ECM as well as the development of mechanical force are known factors affecting embryonic and fetal development of the heart [1–3]. These changes may also be recapitulated during certain disease conditions such as cardiac hypertrophy [1, 3].

The major components of the ECM of the adult heart are known as well as the some of their morphological interactions with the cell surface [4–6]. The ECM consists of three major classes of components: (1) collagens; (2) noncollagenous glyocproteins; and (3) glycosaminoglycans and proteoglycans. The collagens consist primarily of the interstitial collagen types I and III and exist as a three-dimensional network of an epimysium, perimysium and endomysium [4, 6]. The endomysium is formed of bundles of collagen that: (1) connect adjacent myocytes; (2) connect myocytes to capillaries; and (3) forms a weave network that surrounds myocytes. These interstitial collagens exist as mixtures within the bundles of collagen which makes analysis of individual types very difficult [1]. Since the physical properties of the individual collagens are different, subtle changes in the ratio of these individual collagens may affect their function in the heart. In addition, collagen type IV, a component of the basement membrane, surrounds each myocyte and is also found on the basal portion of endothelial cells. The collagen types that have been shown to be important in cardiac cell attachment are the interstitial collagens as well as collagen type IV [5]. The appearance and function of other collagens, however, remains unclear.

While numerous glycoproteins of the ECM are exposed to the myocytes, only a few have been shown to influence attachment and progression in culture [7–10]. Attachment of cardiac myocytes to some glycoproteins such as fibronectin appears to be developmentally regulated [5]. Embryonic, fetal and

* The research described was supported in part by NIH grants HL-24935, HL-42249, HL-404424, and HL-37669.

neonatal cells readily attach to fibronectin, whereas adult cells demonstrate little adhesion (Fig. 1). The adhesion to fibronectin also shows some species specificity with the highest adhesion occurring on fibronectin isolated from the same species as the myocytes [10]. The other major glycoprotein that has been shown to influence adhesion is the basement membrane component laminin. As with the other ECM components, there appears to be some developmental regulation in the ability to recognize laminin. Adhesion assays for rat cardiac myocytes indicate that fetal cells weakly attach to laminin, whereas neonatal, normal adult and hypertrophied adult myocytes attach at the highest rates to laminin (Fig. 1) [1, 10].

The adhesion of cardiac myocytes to proteoglycans is poorly understood. Proteoglycans may regulate cell adhesion by inhibiting rather than promotion cell adhesion [11, 12]. While there is clear morphological evidence of interaction of proteoglycans with the cell surface of cardiac myocytes, their role in adhesion needs further study [13].

The regulation of adhesion to ECM components is associated with the presence of receptors for these components on the cell surface [14, 15]. While the number and type of receptors that may be involved in the regulation of attachment of cells to various substrates is not known, some specific types of ECM receptors have been described. Principally, these receptors belong to the integrin superfamily of adhesion receptors [14, 15]. These transmembrane receptors are composed of alpha and beta subunits in which a class or subfamily of receptors have a common beta subunit non-covalently linked with different alpha subunits [15]. The alpha subunits are thought to impart specificity for the different ECM components.

Investigations on cellular function of cardiac myocytes in vitro has shown that the ECM profoundly influences the structure and function of the cell [16,

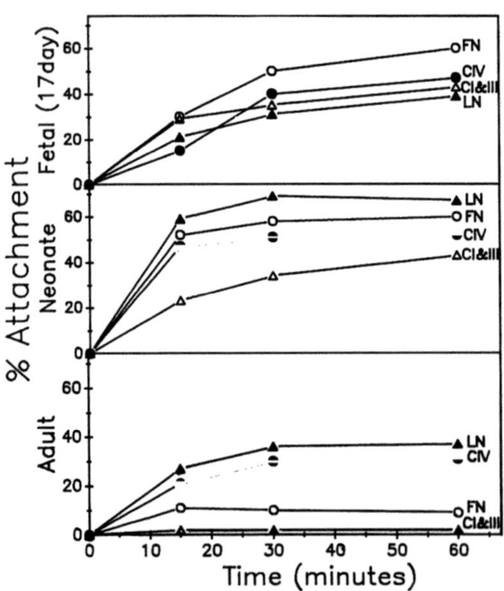

Fig. 1. The percentage of cardiac myocytes from different developmental stages attaching to different extracellular matrix components; fibronectin (*FN*), collagen type IV (*CIV*), collagen type I and III (*CI & III*) and laminin (*LN*). These data indicate a developmental differences in the ability of the cardiac myocytes to recognize various ECM components

17]. The ability to culture cardiac myocytes from different stages of development and from diseased hearts is dependent upon the ECM that the cells are exposed to during the initial stages of culture [10]. During embryonic and fetal development the isolated myocytes attach to ECM components at different rates from those isolated from either neonatal or adult hearts (Fig. 1). These data indicate that the receptors for different components may vary with development as well as in diseases such as cardiac hypertrophy.

The successful attachment of the cardiac myocytes to various types of substrates is dependent upon the method of isolation. Since most methods utilize exposure of the cells to proteases, it is highly likely that this process destroys receptors on the cell surface. The receptors for ECM components may show different susceptibilities to proteases [5]. It is important when using collagenases to include an attachment assay as part of the selection procedure in the determination of an appropriate lot of collagenase for purchase. It is the action of contaminating proteases other than collagenase that are significant in determining viability and survival in culture. These proteases tend to cleave the connection of collagen to its receptor on the cell surface as well as at the regions of cell-cell contact. The relatively slower action of collagenase attacks the helical collagen of the endomysial weave network as well as the collagenous struts. The purpose of this review will be to describe the techniques for the preparation and use of different substrates for attachment of isolated heart cells.

Methods

Substrate Preparation

ECM components can be either purchased commercially or prepared by relatively simple biochemical techniques (Table 1). If commercial substrates are used, they should be checked for purity by sodium dodecyl sulfate polycrylamide gel electrophoresis (SDS-PAGE). This procedure is particularly necessary if either laminin or collagen type IV are being used. Basic SDS-PAGE procedures can be used; however, the samples must be reduced with either

Table 1. Summary of procedures for the purification of extracellular matrix components and commercial sources for ECM components[a]

Component	Reference	Commercial source
Interstitial collagens	Obrink [21]	Collagen Corp.
Collagen type IV	Kleinman et al. [18]	Gibco/BRL, Collaborative Res.
Fibronectin	Weiss and Reddi [29]	Sigma
Laminin	Kleinman et al. [18]	Gibco/BRL
EHS gels	Kleinman et al. [24]	Gibco/BRL

[a] These are the commercial sources used by our laboratory with consistent results. This should not be interpreted as the only source of acceptable attachment factors nor an endorsement on the part of the authors. Please consult supplier list for complete address

dithiothreitol (DTT) or beta-mercaptoethanol [18]. We usually run gradient gels from 5% to 10%. For laminin, two bands at 400 and 220 kDa are readily apparent with either silver or Coomasie blue staining. For collagen type IV where samples are usually in acetic acid, they must first be neutralized with 1.0 M Tris pH 8.0. Again, two distinct bands are apparent at 185 and 175 kDa. Most commercial preparations use either placenta or the Engelbreth-Holm-Swarm (EHS)-tumor. Both of these sources are particularly high in proteases which can cause degradation during the isolation procedure. Most companies will replace "bad" lots, and a simple check by electrophoresis will determine the quality of the substrate and save time and money.

Adhesion Assays

Several procedures can be used to determine the number of cells that are attached to the substrate. The selection of the particular technique is based on whether the cells are freshly isolated or in culture. The simplest, nonlethal method of counting cells is to select five random fields with a phase contrast or differential interference microscope and count all the cells within the field to determine cell density. This method tends to show a fair degree of variability between individuals and should probably be assigned to a single person in the laboratory in order to obtain more consistent results. If the cells have adapted to culture for 7 days or more, it is difficult to separate the cell borders and thus difficult to obtain an exact count. Stains for nuclei tend to be misleading since the number of binucleate cells is variable with the preparation. Removal of the cells with trypsin and counting the number of cells in a hemocytometer or electronic counter again can also give variable results. There are several reasons for this variability, including the failure to remove all the cells from the surface and lysis of the cells. Once cells have been removed from the surface of culture dish, they cannot be replated. This is probably due to the disruption of the transmembrane receptor-cytoskeletal complex.

For acute experiments, the most consistent method has been to use an enzymatic assay for either lactate dehydrogenase or the lysosomal enzymes [10, 19, 20]. These assays obviously are lethal to the cells; yet they are the most accurate. The cells are first washed with phosphate buffered saline (PBS) then solubilized with a known amount of PBS containing 1% Triton X-100 (usually 1–2 ml depending upon the size of the culture dish). After the cells have been extracted for 15 min, the supernatant is removed and assayed, or frozen to be assayed later. The supernatants can be frozen at $-20\,°C$ for up to 3 weeks without loss of activity. Standards are constructed by adding known numbers of cells to a standard volume of the solubilization buffer. The range for the standard should be within the range expected experimentally. Usually this is between 1×10^6 and 1×10^3 cells. The number of cells is determined by regression analysis of the measured absorbance of experimental treatments compared to known standards [20, 21].

The selected references for the purification of various components are listed in Table 1. If experiments involve the culture of cells, then all solutions should be filter sterilized. This step is not necessary for acute experiments.

Regardless of the source of the substrate, the application to the culture dishes is similar. There are two basic surfaces: (1) tissue culture grade; and (2) bacteriological grade. There can be significant differences between culture dishes from different companies, and care should be taken to be consistent in the type of dish and the application of particular ECM component. When starting new experiments, culture dishes from several manufactures should be tested for adhesion after coating with a known concentration of ECM component.

Variables to be considered when coating the surface of culture dishes are: (1) concentration; (2) time of exposure of the substrate with the culture dish; and (3) temperature [22]. Attachment of cardiac myocytes is dependent on concentration of the substrate at amounts usually less than 5 µg/ml. The exception may be the interstitial collagen types I and III which show a saturation at 10 µg/ml. This has been shown to be true for all stages of isolated cardiac myocytes. Exposure of the ECM substrate to the culture surface usually requires a minimum of 30 min at 37 °C for noncollagenous substrates. Exposure for longer periods of time does not seem to alter attachment. Some laboratories coat the substrates overnight at 4 °C. It is important to note that when using *any* type of collagen to coat a planar substrate, the initial attachment should be at 4 °C to avoid denaturing the collagen. Denaturation of the collagen will in many cases cause a reduction in the number of cells attached to the substrate.

The substrate can be applied in a variety of ways. The selected concentration of the ECM component can be made in culture medium or various buffers (e.g., PBS). The medium is then added to the dishes in the minimum volume to completely cover the surface. Alternatively, the correct amount of ECM is placed directly on the culture plate followed by the addition of media or buffer. Both of these procedures result in the adhesion of the component in a uniform layer. Beyond the initial monolayer, the deposition of substrate to the culture dish appears to form multilayers. In addition, some investigators coat the culture dishes with a nonspecific protein such as 1% – 2% bovine serum albumin (BSA) to ensure that all areas of the culture plate have been coated [22]. This may be important if the purpose of the experiment is to determine the effect of ECM substrate on cell function, since it would eliminate any effect of the tissue culture plastic. However, this step also may be superfluous if concentrations above 5 µg/ml are used and the ECM component is soluble in applied vehicle to preclude aggregation of the ECM component.

At least two types of substrates will form gels and can be used as three-dimensional matrices: (1) interstitial collagen gels; and (2) basement membrane extract or Matrigel. Collagen gels can be used in any size culture dish and are easy to construct [23]. The investigator should determine if the gel should be attached or floating prior to gelation. If floating gels are desired, then culture plates should be coated with 2% BSA. For a 24-well dish, each well would receive 2 ml of 2% BSA and be incubated at 37 °C for at least 2 h. The wells would then be washed twice with sterile PBS, taking care not to scratch the plate with the washing pipette. To make collagen gels, 1 ml 10 × minimum essential medium (MEM) is mixed with 1 ml hydroxyethylpiperazine ethanesulfonic acid 0.2 *M* (HEPES) at pH 9.0. Collagen type I, 8 ml, at a concentration of 2.5 to 3.0 mg/ml of purchased (Vitrogen 100, Collagen Corp. Palo Alto,

CA) is rapidly mixed with the MEM, taking care to avoid air bubbles. Whatever procedure is chosen, a $2 \times$ cell mixture with Dulbecco's modified Eagles medium (DMEM) and serum is mixed 1:1 with the above mixture of collagen and polymerized rapidly at 37°C in an incubator. For attached collagen gels, the same procedure can be used but without precoating the culture dish with the 2% BSA.

The second type of gel that can be used for three-dimensional culture is made from an extract of basement membrane of the EHS tumor [24]. This is available commercially and is known as Matrigel (Table 1). This material will polymerize rapidly and care must be taken to keep *all* materials that contact the gel at temperatures as close to 4°C as possible. Cells can either be mixed directly with the gel or added on top of the gel. The gel will polymerize rapidly at 37°C.

Discussion

The use of isolated cells for experimentation is well documented in this volume. However, it is important to note that in vitro methodology offers a window through which we can gaze. An assumption of these techniques is that cells respond to external stimulation and all cells adapt to the culture conditions with time. With this assumption, the choice of substrate is dependent upon a number of factors, including: (1) whether the experiment is acute or chronic; (2) developmental stage of the cardiac myocytes; and (3) parameters to be measured. In acute experiments, in which the isolated cells will be used within hours after isolation, the cheapest method of adhesion would probably be sufficient. In most cases, this would be attaching the cells to 10% serum coated culture dishes [8]. In particular for acute electrophysiological experiments serum coating is enough to provide attached cells, without having to use more expensive substrates.

The developmental stage of the isolated cardiac myocytes is perhaps more critical in terms of providing the proper substrate. Published results clearly indicate that to culture adult cells either laminin or collagen type IV should be used as an attachment substrate [7–9]. Which ECM component(s) works best for neonatal, fetal, or embryonic cells is not as well established. This could be due to our lack of knowledge of the physiological state of cardiac myocytes at different developmental periods. A clear difference between various developmental stages of cardiac myocytes and adult cells is that when freshly isolated, the majority of the viable adult cells are rodshaped, while viable isolated embryonic, fetal, or neonatal cells are mostly round. The principal difference between these stages is the degree of cytoskeletal development that interconnects the forming myofibrillar components. The use of ECM substrates is believed to influence the formation of myofibrils both in vivo and in vitro via specific integrin receptors [25, 26]. The use of different substrates for different times of development may be important in altering the temporal maturation of myocytes.

Myocytes at different stages of development are capable of synthesizing ECM components [27]. To date, only adult and neonatal cells have been investigated for their ability to regenerate their basement membrane [27]. These investigations indicate that laminin and collagen type IV are synthesized and deposited at different rates and in distinctive patterns following culture. Adult cardiac myocytes synthesize both laminin and collagen type IV over the time they spread and adapt to culture, with synthesis plateauing by 7 days in vitro. Neonatal cells also synthesize these basement membrane components while adapting to culture; however, the synthesis data indicates a maximum level by 24 h in vitro. While laminin is still present on freshly isolated myocytes, the collagenase digestion removes the collagen type IV as judged by immunofluorescence with anticollagen type IV antibodies. Data on embryonic and fetal cells will provide important clues on the maturation on cardiac myocytes. Another difference between the developing myocytes and the terminally differentiated adult cells is their ability to migrate on the substrate. Certainly embryonic, fetal, and, to a lesser extent, neonatal cells migrate in vivo and also move on cultured plastic. Which substrates promote adhesion and which promote migration is not known, this, however, is an important area for future investigation.

Three-dimensional culture systems that have been used successfully to demonstrate cell polarity in epithelial cells, have not been extensively used to investigate heart cells. While the three-dimensional cultures have great promise, the required parameters for culture do not appear to be well established. Clearly there are limitations using two-dimensional cultures that are now just becoming known. These include the artificial up or down-regulation of receptor expression by hormones, elevated mRNA expression and protein synthesis rates, and altered phenotype [24, 28]. The changes in cell shape would dramatically affect the functions of the cells, including ion channels, contractility, etc. While these might represent useful parameters to measure, their relation to in vivo function may be tenuous.

In summary, it appears to be important to provide cardiac myocytes with appropriate signals from the extracellular matrix that correspond to particular developmental stages or conditions from which they were isolated. Most cardiac myocytes readily adapt to the in vitro conditions and have the capability of synthesizing the ECM components which are essential in maintenance of viability and survival in vitro. However, for the initial stages of the adaption to culture, it is as important to supply them with the appropriate ECM to provide a proper substrate for attachment. While all the parameters and information that cells derive from the ECM are not known, it is apparent that the ECM does contribute to the homeostasis of the individual cells.

Acknowledgements. The authors wish to thank Dr. Kristofer Rubin for his valuable discussion and criticism on a variety of subjects including those in this chapter.

References

1. Borg TK, Terracio L (1989) Interaction of the extracellular matrix with cardiac myocytes during development and disease. In: Kinne R, Stolte H (eds) Issues in biomedicine: cardiac myocyte–connective tissue interactions in health and disease, Vol. 13. Karger, Basel
2. Terracio L, Tingstrom A, Peters WH, Borg TK (1990) A potential role for mechanical

stimulation in cardiac development. In: Bockman DE, Kirby ML (eds) Embryonic origins of defective heart development. Ann NY Acad Sci, NY, NY

3. Terracio L, Borg TK (1988) Factors affecting cardiac cell shape. Heart Failure 4:114–124
4. Caulfield JB, Borg TK (1979) The collagen network of the heart. Lab Invest 40:354–371
5. Borg TK, Rubin K, Lundgren E, Borg K. Obrink B (1984) Recognition of extracellular matrix components by neonatal and adult cardiac myocytes. Dev Biol 104:86–96
6. Robinson TF, Factor SM, Capasso JM, Wittenberg BA, Blumenfeld OO, Seifter S (1987) Morphology, composition, and function of struts between cardiac myocytes of rat and hamster. Cell Tissue Res 249:247–255
7. Lundgren E, Terracio L, Mardh S, Borg TK (1985) Extracellular matrix components influence the survival of adult cardiac myocytes in vitro. Exp Cell Res 158:371–381
8. Piper HM, Jacobson SL, Schwartz P (1988) Determinants of cardiomyocyte development in long-term primary culture. J Mol Cell Cardiol 20:825–835
9. Haddad J, Decker LM, Hsieh L, Lesch M, Samarel AM, Decker RS (1988) Attachment and maintenance of adult rabbit cardiac myocytes in primary cell culture. Am J Physiol 255:C19–C27
10. Borg TK, Terracio L (1988) Cellular adhesion to artificial substrates and long term culture of adult cardiac myocytes. In: Clark WA, Decker RS, Borg TK (eds) Biology of isolated adult cardiac myocytes. Elsevier, New York
11. Heinegard D, Sommarin Y (1987) Isolation and characterization of proteoglycans. Methods Enzymol 144:319–371
12. Ruoslahti E (1988) Structure and biology of proteoglycans. Ann Rev Cell Biol 4:229–257
13. Funderburg FM, Markwald RR (1987) Conditioning of native substrates by chondroitin sulfate proteoglycans during cardiac mesenchymal cell migration. J Cell Biol 103:2475–2487
14. Hynes RO (1987) Integrins: a family of cell surface receptors. Cell 48:549–554
15. Ruoslahti E, Pierschbacher MD (1987) New perspectives in cell adhesion: RGD and integrins. Science 238:491–497
16. Borg TK, Xuehei M, Hilenski L, Vinson N, Terracio L (1990) The role of the extracellular matrix on myofibrillogenesis in vitro. In: Takao A (ed) Etiology and morphogenesis of congenital heart disease. Aseni, Tokyo
17. Hilenski LL, Terracio L, Sawyer R, Borg TK (1989) Effects of extracellular matrix on cytoskeletal and myofibrillar organization in vitro. Scanning Microsc 3:535–548
18. Kleinman HK, McGarvey ML, Liotta LA, Robey PG, Tryggvason K, Martin GR (1982) Isolation and characterization of type IV procollagen, laminin, and heparan sulfate proteoglycan from the EHS sarcoma. Biochemistry 24:6188–6193
19. Lundgren E, Borg TK, Mardh S (1983) Isolation, characterization and adhesion of calcium tolerant myocytes from adult rat hearts. J Cell Mol Cardiol 6:355–362
20. Landegren U (1984) Measurement of cell numbers by means of the endogenous enzyme hexosaminidase. Applications to detection of lymphokines and cell surface antigens. J Immunol Methods 67:379–388
21. Obrink B (1982) Hepatocyte-collagen adhesion. Methods Enzymol 82:513–529
22. Lundgren E, Terracio L, Borg TK (1985) Adhesion of cardiac myocytes to extracellular matrix components. Basic Res Cardiol 80:69–74
23. Elsdale T, Bard J (1972) Collagen substrate for studies on cell behavior. J Cell Biol 54:626–637
24. Kleinman HK, McGarvey ML, Hassell JR, Star VL, Cannon FB, Laurie GW, Martin GR (1986) Basement membrane complexes with biological activity. Biochemistry 25:312–318
25. Buck CA, Horwitz A (1987) Cell surface receptors for extracellular matrix molecules. Annu Rev Cell Bio 3:319–346
26. Terracio L, Gullberg DL, Rubin K, Craig S, Borg TK (1989) Expression of collagen adhesion proteins and their association with the cytoskeleton in cardiac myocytes. Anat Record 223:62–71

27. Terracio L, Lundgren E, Gullberg D, Borg TK, Terracio MJ, Rubin K (1988) In vitro studies on adult cardiac myocytes: attachment and biosynthesis of collagen type IV and laminin. J Cell Physiol 136:43–53
28. Terracio L, Ronnstrand L, Tingstrom A, Rubin K, Claesson-Welsh L, Funa K, Heldin CH (1988) Induction of platelet-derived growth factor receptor expression in smooth muscle cells and fibroblasts upon tissue culturing. J Cell Biol 107:1947–1957
29. Weiss RE, Reddi AH (1981) Isolation and characterization of rat plasma fibronectin. Biochem J 197:529–534

Microcarrier Culture of Neonatal Heart Cells

P. A. Uusimaa and I. E. Hassinen

Introduction

Cardiac myocytes in culture have been available for three decades [3, 6], but this methodological approach has been in its developmental stage until recent years. The culturing of striated cardiac myocytes even from the neonatal heart is hampered by a low biomass yield because of the limits of cell division [8] in cells derived from a tissue which is already amitotic 17 to 21 days postnatally [14]. Cultured heart cells have many advantages over newly isolated cells, however, because the latter are probably in an abnormal metabolic state after isolation due to reversible damage to the plasma membrane during tissue disaggregation. A marked difference is observed in Ca^{2+} tolerance, for example.

Striated cardiac myocytes account for 80% of the myocardial mass, and their metabolism and its characterization is of the utmost importance for the design of treatment for cardiac disease. There is also a need for experimental models that are better defined and more homogeneous in terms of their cell population than an intact tissue or organ.

Cell culturing on microcarriers has been introduced to provide a means of obtaining high yields of cells [20] and reports have appeared on the cultivation of chicken heart muscle cells in this way [13]. A procedure for the microcarrier culture of mammalian heart cells has recently been devised in this laboratory [19]. The initial goal was to use this method as an aid to investigating subcellular compartmentation, but later on it was found useful for studying the intracellular signaling of the regulation of atriopeptin secretion (Uusimaa et al., in preparation). The method has been refined and found dependable and reproducible. The present communication also aims to indicate useful modifications to the published procedure [19].

Procedure [1]

Solutions and Equipment

The following solutions and equipment are needed for cell culturing:

Disaggregation medium: Phosphate-buffered saline [4] (PBS), Ca^{2+} and Mg^{2+} free (Gibco Europe) containing collagenase CLS II (Worthington, Millipore Corp) 2 g/l and 50 μM $CaCl_2$.

---

[1] Adapted from Uusimaa et al. (1988) Cardiovasc Res 22:291–295 with permission of the publisher.

Culture medium I: RPMI-1640 [11] cell culture medium (Gibco) supplemented with 2 mM glutamine, 10^5 U of benzylpenicillin, 100 mg streptomycin sulfate (Sigma Chemical Co) and 200 ml of fetal calf serum (FCS; Gibco) per liter.
Culture Medium II: RPMI-1640 supplemented with 2 mM glutamine, 10^5 U of benzylpenicillin, 100 mg of streptomycin sulfate and 100 ml of FCS per liter.
Spinner flasks: Spinner flasks (Belco Glass) equipped with a suspended magnet rod for stirring were siliconized using dimethyldichlorosilane (SERVA), air dried, and cured in an oven at $100°-150°C$.
Microcarriers: Cytodex 3 beads (Pharmacia) were preswollen in Ca^{2+} and Mg^{2+}-free PBS overnight, washed once with fresh PBS, and sterilized by autoclaving at $115°C$ for 15 min in PBS. Being sterile, the beads could be stored swollen for 1 month at $4°C$. A suitable aliquot was washed with RPMI-1640 medium before use.
Tissue culture dishes: Polystyrene tissue culture dishes, 10 cm in diameter (Falcon Labware).
Stirrer: Magnetic stirring platform capable of low stirring speeds of less than 30 rpm (Belco μ-Carrier Magnetic Stirrer, Belco).

Initiation and Maintenance of the Cell Culture

The procedure for culturing cells is summarized in Fig. 1. Sprague-Dawley rats 3–5 days old are used as tissue donors, 10–16 animals for each batch of culture. After decapitation, the thorax is opened aseptically, the aorta clamped, the right atrium perforated, and the heart punctured at the apex and

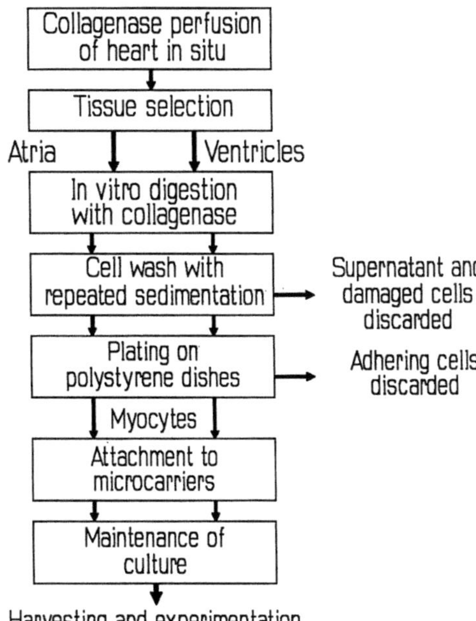

Fig. 1. Flow chart for the isolation and cultivation of neonatal cardiac myocytes on microcarriers. For details, see "Procedure"

slowly perfused with 5–10 ml disaggregation medium. The heart is then excised, placed in disaggregation medium and cut into 2-mm fragments with a surgical scalpel. When initiating the culture of ventricular cells, the apex of the heart is cut out and processed separately. The original procedure has been modified by an addition of 50 μM Ca^{2+} to the disaggregation medium, which is found to increase the cell yield. The mince is transfered to a 50-ml conical flask, incubated at 37 °C for 5 min, and shaken. The supernatant is filtered through a double layer of 160-μm nylon mesh and supplemented with culture medium I. Fresh disaggregation medium is added to the original tissue mince, which is mixed by repeated suction into a Pasteur pipette. This cycle of incubation and shaking in fresh disaggregation medium, suction into a pipette, and collection of the supernatant is repeated until most of the tissue has disintegrated.

The combined supernatants are centrifuged at 175 g for 5 min at room temperature. The sediment is resuspended in fresh culture medium 1 and washed twice by centrifugation. To eliminate most of the fibroblasts by adhesion, the final suspension in culture medium I (15 ml) is transfered to a culture dish and kept for 30 min at 37 °C in a humidified atmosphere of 5% CO_2 in air. After gentle shaking, the cell suspension is transfered to the spinner flask containing preswollen Cytodex 3 beads (bead volume approx. 7 ml) and the final volume adjusted to 50 ml by adding culture medium I. The microcarriers must be incubated in culture medium for several hours before addition of the cells. This has been found necessary in order to enhance the attachment of the cells, probably due to coating of the carriers by the serum proteins. The pH is adjusted to 7.2 by gassing with CO_2 prior to adding the cells. The gas phase of the spinner flask is then once again adjusted to approx. 5% CO_2 in air by introducing CO_2 until the color of the phenol red in the culture medium turns orange. To ensure attachment of the cells to the carriers, the contents of the spinner flasks (for 30 s) are stirred intermittently during the first 6 h, at 30-min intervals at first and ending with a 1½-h interval. Stirring then proceeds continuously at 25 rpm on a magnetic stirring platform. The culture is maintained by adding 50 ml of fresh culture medium II. After 36 or 48 h, depending on growth and the purpose of the experiment, half of the medium is replaced by fresh culture medium II. This means that the FCS concentration is lowered in a stepwise manner from 20% to 12.5%.

When atrial cells are cultivated, the growth is usually faster, and since the number of the nonmyotypic cells is higher, shorter cultivation times (2–3 days) are preferable. For optimum yield, culture of ventricular cells can be continued for 3–4 days. After that time problems are encountered with cell adhesion.

Comments on Alternatives and Design of the Procedure

The present method employs crude collagenase for tissue dispersion, whereas most researchers working with cultured heart cells until now have used trypsin [see 7, 17]. Collagenase digestion has become the standard procedure in experiments with freshly isolated cardiac myocytes, however [see e.g. 5, 21], and in the present case the crude collagenase produced higher cell yields than trypsin and was adopted for this reason. No systematic research has been done on the effects of extracellular proteolytic enzymes on receptor proteins in the plasma membrane, but it is plausible that a peptidase with a higher specificity is less liable to produce random processing of intrinsic membrane proteins, as has been noticed in organelle isolation procedures employing bacterial proteinases [15].

Both Biosilon (Nunc) and Cytodex 3 microspheres have been tested as carriers. The former are made of surface-treated Nunclon plastic and have been found to be impermeable to aqueous solutions, but tests involving heart cell cultivation on Biosilon have not been successful in our laboratory. Cytodex 3 (type-I collagen-coated Sephadex) microspheres have been found suitable as a cultivation substratum, and no other types of Cytodex carrier have been tested.

In the first trials the flasks were filled with an appropriate mixture of air and CO_2 to adjust the pH of the culture medium to the colour change point of phenol red. Alkalinity was occasionally noticed in the culture, and continuous gassing with an air/CO_2 mixture adjusted with a precision rotameter was therefore adopted. It was later found that adequate pH values could be maintained for the lengths of time needed for initial gas filling, provided that the lids of the flasks were gastight. The bulkiness of the flasks and the need for stirring makes the use of conventional incubator cabins awkward or impossible when using spinner flasks, and it has been found that the most convenient arrangement for temperature regulation and stirring is a thermostated water bath on top of a magnetic stirring platform.

Morphology

Cardiac myocytes beat spontaneously when cultured on microspheres. Single beating cells can be observed on the 1st day, and beating in groups of cells becomes more pronounced on the 2nd day and is more prominent when the culture reaches confluence.

The morphology of the cells has been investigated by light and electron microscopy in the transmission and scanning mode [19]. Taking a high density population of mitochondria and myofilament structure as the criteria for myotypy, 63% of the cells were myocytes 24 h after inoculation, a proportion which decreased to 48% at 48 h and 44% at 72 h.

Some Useful Experimental Designs with the Culture on Microcarriers

The ease of separation of the cells from the incubation medium can be exploited when studying the excretion of substances from the cells. Moreover, for continuous monitoring of the production of extracellular substances, perifusion of the cells can be performed in a chromatographic column, as has been shown previously with endothelial cells [2, 10, 18]. The heart myocyte cultures

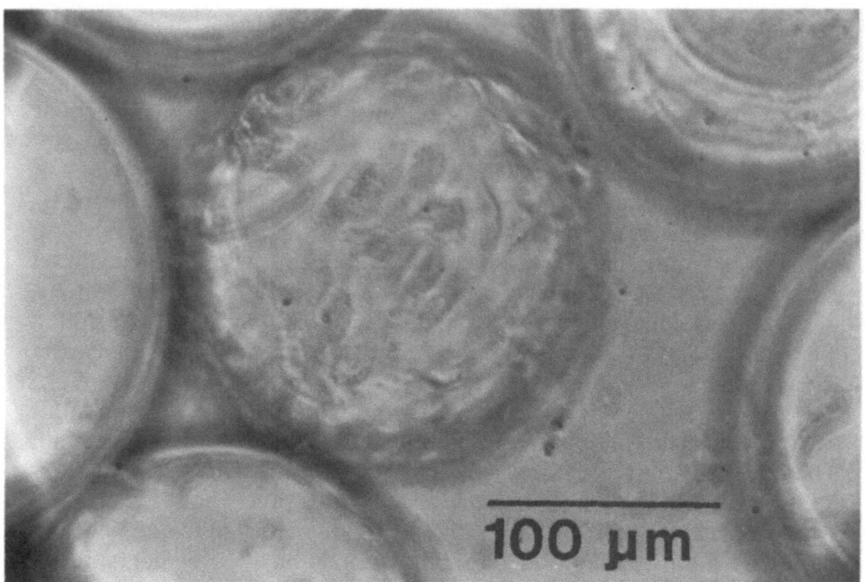

Fig. 2. Microcarrier cell culture of neonatal heart cells after harvesting, perifusion for 2½ h with Krebs-Ringer bicarbonate solution in a chromatographic column (0.9 × 3 cm) and its disassembly. Phase contrast photomicrograph of an unstained specimen (microcarriers in PBS on a tissue culture dish. Beating of the cells was observed. The *scale bar* represents a length of 100 µm

have been perifused in a column of 1.5-ml bed volume for 2–3 h at flow rates of 1 ml/min. When the cells are subsequently observed by light microscopy with phase contrast optics after disassembly of the column, they seem to be viable as judged by their appearance and spontaneous beating (Fig. 2).

Immobilization of the cells in the column also allows other means of observation of the cells, e.g., by optical methods such as fluorometry or spectrophotometry of endogenous chromophores (nicotinamide nucleotides, flavins, and cytochromes) or exogenous optical probes. Thus the cell damage which occurs during continuous stirring of free cell suspensions can be avoided. Chromatography columns of the dimensions of a photometric cuvette are commercially available, and an optical readout can be easily arranged with commercial or dedicated instruments. Excellent performance has been observed in monitoring cytosolic free calcium concentrations by means of Fura-2 (Molecular Probes, Inc.) in cultivated heart muscle cells immobilized on Cytodex 3 (Uusimaa et al. in preparation)

Comparison with Alternative Procedures

There are few reports available on the cultivation of heart myocytes on microcarriers. A chick embryo heart cell model has been reported by Norrgren et al.

[13]. This was used in preliminary experiments with dish cultures, leading to the development of the present procedure. Initation of a neonatal rat heart cell culture was found dependable and at least as productive as the chick embryo cells. Future experiments on hormonal regulation, for example, call for serum-free defined media, but they have not yet been tested on microcarrier cultures of mammalian neonatal cardiomyocytes.

The main issue in introducing microcarriers as a growth substratum is their large total surface and concomitantly increased cell yield. This is a real advantage when immortal cell lines are being used. No case of the development of transformed heart cell lines for permanent culture has been reported up to now [9]. The surface area of the growth substratum is no longer a crucial point in primary cultures of a cell type with a low mitotic capacity, and the emphasis in method selection must be placed on the technical aspects of experimental design. The microcarrier culture of heart muscle cells certainly has a potential of its own in this respect.

One criterion for the integrity of the cell is cytosolic Ca^{2+} homeostasis. Preliminary experiments have given $147 \pm 15\,nM$ (mean $\pm$ SE) for the free cytosolic $[Ca^{2+}]$ of heart ventricular myocytes cultured on Cytodex 3 and perfused in a chromatographic column. This value can be compared with figures of 89 ± 1.6 [1], 181 ± 18 [16] and $129\,nM$ in freshly isolated rat heart myocytes and $126 \pm 14\,nM$ in cultured chick heart cells [7]. All this indicates that microcarrier cultures of cardiac myocytes can be manipulated without undue damage to the cells.

References

1. Beekman RE, van Hardeveld C, Simonides WS (1988) Effect of thyroid state on cytosolic free calcium in resting and electrically stimulated cardiac myocytes. Biochim Biophys Acta 969:18–27
2. Busch C, Owen WG (1982) Identification in vitro of an endothelial cell surface cofactor for antithrombin III. Parallel studies with isolated perfused rat hearts and microcarrier cultures of bovine endothelium. J Clin Invest 69:726–729
3. Cavanaugh MW (1955) Pulsation, migration and division in dissociated chick embryo heart cells in vitro. J Exp Zool 128:573–589
4. Dulbecco R, Vogt M (1954) Plaque formation and isolation of pure lines with poliomyelitis viruses. J Exp Med 99:167–182
5. Hansford RG (1987) Relation between cytosolic free Ca^{2+} concentration and the control of pyruvate dehydrogenase in isolated cardiac myocytes. Biochem J 241:145–151
6. Harary I, Farley B (1960) In vitro studies of single isolated beating heart cells. Science 131:1674–1675
7. Kim D, Okada A, Smith TW (1987) Control of cytosolic calcium activity during low sodium exposure in cultured chick heart cells. Circ Res 61:29–41
8. Klein I, Daood M, Whiteside T (1985) Development of heart cells in culture: studies using an affinity purified antibody to a myosin light chain. J Cell Physiol 124:49–53
9. Lieberman M, Hauschka SD, Hall ZW, Eisenberg BR, Horn R, Walsh JV, Tsien RW, Jones AW, Walker JL, Poenie M, Fay F, Fabiato F, Ashley CC (1987) Isolated muscle cells as a physiological model. Am J Physiol 253:C349–C363
10. Lückhoff A, Busse K, Winter I, Bassenge E (1987) Characterization of vascular relaxant factor released from cultured endothelial cells. Hypertension 9:295–303

11. Moore GE, Gerner RE, Franklin HA (1967) Culture of normal human leukocytes. JAMA 199:519–524
12. Moreno-Sánchez R, Hansford R (1988) Relation between cytosolic free calcium and respiratory rates in cardiac myocytes. Am J Physiol 255:H347–H357
13. Norrgren G, Ebendal T, Gebb C, Wikström H (1984) The use of cytodex® 3 microcarriers and reduced-serum media for the production of nerve growth promoters from chicken heart cells. Dev Biol Stand 55:43–51
14. Overy HR, Priest RE (1966) Mitotic cell division in postnatal cardiac growth. Lab Invest 15:1100–1103
15. Pande SV, Blanchaer MC (1970) Preferential loss of ATP-dependent long-chain fatty acid activating enzyme in mitochondria prepared using Nagarse. Biochim Biophys Acta 202:43–48
16. Sheu SS, Sharma VK, Banerjee SP (1984) Measurements of cytosolic free calcium concentration in isolated rat ventricular myocytes with Quin2. Circ Res 55:830–834
17. Shields PP, Dixon JE, Glembotski CC (1988) The secretion of atrial natriuretic factor-(99-126) by cultured cardiac myocytes is regulated by glucocorticoids. J Biol Chem 263:12619–12628
18. Shikano K, Berkowitz BA (1987) Endothelium-derived relaxing factor is a selective relaxant of vascular smooth muscle. J Pharmacol Exp Ther 243:55–60
19. Uusimaa PA, Hiltunen JK, Sormunen RT, Hassinen IE (1988) Microcarrier culture of neonatal cardiac myocytes in metabolic studies. Cardiovasc Res 22:291–295
20. Van Wezel AL (1967) Growth of cell strains and primary cells on microcarriers in homogenous culture. Nature 216:64–65
21. Wittenberg BA, Robinson F (1981) Oxygen requirements, morphology, cell coat and membrane permeability of calcium-tolerant myocytes from hearts of adult rats. Cell Tissue Res 216:231–251

Endothelial Cells

Macrovascular Endothelial Cells from Porcine Aorta

A. M. Rosenthal, and A. I. Gotlieb

Introduction

The ability to grow pure cultures of large vessel endothelial cells has resulted in a marked increase in our understanding of the function of these cells. For a long time endothelial cells were thought to be rather inert cells forming a physical barrier on the inside of blood vessels. There was very little appreciation of the complex metabolic activity that these cells carry out. It is now clear that in addition to actively participating in the regulation of the two main functions of the endothelium, that of a macromolecular barrier and of a thromboresistant surface [16], large vessel endothelial cells have a wide range of metabolic activities. Culture studies have also shown that the endothelial cells secrete components of the wall and substances that regulate smooth muscle cell function.

In the 1970s, the ability to harvest and grow pure cultures of large vessel endothelial cells was reported. The culture of human umbilical vein cells [8, 13], bovine aortic cells [1], porcine aortic endothelial cells [20], and pulmonary artery cells [17] were described.

A variety of laboratories choose the porcine aortic endothelial cell to work with [2, 9–11, 17, 25] since the porcine cardiovascular system has many similarities to that of humans [4]. With respect to atherosclerosis, older pigs develop spontaneous aortic atherosclerotic plaques [3, 12] and the pig is an excellent model for studying experimental atherosclerosis [5, 6, 18, 21].

Materials

This section summarizes (a) preparation of medium and solutions, (b) materials for the slaughterhouse, and (c) materials for the tissue culture laboratory.

Preparation of Medium and Solutions [1]

1. *Medium 199 with 1% antibiotics and 5% FBS (M199, 1% PSF, 5% FBS);* (100 ml)

---

[1] *Abbreviations:* M199 – Media 199; PSF – Penicillin-streptomycin-fungizone; FBS – Fetal bovine serum; DPBS – Dulbecco's phosphate buffered saline; EDTA – Ethylenediamine tetracetic acid; HBSS – Hank's balanced salt solution; Hepes – *N*-2-hydroxyethylepiperazine-*N*-2 ethanesulfonic acid; Dil-AC-LDL – 1,1-dioctadecyl 1-3,3,3,3-tetramethyl-indocarbocyanine-perchlorate low density lipoprotein

Medium 199 with 25 mm Hepes buffer, Earles salts,
and L-glutamine, (e.g., Gibco #380-2340)   93 ml
Penicillin (5000 units/ml)-streptomycin (5000 mcg/ml)   1 ml
   (Gibco #600-5070)
Fungizone (amphotericin B 250 mcg/ml)   1 ml
   (Gibco #600-5295AE)
FBS (Gibco #240-6000) (Stored at 4°–8°C)   5 ml

FBS is supplied in 500 ml bottles. The serum is shipped and stored frozen.
To use thaw a bottle and aliquot it into 5×100 ml bottles which may be
refrozen once.

2. *Dulbecco's Phosphate Buffered Saline (without Ca^{++} and Mg^{++}) [DPBS]*
   (to make 1000 ml)

	gm/l
KH_2PO_4	0.20
KCl	0.20
$NaCl$	8.00
$Na_2HPO_4 \cdot 7\,H_2O$	2.16

Sterilized by filtration through 0.2 μm filter

3. *DPBS (2% antibiotics)* (to make 500 ml)

DPBS	480 ml
Penicillin-streptomycin	10 ml
Fungizone	10 ml

Prepare 1.5 l within 24 h of use and store at 4°–8°C

4. *Collagenase (Worthington, class II)* (to make 0.75 mg/ml)

Collagenase	15 mg
DPBS (2% antibiotics)	20 ml

The collagenase is dissolved in the DPBS, in a sterile 50 ml test tube, filtered
through the 0.2 μM sterile filter, decanted into a second sterile 50 ml test
tube, and wrapped in aluminium foil to protect it from exposure to light.
This solution should be prepared immediately prior to use, warmed to 37 °C
and used. It should not be stored for an extended period of time. Each lot
of collagenase will have to be tested to fine adjust the length of time of
incubation.

5. *Trypsin-EDTA (1 ×)*
   0.5 g Trypsin (1:250) and 0.2 g EDTA/l of $1 \times$ HBSS without Ca^{++} and
   Mg^{++} (Gibco #610-5300)
   A 100 ml bottle may be aliquoted into sterile plastic test tubes and refrozen.
   This maintains more stable enzymatic activity.

Materials for the Slaughterhouse

2 Sterile fleakers (500 ml; Corning Glass Works)
2 Sterile fleaker caps
1 Sterile surgical scissors
1 Sterile long blunt forceps
1 Pair of latex medical gloves
1 Hair net, beard net, clean white laboratory coat and hard hat (as per slaughterhouse regulations)
600 ml of DPBS (2% antibiotics)

Materials for the Tissue Culture Laboratory

1 Sterile fleaker (500 ml)
1 Sterile fleaker cap
2 Sterile beakers (400 ml)
1 Sterile long blunt forceps
1 Sterile fine surgical scissors
1 Sterile surgical scissors
1 Sterile fine forceps
1 Sterile hemostat for each aorta to be used
1 Sterile plastic test tube (17×100 mm) with cap for each aorta to be used
1 Test tube rack for 17×100 mm test tubes
1 Sterile tissue culture dish (150×15 mm)
1 Sterile tissue culture dish (100×20 mm)
3–5 Sterile tissue culture dishes (35×10 mm) for each aorta to be used
2 Sterile test tubes (50 ml)
1 Sterile filter unit (0.2 µm, 30 ml capacity)
1 Pipetting device (10 ml), e.g., "pi-pump", Glasfirn, FRG
2 Sterile disposable pipettes (10 ml)
2 Sterile disposable pipettes (5 ml)
1 Timer (minutes)
2 Sterile cannisters of Pasteur pipettes (230 mm)
1 Roll aluminium foil
1 Garbage bag (for discarded animal tissue)
5 Sterile glass coverslips (22×22 mm) for each aorta used
1 Plastic wash bottle containing 70% ethanol
1 Sterile bottle containing 500 ml of DPBS (2% antibiotics)
The fleakers, fleaker caps, scissors, forceps, hemostats, beakers, and coverslips are sterilized (individually) by autoclaving.

The sterile glass cover slips are kept in a 100-mm glass Petri dish. The coverslips are spread individually on filter paper. Several layers of filter paper may be used per dish. The filter paper prevents coverslips from becoming stuck to each other during autoclaving. If kept sterile, the dish need not be resterilized between uses. The "pi-pump" is rinsed with 70% ethanol and rinsed in sterile water.

Some investigators do not use the glass coverslips. Instead they plate the harvested cells directly onto the surface of the tissue culture dish.

Methods

Obtaining Aortas from the Slaughterhouse

At the abattoir aortas are taken by the investigator from the heart-lung block of organs passing along in the assembly line tray. Tissue that has come into contact with other tissues such as gastrointestinal organs may be more readily contaminated by organisms and is not taken. The thoracic aorta has been previously cut at the level of the diaphragm to remove it from the carcass. The aorta is cut just distal to the aortic arch by the investigator with a sterile surgical scissors and immediately placed into one of the sterile 500 ml fleakers containing 300 ml of Dulbecco's phosphate buffered saline (DPBS, 2% antibiotics). The fleaker is capped after 6-8 aortas have been collected and placed in the fleaker. Within 5 min, the aortas are moved to a room away from the slaughterhouse floor and are transfered to the second 500 ml fleaker containing 275–300 ml of DPBS (2% antibiotics). The aortas are transported to the laboratory within 1 h at ambient temperature. Each aorta provides a quantity of endothelial cells to seed 3–5 tissue culture dishes (35 mm). Some investigators choose to transport the aortas at 4°C; however, we have not found any advantage to cooling the tissue during transportation.

Obtaining Endothelial Cells from the Aorta

Upon returning to the laboratory, the surface of the laminar flow biosafety hood, which is on, is wiped with 70% ethanol and left for 5 min. The outside surface of the fleaker is washed thoroughly with 70% ethanol and placed in the hood. All the materials listed in "Materials for the Tissue Culture Laboratory" except for the wash bottle of 70% ethanol, the garbage bag, and the aluminium foil are placed in the hood.

1. Pour 275–300 ml of DPBS (2% antibiotics) from its bottle into the sterile fleaker.
2. Wash the aortas by transfering them with the sterile long blunt forceps to the fleaker of DPBS prepared in step 1. One must handle the aorta with care whenever it is picked up with forceps so as not to damage the endothelial surface of the vessel.
3. Remove one aorta from the wash fleaker with the long blunt forceps. Open the 150 mm culture dish and place the aorta into the bottom half.
4. With the sterile surgical scissors, remove all large pieces of nonaortic tissue and place them into the top half (lid) of the 150 mm dish for future disposal.
5. Trim the adventitia of the aorta thoroughly with the sterile fine scissors and fine forcep and place it in the top half of the dish for disposal as well. One must be careful in step 4 and 5 not to cut the vessel wall.
6. Place the trimmed aorta in a 400 ml beaker containing 200 ml of DPBS (2% antibiotics). If the vessel is very long and has many branch points, then cut away the lower portion and retain the upper section (i.e., aortic arch with few branch points).

7. Locate the first branch point distal to the aortic arch and clamp the vessel with a sterile hemostat just above this point. Cut the vessel just below the clamp and discard this lower section. This should result in the maximum length of vessel for the collagenase dispersion of the endothelial cells. This also avoids the need to tie off branches with sterile sutures.

8. Place the hemostat across the bottom half of a 100 mm culture dish so that the clamped aorta stands upright.

9. Wash the lumen of the aorta by gently pipetting DPBS (2% antibiotics) into the clamped vessel. Check that there is not any leakage of fluid and that the aorta is completely sealed at the site of clamping. If necessary reclamp the vessel. Check that the aorta remains stable in the upright position when filled with fluid. If it is not stable, then adjust the position of the hemostat on the dish, or use a sterile beaker as a support for the aorta.

10. Decant the DPBS into a sterile 400 ml beaker and fill the vessel once again with DPBS. This will gently wash out any blood cells from the vessel.

11. Repeat steps 3–10 with a second aorta.

12. Decant the DPBS from the vessel into the sterile 400 ml beaker. Fill the aorta to just below the top with the warm collagenase solution (4. in "Preparation of Medium and Solutions"), let it incubate for 8 min, and then gently decant the collagenase solution into the 400 ml beaker.

13. In order to remove endothelial cells from the wall, start at the bottom of the vessel and progressing around its circumference and moving toward the top, use the pasteur pipette attached to the pipetting device (e.g., pi-pump), to gently wash the intimal surface with medium (1. in "Preparation of Medium and Solutions"). Complete the washing of the wall with additional medium as needed. Do not move back and forth over the surface and do not touch the vessel itself with the pipette tip since these actions may dislodge smooth muscle cells as well.

14. Aspirate the wash medium from the clamped aorta with the Pasteur pipette and place it in a sterile plastic test tube.

15. Repeat steps 13–15 with the second aorta.

16. Estimate the total volume of cell suspension in the test tube. Use one 35 mm tissue culture dish for each ml of cell suspension. With a sterile fine forcep place a sterile glass coverslip (22 × 22 mm) in each 35 mm tissue culture dish.

17. Pipette 1 ml of medium (M199, 1% PSF, 5% FBS) into each dish.

18. Aspirate the cell suspension up and down a few times and add 1 ml of cell suspension to each dish until all of the cell suspension is plated.

19. View the dishes under an inverted microscope with good quality phase contrast optics. Approximately 4–10 clumps of 10–100 rounded cells should been seen under a 10× objective (Fig. 1).

20. Incubate the primary endothelial cell cultures in a tissue culture incubator with humidification and set at 5% CO_2 and 37 °C.

21. The cultures should be left undisturbed for 2–3 days. To feed the cells, the medium is aspirated, leaving about 25% of the "conditioned medium" in the dish. Fresh M199, 1% PSF, 5% FBS is added to a total volume of 2–2.5 ml. The cultures are fed twice a week until they are nearly confluent

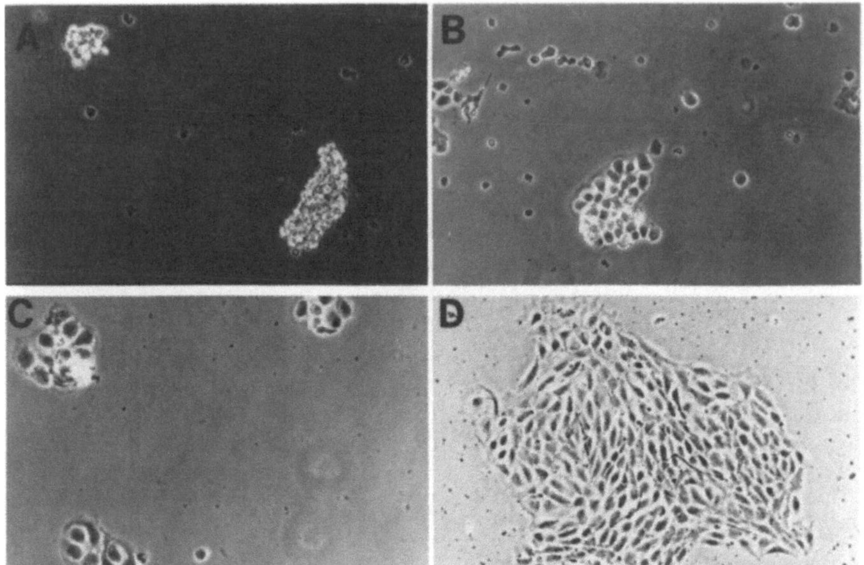

Fig. 1A–D. Phase contrast photomicrograph of porcine aortic endothelial cells at various times after being harvested from the aorta and plated: **A** 2 h, **B** 4 h, **C** 24 h, and **D** 1 week post plating. (Magnification **A, D** ×75, **B, C** ×150)

(3–4 weeks after the initial plating). The cultures should be closely monitored during this period for signs of bacterial or fungal contamination. Contaminated dishes should be discarded immediately in an appropriate fashion.

Passaging Endothelial Cell Cultures

The primary cell cultures are best passaged before they reach confluency. Subsequent subcultures are passaged from 1–4 days post confluency.

Cell cultures are passaged by washing with DPBS (2. in "Preparation of Medium and Solutions"), incubation with a trypsin ethylenediamine tetraacetic acid (EDTA) solution and dispersion in M199, 1% PSF, 5% FBS. The step-by-step procedure is as follows for cells in a 35 mm dish. For later subcultures which may be in a 60 or 100 mm dish, the volumes of the solutions need to be increased accordingly.

1. Aspirate all the media from the 35 mm culture dish.
2. Wash the cells in the dish with 2 ml of DPBS (without $Ca^{++} + Mg^{++}$). Repeat this wash.
3. Add 1 ml of warm (37 °C) trypsin-EDTA solution (5. in "Preparation of Medium and Solutions")
4. Observe the cell culture under the microscope. The cells begin to round up within 30 s (Fig. 2). When the cells are rounded up (usually ½–1 minute,

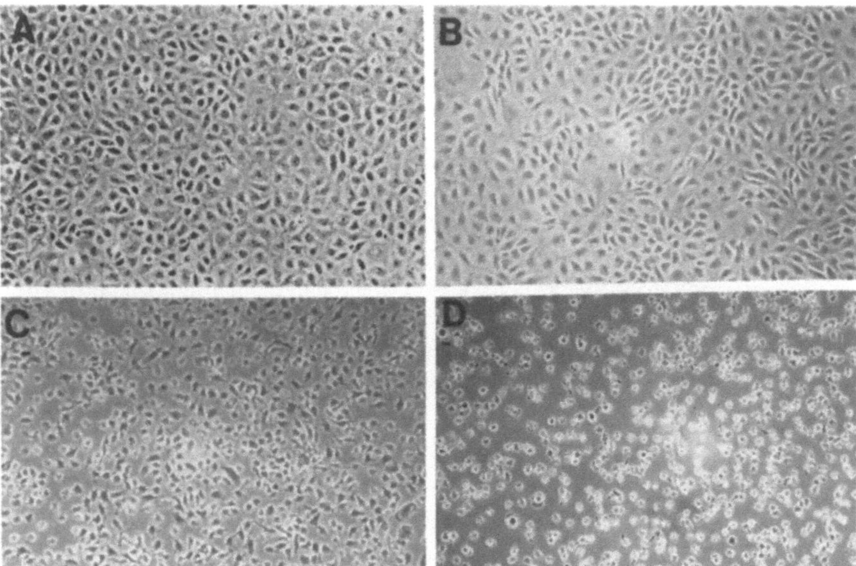

Fig. 2A–D. Phase contrast photomicrograph of a confluent culture of porcine aortic endothelial cells at various times after the addition of the trypsin-EDTA solution: **A** after 15 s (There is no changes from normal), **B** 30 s, **C** 30–45 s, and **D** 60 s incubation with trypsin-EDTA, ×75

Fig. 2C), aspirate the trypsin solution and add 1–2 ml of M199, 1% PSF, 5% FBS. Take care not to over expose the cells to the trypsin solution. This may injure the cells or cause them to come off the dish surface. If the cells have come off the dish, do not aspirate the trypsin solution. Go to step 5a.

5. Disperse the cells by repeated flushing of the dish surface with the media. When fully dispersed (single cells floating in the media as seen under the microscope), pipette the cells into a 10 ml test tube and bring the volume up to 5 or 10 ml.

   a. Disperse the cells as in 5., but use the trypsin-media mixture. Pipette the dispersed cells into a 50 ml plastic disposable centrifuge tube, add 20 ml of media, and centrifuge at 1000 rpm for 6–8 min. Remove the supernatant and resuspend the cells in 5 or 10 ml of media.

6. Mix the cell suspension by inverting the test tube. For the primary cultures grown in a 35 mm dish, the cell suspension is divided by pipetting equal volumes into fresh 35 mm dishes (four to five dishes) or 60 mm dishes (one to two dishes). For later subcultures, the cell suspension is pipetted into fresh dishes to give a cell splitting ratio of 1:5 or 1:10. Cell cultures passaged in this manner grow to confluency in 4–7 days. Cell should be fed twice a week.

Discussion

The porcine aorta is a useful source of aortic endothelial cells. The method of harvesting described in this chapter is not complicated, and the cells grow well without the need to add aditional growth factors.

There are a few findings in the porcine cultures to be aware of. Confluent cultures of endothelial cells will show occassional large endothelial cells, sometimes with more than one nucleus, which are thought to be senescent cells (Fig. 3). These are not smooth muscle cells. In addition, the cultures show sprouting of some endothelial cells which are located on the surface of the confluent endothelial cell monolayer. The cells are not smooth muscle cells and do not overgrow the culture. Smooth muscle cells contaminating the culture can be identified using techniques to localize smooth muscle specific actin (Fig. 4).

Special Procedures

With human and bovine large vessel endothelial cells, the presence of Factor VIII related antigen (VIII R:AG) and Weibel-Palade bodies can be used to establish the identity of cultured cells. However, porcine aortic endothelial cells in cell culture do not exhibit Factor VIII or Weibel-Palade bodies [7, 26].

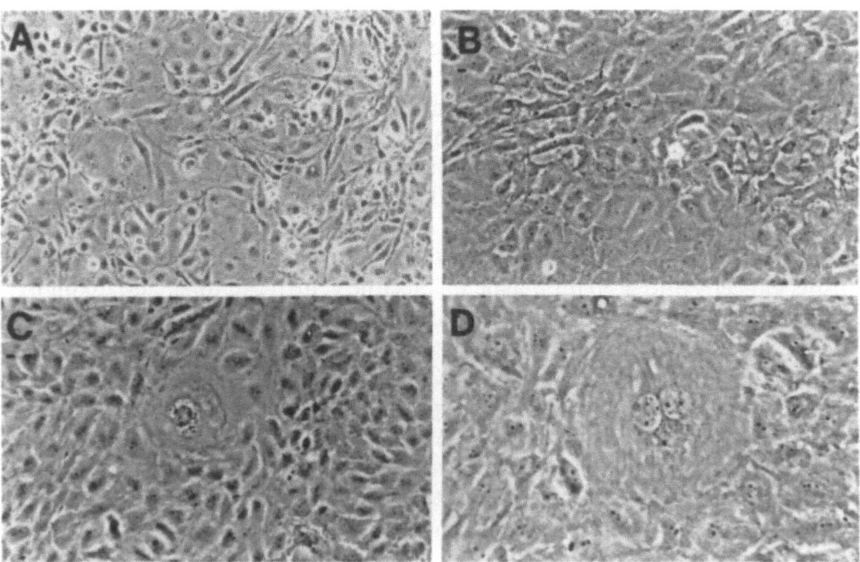

Fig. 3A–D. Phase contrast photomicrograph of confluent porcine aortic endothelial cell cultures, **A** endothelial cells which have lost cobblestone morphology. These cultures should be discarded: **B** sprouting endothelial cells located on the surface of a monolayer of endothelial cell, **C, D** endothelial cell monolayer with occassional large endothelial cells in the monolayer, **A** ×75; **B, C, D** ×150

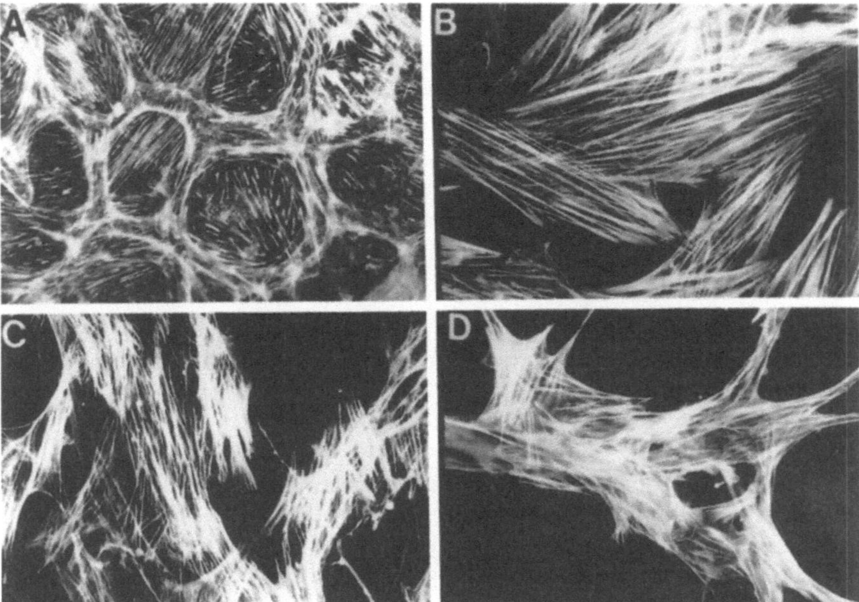

Fig. 4A–D. Fluorescence photomicrograph of the actin staining of porcine aortic endothelial and smooth muscle cell cultures. **A** Confluent porcine aortic endothelial cell monolayer stained for actin with rhodamine-phalloidin; **B** porcine aortic smooth muscle cells stained for actin with rhodamine phalloidin; **C** porcine aortic smooth muscle cells growing on the surface of a confluent monolayer of porcine aortic endothelial cells and stained for smooth muscle actin with an anti-α smooth muscle actin antibody (#A2547 Sigma); **D** same as **C** but stained for muscle actin with an anti-muscle actin antibody (HHF35), ×400

Therefore, in addition to documenting the cobblestone morphology of the confluent cultures of endothelial cells, we have used the following procedures to verify that the cells in culture obtained from porcine aortas are endothelial cells and are not contaminated with smooth muscle cells.

Acetylated Low Density Lipoprotein Uptake

Uptake of acetylated low density lipoprotein (LDL) (Dil-Ac-LDL #BT-902 Biomedical Technologies Inc.) by porcine endothelial can be used to verify that the cells are endothelial and not vascular smooth muscle cells [14, 24]. This test is carried out by incubating cell cultures on glass coverslips with acetylated low density lipoprotein labelled with 1,1-dioctadecyl 1-3,3,3,3-tetramethyl-indo-carbocyanine-perchlorate (DIL-Ac-LDL) at a concentration of 10 µg/ml in M199 for 4 h at 37 °C. Cells are washed well three times with standard medium, three times with PBS and fixed with 3% paraformaldehyde in PBS for 20 min at room temperature. The cells are subsequently washed with PBS for 10 min, rinsed for 5 s with distilled water, and mounted on glass slides cell-side down

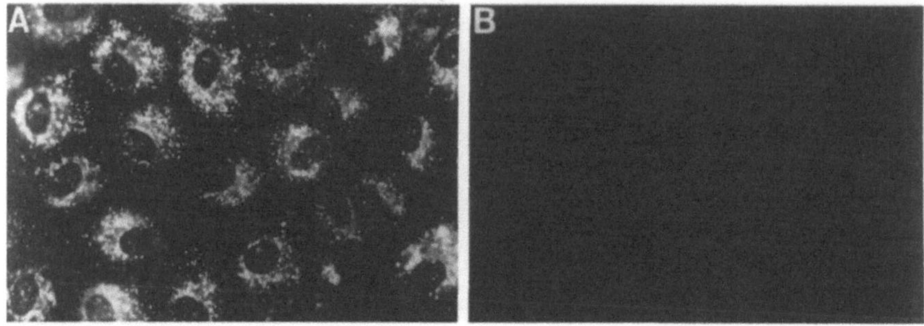

Fig. 5A, B. Fluorescence photomicrograph of DiL-Ac-LDL uptake by porcine aortic endothelial and smooth muscle cell cultures. **A** Porcine aortic endothelial cells; **B** porcine aortic smooth muscle cells, × 300

over glycerol-PBS (1:1). Using a fluorescence photomicroscope (Zeiss) equipped for epi-illumination, the fluorescence of the DIL-Ac-LDL taken up by the cells is visualized using standard rhodamine excitation and emission filters (Zeiss filter set #14) and photographed with Kodak Tri-X-Pan film (Fig. 5).

Actin and Smooth Muscle α Actin and Muscle Specific Actin Labeling of Tissue Culture Cells

Endothelial cells and smooth muscle cells contain actin filaments (stress fibers) that can be visualized by staining with a fluorescent probe. The pattern of these actin filaments when stained with rhodamine-phalloidin differ between endothelial and smooth muscle cells. In addition, employing an antimuscle α, γ-actin antibody [22, 23] or an anti-α smooth muscle actin antibody which binds to smooth muscle cell α-actin [19] and not to endothelial cell actin allows for the verification of the purity of the cell cultures. Cells are grown on glass coverslips under standard conditions and stained for actin as follows.

Fixation and Permeabilization of Cells for Staining with Rhodamine-Phalloidin or Anti-α-Smooth Muscle Actin

1. Wash the cells three times with PBS. Do not let the cells dry. They must be kept wet at all times.
2. Fix the cells for 20 min with 3% paraformaldehyde in PBS.
3. Wash the cells for 15 min with PBS (four changes of PBS).
4. Permeabilize the cells with Triton X-100 (0.1% in PBS) for 3.5 min.
5. Wash the cells with PBS for 15 min (four changes of PBS).

Fixation and Permeabilization of Cells for Staining with HHF 35 Anti-Muscle α,γ-Actin

1. Wash the cells three times with PBS.
2. Add cold methanol ($-20\,°C$) and transfer to $4\,°C$ for 5 min.
3. Transfer the coverslips to cold acetone ($-20\,°C$) for 2 min in an acetone resistant dish (e.g., glass dish).
4. Wash the cells with PBS for 10 min with three changes of the PBS.

Staining Chamber

Prepare a staining chamber by placing a piece of filter paper in the bottom half of a 150 mm Petri dish. Moisten the filter paper with water. Place two wooden rods across the filter paper, and place a glass staining slide across the two rods. Close the Petri dish by replacing the lid (top half) to maintain a moist environment. To prevent contamination, change the filter often.

Prepare glass staining slides by scoring a glass coverslip (22×22 mm) with a diamond knife to obtain pieces approximately 5×22 mm. Cement one piece each on opposite sides of a standard glass slide (25×75 mm) at one end of the slide. To stain, the staining solution is placed with a calibrated micropipettor as a drop between the two pieces of cemented glass. The coverslip to be stained is then inverted and placed over the drop of fluid to rest on the two cemented pieces of glass. Following staining, the staining slides are washed well, dried, and reused.

Actin Staining

Rhodamine-Phalloidin (#R-415 Molecular Probes, Inc.)

1. For each coverslip to be stained, evaporate a 10 µl aliquot of the rhodamine-phalloidin solution in a small test tube, and redissolve in 200 µl of PBS.
2. Remove the lid of the staining chamber and place 150 µl of this solution on the glass slide in the staining chamber.
3. Carefully place the coverslip with cells to be stained (cell-side down) on the rhodamine-phalloidin solution. Replace the lid of the staining chamber.
4. Cover the staining chamber with aluminum foil to protect the fluorescent probe from room light and allow the cells to stain for 20 min at room temperature.
5. Wash the coverslip for 15 min with three changes of PBS. Rinse the side with no cells only with distilled water and mount the coverslip on a slide with the cell-side down in a 1:1 solution of PBS and glycerol. The slide should not be sliding on the mounting media. Extra media can be removed by capillary action, using a tissue.
6. Actin filaments are observed by epifluorescence microscopy using a Zeiss photomicroscope III or equivalent with filter set #14 for rhodamine excitation and emission.

α,γ-Muscle Actin and α Smooth Muscle Specific Actin Staining

1. For each coverslip to be stained, place 150 µl of the primary antibody (HHF 35 #MA 931 ENZO Biochem or anti-α smooth muscle actin #A2547, Sigma), diluted according to the suppliers instructions, on a glass slide in the staining chamber.
2. Place the coverslip (cell-side down) on the antibody solution, replace the lid of the staining chamber, and allow the cells to stain for 20 min at room temperature.
3. Wash the coverslip with PBS for 15 min (with three changes of PBS).
4. Prepare 150 µl of secondary antibody diluted 1:20 with PBS and place on a glass slide in the staining chamber. Fluorescein or rhodamine conjugated affinity purified anti-mouse IgG (H&L) can be used as the secondary antibody.
5. Place the coverslip cell-side down to be stained on the solution of secondary antibody. Replace the lid of the staining chamber, cover the staining chamber with aluminium foil, and allow the cells to stain for 20 min at room temperature.
6. Wash the coverslip with PBS for 15 min (change the PBS two to three times). Rinse the side with no cells only with distilled water.
7. Mount the coverslip (cell-side down) over PBS:glycerol (1:1) on a glass slide.
8. The fluorescence of the actin filaments stained by the antibodies is observed with the epifluorescence microscope (for a the Zeiss photomicroscope III fitted with filter set #14 for rhodamine or filter set #09 for fluorescein).
9. Photomicrographs may be taken using Kodak Tri-x-pan film, which is developed with Kodak microdol X-100 as per manufacturers instructions.

References

1. Booyse FM, Sedlak BJ, Rafelson ME (1975) Culture of arterial endothelial cells: characterization and growth of bovine aortic endothelial cells. Thromb Diath Haemorrh 34:825–839
2. Dickinson ES, Slakey LL (1982) Plasma-derived serum as a selective agent to obtain endothelial cultures from swine aorta. In Vitro 18:63–70
3. French JE, Jennings MA, Poole JCF, Robinson DS, Florey H (1963) Intimal changes in the arteries of ageing swine. Proc R Soc Lond [Biol] 158:24–42
4. French JE, Jennings MA, Florey HW (1965) Morphological studies on atherosclerosis in swine. Ann NY Acad Sci 127:780–799
5. Gerrity RG (1981) The role of the monocyte in atherogenesis. II. Migration of foam cells from atherosclerotic lesions. Am J Pathol 103:191–200
6. Gerrity RG, Naito HK, Richardson M, Schwartz CJ (1979) Dietary induced atherogenesis in swine. Morphology of the intima in prelesion stages. Am J Pathol 95:775–792
7. Giddings JC, Jarvis AL, Bloom AL (1983) Differential localization and synthesis of porcine factor VIII related antigen (VIII R:AG) in vascular endothelium and in endothelial cells in culture. Thomb Res 29:299–312
8. Gimbrone MA, Cotran RS, Folkman J (1974) Human vascular endothelial cells in culture: growth and DNA synthesis. J Cell Biol 60:673–684

9. Gotlieb AI, Spector W (1981) Migration into an experimental wound: a comparison of porcine aortic endothelial and smooth muscle cells and the effect of culture irradiation. Am J Pathol 103:271–282

10. Gotlieb AI, Subrahmanyan L, Kalnins VI (1983) Microtubule organizing centers and cell migration. Effect of inhibition of migration and microtubule disruption in endothelial culture. J Cell Biol 96:1266–1272

11. Gotlieb AI, Spector W, Wong MKK, Lacey C (1984) In vitro reendothelialization: microfilament bundle reorganization in migrating porcine endothelial cells. Arteriosclerosis 4:91–96

12. Gottlieb H, Lalich JJ (1954) The occurrences of arteriosclerosis in the aorta of swine. Am J Pathol 30:851–853

13. Jaffe EA, Nachman RL, Becker CG, Mimick CR (1973) Culture of human endothelial cells derived from umbilical veins. J Clin Invest 52:2745–2756

14. Koo EWY, Gotlieb AI (1989) Endothelial stimulation of intimal cell proliferation in a porcine aortic organ culture. Am J Pathol 134:497–503

15. Magargal WW, Dickinson ES, Slakey LL (1978) Distribution of membrane marker enzymes in cultured arterial endothelial cells and smooth muscle cells. J Biol Chem 253:8311–8318

16. Ryan US (ed) (1988) Endothelial cells, vols 1–3. CRC, Boca Raton

17. Ryan US, Clements E, Habliston D, Ryan JW (1978) Isolation and culture of pulmonary artery endothelial cells. Tissue Cell 10:535–554

18. Scott RF, Thomas WA, Lee WM, Reiner JM, Florentin RA (1979) Distribution of intimal smooth muscle cell masses and their relationship to early atherosclerosis in the abdominal aortas of young swine. Atherosclerosis 34:291–301

19. Skalli O, Ropraz P, Trzeciak A, Benzonana G, Gillessen D, Gabbiani G (1986) A monoclonal antibody against α-smooth muscle actin. A new probe for smooth muscle differentiation. J Cell Biol 103:2787–2796

20. Slater DN, Sloan JM (1975) The porcine endothelial cell in culture. Atherosclerosis 21:259–272

21. Thomas WA, Reiner JM, Florentin RA, Lee KT, Lee WM (1976) Population dynamics of arterial smooth muscle cells. V. Cell proliferation and cell death during initial 3 months in atherosclerotic lesions induced in swine by hypercholesterolemic diet and intimal trauma. Exp Mol Pathol 24:360–374

22. Tsukada T, Rosenfeld M, Gown A (1986) Immunocytochemical analysis of cellular component in atherosclerotic lesions: use of monoclonal antibodies with the Watanabe and fat-fed rabbit. Arteriosclerosis 6:601–613

23. Tsukada T, Tippens D, Gordon D, Ross R, Gown A (1987) A muscle-actin-specific monoclonal antibody: I Immunocytochemical and biochemical characterization. Am J Pathol 127:51–60

24. Voyta JC, Via DP, Butterfield CE, Zetter BR (1984) Identification and isolation of endothelial cells based on their increased uptake of acetylated-low density lipoprotein. J Cell Biol 99:2034–2040

25. Wong MKK, Gotlieb AI (1986) Endothelial cell monolayer integrity. I. The characterization of the dense peripheral band of microfilaments. Arteriosclerosis 6:212–219

26. Wu QY, Drouet L, Carrier JL, Rothschild C, Berard M, Rouault C, Caen JP, Meyer D (1987) Differential distribution of von Willebrand factor in endothelial cells: comparison between normal pigs and pigs with von Willebrand disease. Arteriosclerosis 7:47–54

Microvascular Endothelial Cells from the Lungs *

U.S. Ryan

Introduction

The overwhelming majority of all the endothelial cells in the body are found in the pulmonary microcirculation. However, despite early recognition that many of the most important functions attributable to the pulmonary vasculature occur at the level of the small vessels of the lungs [1, 2], relatively few in vitro studies have used endothelial cells derived from the pulmonary microvasculature. In part this may be due to the relatively greater difficulties encountered in their isolation, identification, and culture.

A number of techniques have been developed for obtaining capillary endothelium from a variety of sources, using mincing techniques followed by enzymatic digestion and selection of cells via buoyant density gradients or cell sorting techniques. By and large these preparations require constant addition of growth factors and have not been applied successfuly for obtaining long-term cultures of lung microvascular endothelium.

We have been able to develop techniques that allow collection of microvascular endothelial cells from pre- and postcapillary vessels [3–6]. Originally, these were described as separate experimental protocols [3, 5], but it is now possible to harvest both arterial and venous endothelial cells from the same preparation [4].

The specific protocols that follow apply equally well to obtaining microvascular endothelial cells from a lobe, or part of a lobe of lung, of a large animal (such as cow or human), or from the lungs of small laboratory animals such as rats or rabbits.

Methods

Comprehensive step-by-step descriptions follow which will enable undertaking the complete isolation procedure from intact animal to continuous culture.

* This work was supported by grants HL 21 568 and HL 33 064 from the National Heart Lung and Blood Institute.

Preparation of Medium and Saline Solutions

Puck's Saline F with 3 × Antibiotics

To 500 ml of Puck's saline F 30 ml penicillin (5000 IU/ml)-streptomycin (5000 µg/ml) and 1.5 ml (50 mg/ml) gentamicin sulfate are added; this is three times the normal concentration of antibiotics. A total of 2-l are required for the isolation procedure. Puck's saline should be stored at 4°C.

Medium 199 with 1 × Antibiotics

Dry medium 199 (M199) powder is placed into a 1-l flask, 4.35 g sodium bicarbonate and 1000 ml Milli-Q purified water are added. Aliquots are filter sterilized and stored at 4°C. At the time of use, 2 ml penicillin-streptomycin and 0.1 ml gentamicin sulfate are added per 100 ml.

Ryan Red Medium

Dry M199 powder is placed in a 5-l flask, 11.0 g sodium bicarbonate, 4500 ml Milli-Q purified water, 0.012 g thymidine (final concentration $= 10^{-5}$ M), 250 ml pretested bovine calf serum, and 250 ml pretested fetal bovine serum are added and mixed well. The medium is then filtered into sterile 500-ml bottles, capped, and stored at 4°C. Sterility is tested as described [7]. At time of use, (per 500 ml) 10 ml penicillin-streptomycin and 0.5 ml gentamicin sulfate are added.

Preparation of Microcarriers

Dry microcarriers, 1 g, are placed in 50 ml of 70% ethanol for 4 h and then rinsed three times (20 ml for each rinse) with sterile calcium and magnesium free Dulbecco's phosphate buffered saline (PBS) containing antibiotics (penicillin at 100 IU/ml and streptomycin at 100 µg/ml).

The last PBS rinse is aspirated off and 50 ml sterile serum is added to coat the positively charged microcarrier beads which can then be stored in a refrigerator.

A count is performed on the microcarriers. They are used at a concentration of 600 microcarriers/ml of perfusion buffer.

Harvesting Arterial and Venous Microvascular Endothelial Cells: Preparation of Small Animal Lung

A small laboratory animal (e.g., rat or rabbit) is weighed, and an ear is shaved to expose the marginal ear vein. Then the needle of a butterfly infusion set

(23 × 3/4, 12″ tubing, Abbott Hospitals Inc.) is inserted into the vein and secured with tape. Nembutal diluted 50 : 50 with PBS is administered slowly (IV), until the corneal reflex is abolished (up to a dose of 5 mg/100 g body weight). Then heparin is administered (IV) at 100 units/200 g body weight.

The fur of the abdomen and neck area are sterilized with aerosolized 70% ethanol. An abdominal incision is made to expose the viscera, which are deflected to one side so that the descending aorta is exposed. Then the trachea are exposed for tracheal cannulation and the tracheal cannula is tied in place. The diaphragm is punctured carefully and the lungs are allowed to collapse. The sides of the rib cage are cut down to expose the lungs and heart. The thymus and fat tissue are removed to expose the pulmonary artery and aorta. Next a pair of blunt forceps are inserted under the aorta and pulmonary artery to pick up the suture which will secure the cannula in place.

An incision is made through the wall of the right ventricle. A cannula ["suction catheter" #1761 (Bardic Brand) C.R. Bard, Inc. modified to flare tip] is inserted into the pulmonary artery via the slit in the right ventricle. The pulmonary arterial cannula is secured with a ligature around the pulmonary artery and aorta.

An incision is made through the wall of the left ventricle. A pair of blunt forceps is inserted via the slit in the right ventricle through the ventricle wall on the dorsal side of heart. A suture can then be taken through the right ventricle and placed around the heart to secure a cannula within the left atrium. This suture is prevented from slipping by making a small nick in the left ventricle wall and locating the suture within it. A flared-tip cannula is inserted into the left atrium via the slit in left ventricle. The left atrial cannula is secured by clamping heart muscle around it with the previously placed suture.

Approximately 20 ml of air are injected into the lungs via the tracheal cannula. The lungs are then perfused blood free with warm (37 °C) PBS containing 0.1 mg/ml gentamycin at a flow rate of approximately 15 ml/min (Fig. 1a). The direction of perfusate flow is alternated several times to assist in clearance of blood from the lungs.

When the lungs are blood free, the tubing is switched to a reservoir containing cold calcium and magnesium free (CMF) PBS, 0.02% ethylendiamine-tetraacetate (EDTA) (0.04% for guinea pig lungs), 0.1 mg/ml gentamycin and approximately 600 beads/ml Jacobson-Ryan microcarriers 40–80 μm [8] (Fig. 1b).

The solution containing microcarriers is pumped into the arterial side at a flow rate of approximately 15 ml/min until the effluent is cold. The flow direction is reversed, and the perfusate from the arterial side is collected (10–15 ml; Fig. 1c).

The direction of flow is reversed again, and the flow is increased to approximately 16 ml/min. The perfusate is collected from the venous side (again 10–15 ml). These steps are repeated until 8–10 centrifuge tubes are collected, increasing the input flow rate with each tube.

The tubes are centrifuged at 250 g for 15 min at 4 °C. PBS is aspirated off and 2 ml Ryan red medium are added. In each tube the microcarriers are resuspended and recentrifuged.

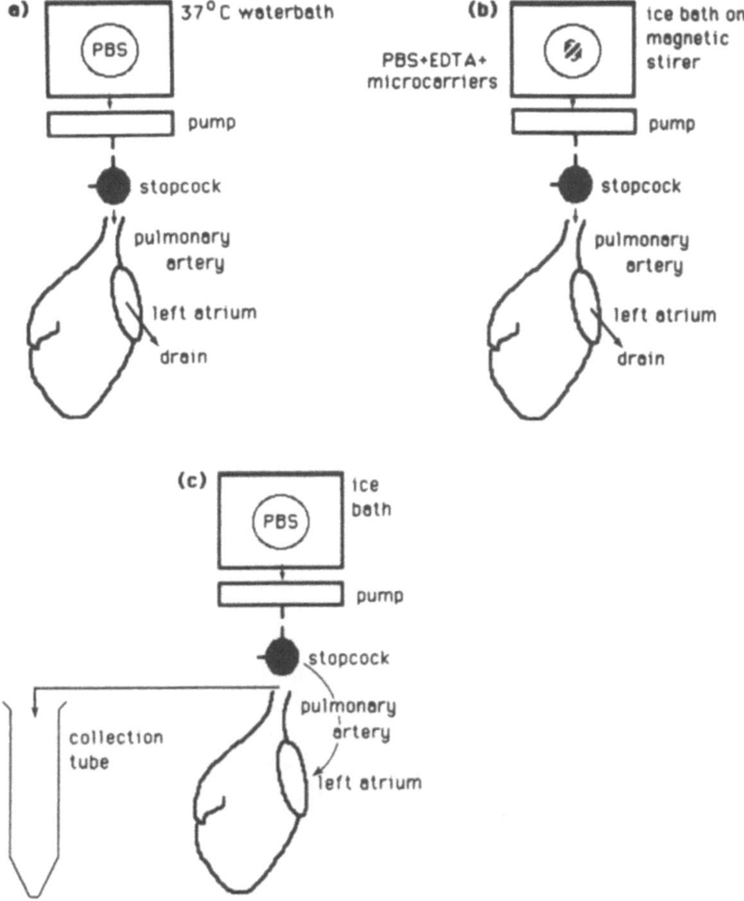

Fig. 1. Isolation of microvascular endothelial cells using microcarriers [from 4]. *PBS,* phosphate buffered saline

Each tube of microcarriers is seeded into a separate T25 flask; these are then placed in a 37 °C incubator with 5% CO_2 and left undisturbed for 1 week. This allows the cells on the microcarriers to attach to the flask.

After 1 week, approximately ½ of the old medium is aspirated off and new medium is added to each flask, keeping the flasks horizontal during the feeding operation. When the cells are attached and dividing, they can be fed twice weekly.

Preparation of Lobe of Lung of Large Animal

Similar methods to those described for the lungs of small animals can be used to obtain microvascular endothelial cells from a lung lobe of a larger animal,

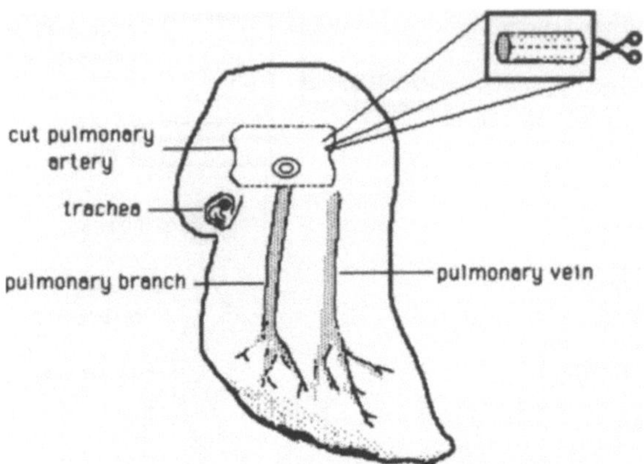

Fig. 2. Calf lung lobe perfusion. The pulmonary artery is sliced open longitudinally revealing the branches. A flap is cut around the opening of one of the branches that serves the lower lobe and a suture tied in place. The perfusion is started and a second catheter is placed in the vein where the perfusate exits [from 4]

such as cow, or of a human. A small lobe of lung is selected, a ligature (1-0 silk) is placed around a pulmonary artery, and a catheter (sterile Tygon catheter tubing, 1/16″ ID × 1/8″ × 1/32″ wall, with flared end) is inserted into the artery and securely tied in place (Fig. 2). A catheter is gently inserted into the corresponding pulmonary vein and secured with a ligature.

The lobe is perfused via a pulmonary artery (19–25 ml/min) with sterile phosphate buffered saline (PBS) 37 °C, containing 0.1 mg/ml gentamycin, until free of blood. The lobe is then placed on a plastic bag of ice and perfused with cold (4 °C) CMF PBS (containing EDTA, 0.02%, and polystyrene microcarriers [8]). Microcarriers (40–80 μm diameter) previously soaked overnight in 100% calf serum (CS) are perfused at a concentration of 600/ml. When the effluent emerging is cold, the direction of flow is reversed periodically to loosen the cells, and the bead-cell harvest is collected first from the arterial then from the venous side (at least 25 ml). The process is repeated as described in the previous section.

Cold shock and EDTA cause the endothelial cells to detach from the vessels under conditions such that the cells remain attached to the microcarriers. The

Fig. 3. a Initial isolate of microcarriers and cells from rabbit lung perfusion approximately 2 h after seeding into culture flasks. Most cells are rounded up and still attached to beads, some are beginning to flatten and to migrate onto the flask. **b** A primary isolate similar to that shown in **a**, but approximately 12 days after seeding. Monolayer patches are formed as a result both of migration from beads and of division. **c** Monolayer of rabbit microvascular endothelial cells in the seventh passage. All beads have been washed away during medium changes. **d** As shown in the sequences **a–d** endothelial cells isolated on microcarriers can be introduced into flasks, allowed to form stationary monolayer cultures, and then seeded back on to microcarriers for scale-up in roller bottles. This sequence provides better quantitation of cultures and allows easier characterization (monolayer growth, immunocytochemical localization) than does placing an isolate such as that shown in **a** directly into a roller bottle

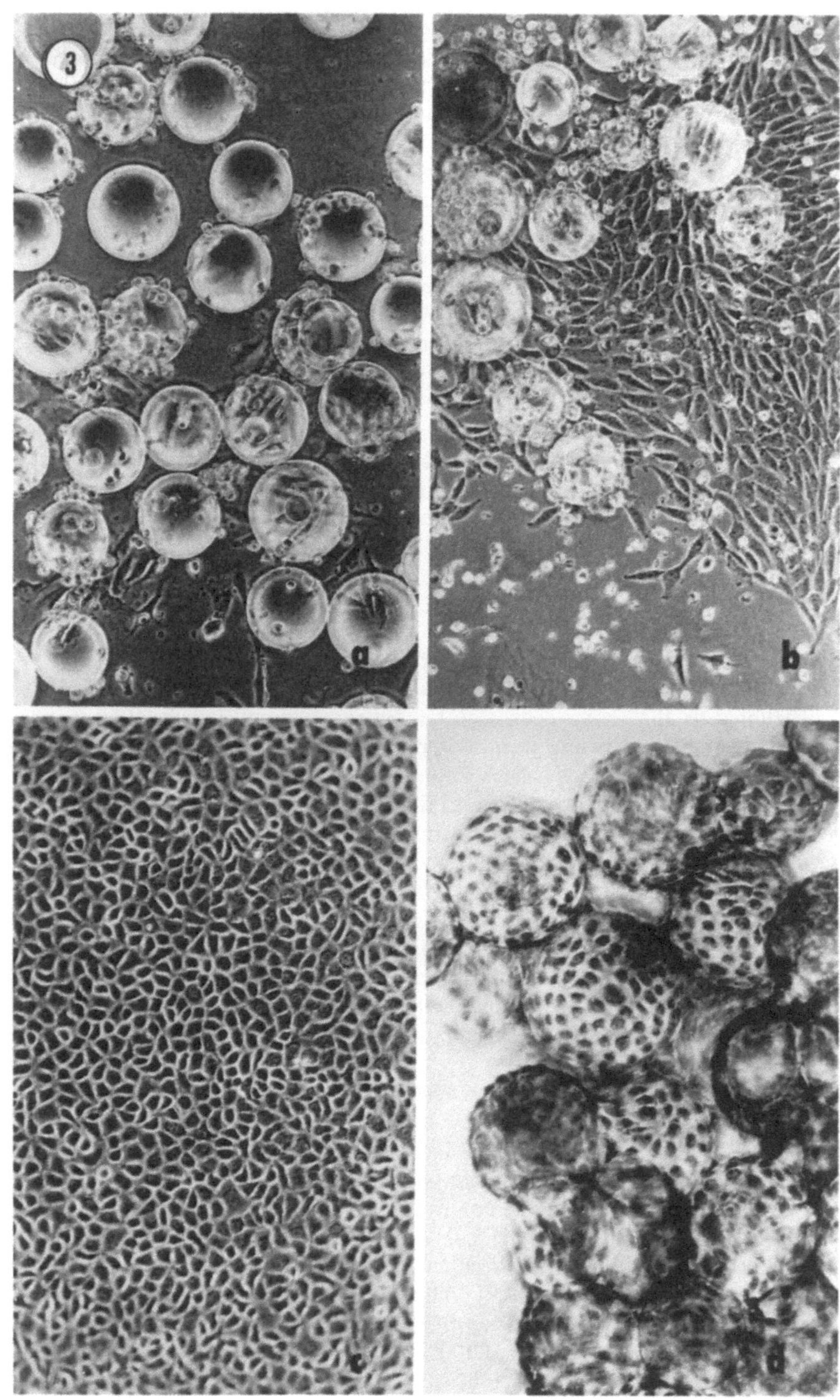

selective attachment to microcarriers is apparently aided by the tight fit of the beads within vessels of the same diameter. Beads do not emerge on the arterial side, all being trapped at the immediate postcapillary level.

Endothelial cells recovered on microcarriers are spun at 250 g for 15 min at 4°C and then allowed to migrate from the microcarriers on to Primaria flasks (25 cm^2) in 2 ml of M199 containing 10% fetal bovine serum (FBS), 5% CS, 100 IU/ml penicillin, 100 µg/ml streptomycin, and 0.05 mg/ml gentamycin (Fig. 3a, b).

The cells are maintained at pH 7.4 and are not fed or moved for at least 1 week. After 1 week, half of the medium is gently replaced while the flask remains horizontal. In the 2nd week, the medium is replaced once.

Thereafter, dividing cultures are fed twice weekly with 5 ml medium. The number of cells obtained in the initial isolate is small, and primary cultures of microvascular endothelial cells harvested from the venous side of the microvasculature appear to grow more slowly than those harvested from the arterial side. By the above procedure, cell lines obtained from both the arterial and from the venous side of the pulmonary microvasculature, which have never been exposed to proteolytic enzymes, can be established and the cultures can subsequently be scaled up on 150–200 µ microcarriers [9, 10] or maintained as monolayers in flasks (Fig. 3c, d).

Long-Term Culture on Microcarriers

Endothelial cells that have been isolated by mechanical means [10] can be passaged mechanically using a rubber policeman [11]. However, to obtain large-scale cultures, methods have been developed for seeding endothelial cells onto polyacrylamide microcarriers and maintaining them in roller bottle cultures [5, 9, 10]. Endothelial cells can be transfered onto fresh microcarriers, or from microcarriers back onto flasks (or any planar surface), and vice versa, without the necessity for exposure to trypsin. Microcarrier methods provide a system for scale-up and efficient use of nutrients in medium that conserves laboratory space and is ideal for shipping. Microcarriers also provide a means of continuous culture that is well suited for experiments involving interactions of fluid phase substrates or agonists with cell-bound enzymes or receptors. They have been particularly useful in studies of endothelium derived vasoactive substances [12].

Microcarriers (Cytodex-3, Pharmacia) are prepared by following the instructions in the pamphlet provided with the microcarriers: Essentially, the beads are hydrated, swollen, washed, autoclaved, and stored at 4°C in Ca^{++}- and Mg^{++}-free phosphate buffered saline (CMF PBS).

To set up roller bottle cultures of endothelium avoiding exposure of cells to enzymes, plastic pipettes must be used for all bead suspensions. Roller bottles must be siliconized and autoclaved. Cytodex-3 bead stock solution is added to a sterile siliconized 100-ml roller bottle. The bottle is tilted to let the beads settle, and PBS is carefully withdrawn.

The cells are detached from a T75 flask with a rubber policeman (Fig. 4), aspirated well to break up clumps, and seeded into a roller bottle in a total

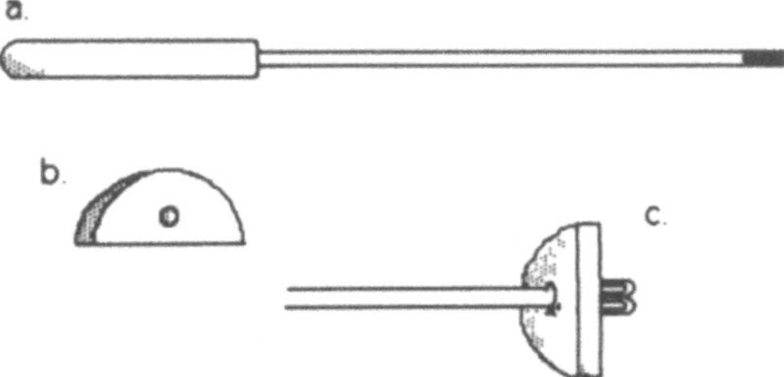

Fig. 4a–c. This rubber policeman can be bent to accommodate flask shapes and may be autoclaved repeatedly. The rubber stopper piece can be shaped to accommodate the individual needs of the investigator. **a** Copper wire is cut with a jewelers saw (blade no. 8/0 no. 54478, Ted Pella, no. 54470). **b** A piece of rubber stopper is trimmed smoothly with a sharp razor blade. **c** The split end of the wire is filed and inserted through the piece of stopper. The cut ends are curled back to retain the stopper piece [from 11]

volume of 13 ml Ryan red medium. The bottle is gassed for 60 s with 5% CO_2 in air, closed tightly, and taped securely. The bottles are placed in a 37 °C roller bottle incubator set for 1.5 rpm.

The microcarrier cultures are fed twice weekly by tilting the bottle and allowing the beads to settle, carefully drawing off old medium, replenishing with fresh medium, gassing as before, and transfering to an incubator at 37 °C.

The bead concentration may vary. As a rule of thumb, 1.3×10^5 beads + 10^7 endothelial cells (1 T75) yields confluence on each bead in 2 days.

Concluding Comments

There are a number of structural and physiological differences between endothelium of large vessels and endothelium of the microvasculature that indicate a need to study endothelial cell cultures derived from various levels of the circulation. The methods described in this chapter allow collection of pre- and postcapillary endothelia without using enzymic digestion. Light and electron micrographs of lungs into which microcarriers had been pumped, but from which the cell-bead harvest had not been collected show that the beads lodge in vessels of approximately the same diameters as those of the beads, and that the endothelial cells attach closely onto the bead [3]. Beads of appropriate dimensions (40–80 µm) consisting of a polystyrene core with a special coating and positive charge have been synthesized [8] and are available to interested researchers. Once monolayers of endothelial cells are established on flasks they can be scaled up, using 175 µm diameter microcarriers in roller bottles as described for large vessel endothelium [9, 10].

Interestingly, although the microcarrier perfusion techniques can be adapted successfully for obtaining both pre- and postcapillary microvascular endothelial cells, when we attempted to collect pulmonary capillary endothelial cells by perfusion with 5–10 µm beads, no cells were harvested. Subsequent examination of the lungs indicated that the 5–10 µm beads had been phagocytosed by the endothelial cells. Arising from this finding, we have recently been examining the phagocytic properties of endothelium using endothelial cells in culture [13–15].

Previous methods for collection of endothelial cells from the pulmonary microvasculature involved retrograde perfusion with collagenase [16] or mincing followed by enzymic digestion modification of [17]. For many reasons it is frequently desirable to isolate and cultivate endothelial cells without exposing them to enzyme treatments. Methods for mechanical harvest and culture of endothelial cells from large vessels, avoiding exposure to proteolytic enzymes, have been described [10, 18]. However, because it is not possible to scrape endothelial cells from the microcirculation using a scalpel, a different approach had to be devised [3–6].

The methods described take advantage of certain aspects of the previously established techniques. For example, we had shown previously [16] that endothelial cells loosened by enzymic digestion could not be collected by anterograde perfusion (artery to vein) but were, on account of their size, trapped by the sieving action of the pulmonary capillary bed. However, the cells could be collected if the direction of flow was reversed and cells were collected from the inflow (arterial) cannula.

Similarly, we had shown [10, 18] that endothelial cells have a great propensity for migrating onto and off beads (confluent microcarrier to an empty flask or from a confluent monolayer to a naked microcarrier). Here we describe techniques that combine nonenzymatic loosening of cells (cold shock and EDTA) with both anterograde and retrograde perfusion of the lungs. We also exploit the ability of endothelial cells to attach to microcarriers, specifically in those vessels of similar or slightly lesser diameter than that of the microcarriers, i.e., where there is close adhesion of the bead against the vessel wall [3]. Endothelial cells are not collected from larger vessels where unrestricted flow can occur, nor from the capillaries. Inasmuch as no microcarriers transverse the capillary bed, collection of precapillary arterial microvascular cells can be achieved by introducing microcarriers into the arterial circulation, and endothelium of postcapillary venules can be collected by introducing microcarriers on the venous side of the perfusion [4].

References

1. Ryan US (1986) Metabolic activity of pulmonary endothelium: modulation of structure and function. Ann Rev Physiol 48:263–277
2. Ryan US (1989) Endothelial processing of biologically active materials. In: Fishman AP (ed) The pulmonary circulation; normal and abnormal. University of Pennsylvania Press, Philadelphia
3. Ryan US, White L, Lopez M, Ryan JW (1982) Use of microcarriers to isolate and culture pulmonary microvascular endothelium. Tissue Cell 14:597–606
4. Ryan US, White L (1986) Microvascular endothelium, isolation with microcarriers: arterial, venous. J Tissue Culture Methods 10:9–13
5. Ryan US (1984) Culture of pulmonary endothelial cells on microcarrier beads. In: Jaffe EA (ed) Biology of the endothelial cell. Njihoff, Leiden, pp 34–50

6. Ryan US, Ryan JW (1984) Inflammatory mediators, contraction and endothelial cells. In: Courtice FC, Garlick DG, Perry MA (eds) Progress in microcirculation research II. University of New South Wales, Sydney, pp 424–438

7. Cour I, Maxwell G, Hay RJ (1979) Tests for bacterial and fungal contaminants in cell cultures as applied at the ATCC. TCA Manual 5:1157–1160

8. Jacobson BS, Ryan US (1962) Growth of endothelial and HeLa cells on a new multipurpose microcarrier that is positive, negative, or collagen coated. Tissue Cell 14:69–83

9. Ryan US, Maxwell G (1986) Microcarrier cultures of endothelial cells. J Tissue Culture Methods 10:7–8

10. Ryan US, Mortara M, Whitaker C (1980) Methods for microcarrier culture of bovine pulmonary artery endothelial cells avoiding the use of enzymes. Tissue Cell 12:619–635

11. Ryan US, Hart MA (1986) Electron microscopy of endothelial cells in culture: III. freeze-fracturing. J Tissue Culture Methods 10:37–40

12. Johns A, Khalil RA, Ryan US, van Breemen C (1988) Endothelium-derived relaxing factor. In: Ryan US (ed) Endothelial cells vol III. CRC Press, Boca Raton, pp 50–60

13. Ryan US (1988) Phagocytic properties of endothelial cells. In: Ryan US (ed) Endothelial cells, vol III. CRC Press, Boca Raton, pp 33–49

14. Ryan US (1988) The macrophagelike properties of endothelial cells. NIPS 3:93–96

15. Ryan US, Vann JM (1988) Endothelial cells: a source and target of oxidant damage. In: Simic MG, Taylor KA, Ward AF (eds) Oxygen radicals in biology and medicine. Plenum, New York, pp 963–968

16. Habliston DL, Whitaker C, Hart MA, Ryan US, Ryan JW (1979) Isolation and culture of endothelial cells from the lungs of small animals. Am Rev Respir Dis 119:853–868

17. Del Vecchio PJ, Ryan JW, Chung A, Ryan US (1980) Capillaries of the adrenal cortex possess aminopeptidase A and angiotensin-converting enzyme activities. Biochem J 186:605–608

18. Ryan US, Maxwell G (1986) Isolation, culture and subculture of bovine pulmonary artery endothelial cells: mechanical methods. J Tissue Culture Methods 10:3–5

Microvascular Endothelial Cells from Brain*

P.D. Bowman, M. du Bois, K. Dorovini-Zis, and R.R. Shivers

Introduction

In addition to providing a nonthrombogenic surface, synthesizing factor VIII, and participating in vascular tone regulation, most endothelial cells of the vasculature of the brain participate in the blood-brain barrier. The blood-brain barrier severely restricts the entry of macromolecules into nervous tissue and provides a selective permeability barrier to a variety of other solutes [5, 23]. The site of this barrier is the brain endothelial cells which exhibit continuous tight junctions, very low amounts of transcytosis, and a polar distribution of transport carriers between luminal and antiluminal plasma membranes.

The relative lack of connective tissue and high ratio of microvessels to large vessels make neural tissue a popular source of microvessels for endothelial cell isolation. Additionally, the microvessels are surrounded by a relatively tough basement membrane, and this protects them from mechanical damage which destroys the surrounding tissue. Microvessels can be separated on various gradients or glass beads or by sieving on mesh. Treatment with collagenase generally follows separation to remove the basement membrane and allow separation of pericytes.

Progress in cultivating brain microvessel endothelial cells has been contemporaneous with isolating and cultivating microvessel endothelial cells from other tissues. Shortly after Wagner and Matthews [62] reported that collagenase digestion could be used to isolate microvessels from epididymal fat pad, Panula et al. [37] reported that endothelial cells from the brain could be maintained in culture. About the same time, De Bault and Cancilla [15] and Bowman et al. [6] reported the sustained growth of brain microvessel endothelial cells in culture. During the 1980s, a variety of reports have confirmed that isolating and cultivating microvessel endothelial cells from the brain can be accomplished without great difficulty [7, 9, 11, 14, 16, 17, 25, 41, 52, 57, 58]. In addition to providing a system for studying the unique cell biology of this cell type, brain microvessel endothelial cells are readily available for studying general microvessel endothelial cell properties.

---

* The opinons and assertions contained herein are the private views of the authors and are not to be construed as official nor do they reflect the views of the Department of the Army or the Department of Defense (AR 360-5)

Description of Methods and Materials

The basic techniques for isolating microvessels from brain are derived from the method of Siakotos [49, 50] or Joó and Karnushina [30]. Siakotos first demonstrated the feasibility of obtaining metabolically active microvessels on a large scale by homogenization, differential centrifugation, and glass-bead column filtration. In Joó's micromethod, more appropriate for small-scale isolations, brain microvessels were mechanically dissociated by forcing minced brain tissue through nylon meshes of different pore sizes followed by differential centrifugation. The following is a description of materials and methods that, in our experience, have yielded pure populations of brain microvessel endothelial cells.

Source Material

Figure 1 outlines a general method for isolating and cultivating microvessel endothelial cells from the brain. Starting material from rat, murine, bovine, porcine, or human sources has been used successfully (step 1). Rodents are killed by cervical dislocation or decapitation in a guillotine. Brains of larger animals obtained from a slaughterhouse are transported on ice in plastic bags. Brains are quickly removed, dura and arachnoid membranes removed, and cerebrum separated from cerebellum and brain stem. Regardless of starting species, gray matter is usually cut from cortices and serves as starting material since it contains the highest density of microvessels [33].

Isolation of Microvessels

All further work is performed in a laminar flow hood under conditions as sterile as possible. The gray matter is placed in a Petri dish and minced to approximately 1 mm^3 with crossed scalpel blades or a razor blade (step 2). A razor blade held in a window scraper and changed frequently is convenient when large amounts of gray matter, such as from bovine brains, are being prepared.

Next, the cortical material is homogenized (step 3). During this procedure, the nervous tissue is destroyed and microvessels are largely freed of neuropil. One volume of minced brain is added to one volume of medium 199 and homogenized with 10–20 strokes of a loose fitting (0.25-mm clearance) teflon pestle. For large volumes, this is best accomplished with a motor drive at approximately 400 rpm. Medium 199 with Hank's salts (GIBCO) is used in all steps of the isolation procedure. It can be prepared 10× and filter sterilized. The pH is not adjusted until it is diluted to 1× with sterile distilled or deionized water and the bicarbonate replaced by adding sterile 1 M HEPES buffer to a final concentration of 15 mM. Penicillin (50 units/ml), streptomycin (50 µg/ ml), and amphotericin B (5 µg/ml) are added to all solutions. From this point, all steps should be monitored under the microscope to assess the degree of microvessel separation from nervous tissue and to make adjustments accord-

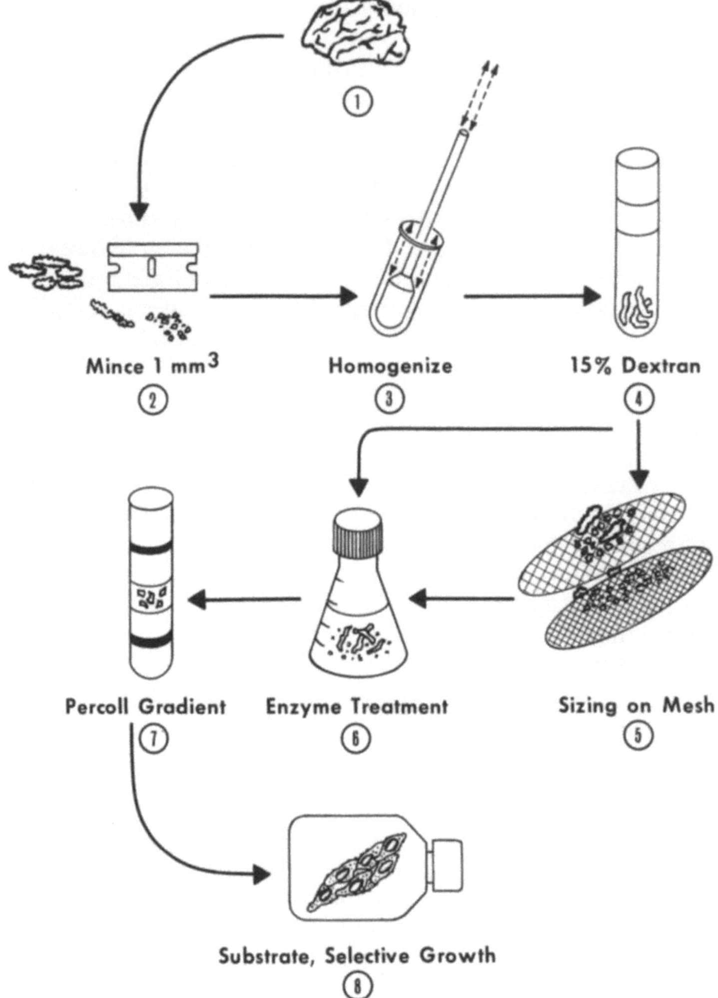

Fig. 1. General scheme for isolating and cultivating brain microvessel endothelial cells; see text for detailed description

ingly. For example, brains from older animals may require more homogenization.

While treatment with collagenase or trypsin in lieu of homogenization is effective in isolating microvessels from the minced tissue [9, 39, 64], some structural features, such as the basement membrane, can be lost. This makes separation of large and small vessels difficult [41], thus contamination by pericytes or cells from larger vessels becomes a greater problem. We have used the protease dispase at 0.5% with good results: viability is increased over homogenization, and it does not attack the basement membrane significantly [9]. However, if larger vessels are present, it may release clumps of these cells which will contaminate the final culture. Therefore, if purity of the final culture is important, it is better to treat with enzymes after the intact microvessels have been isolated.

After homogenization, the homogenate is mixed with a volume of 27.3% dextran to yield a final dextran concentration of 15%, then centrifuged at $5800 \times g$ at $4\,°C$ in a swinging bucket rotor to separate microvessels from the bulk of the damaged neuropil (step 4). Dextran (clinical grade, average molecular weight 70 000; Sigma) is prepared as a 30% solution using double distilled or deionized water, then autoclaved. Before use, it is made $1 \times$ in medium 199 by adding $10 \times$ medium 199. It is then diluted to 15% with medium 199 containing $15\,\mathrm{m}M$ HEPES.

The pellet containing microvessels, red blood cells, nuclei, and other debris is resuspended in medium 199 and examined using phase microscopy. If segments of larger vessels are present, sizing on nylon mesh (step 5) can be incorporated, with some loss of microvessels. By using $149-249\,\mu\mathrm{m}$ nylon mesh in tandem with $20-53\,\mu\mathrm{m}$ mesh, larger vessels are removed on the former while microvessels are trapped on the latter. Some larger extended microvessels will be lost at this time, however. If step 5 is used, the microvessels are recovered by inverting the mesh and washing with medium 199 into a Petri dish.

Removal of Basement Membrane and Pericytes

The pellet from step 4 or microvessels from step 5 are then treated with protease to remove the basement membrane and pericytes, which invest the capillary tube, and free the endothelial cells (step 6). In our experience, a mixture of collagenase and dispase from Boehringer/Mannheim has given good results. A 0.1% solution prepared in medium 199 and filtered ($0.22\,\mu\mathrm{m}$) generally removes the basement membrane without apparently gaining access to the endothelial cells; they tolerate it for as much as 16 h. Electron microscopic examination has demonstrated that most of the basement membrane is gone, and pericytes are not obvious, after 5 h treatment. However, the dissociation is allowed to proceed for 16 h because steps $1-5$ usually are not accomplished until late in the day and continuing the dissociation until the next morning has not proven detrimental.

Separation of Pericytes from Endothelial Cells

The enzyme treatment is terminated by washing with medium 199 and pelleting at $800 \times g$. At this point there is considerable debris from the dissociation procedure and it is useful to separate endothelial cells by Percoll gradient centrifugation (step 7). Stock Percoll (Sigma) is prepared by mixing one part $10 \times$ medium 199 to nine parts Percoll, making it $15\,\mathrm{m}M$ in HEPES buffer, and adjusting the pH to 7.4. Prepare 45% Percoll by mixing 55 parts medium 199 and 45 parts stock Percoll. This is dispensed into sterile polycarbonate centrifuge tubes, capped, and spun at $30000 \times g$ for 40 min to establish the

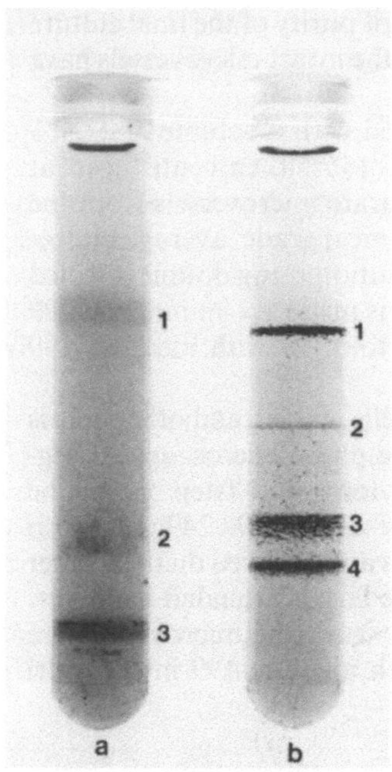

Fig. 2a, b. Percoll gradient separation of microvessel endothelial cells. **a** *Band 1* contains cell debris and damaged cells; *band 2* contains microvessel segments and some single endothelial cells; *band 3* is comprised of red blood cells that have been freed from the capillary lumen. The position of *band 2* is not always at this density, most likely due either to differences in numbers of red blood cells trapped with the endothelial cells or to incomplete removal of basement membrane. **b** Density gradient markers: (*1*) 0.018 g/ml; (*2*) 1.033 g/ml; (*3*) 1.049 g/ml; (*4*) 1.062 g/ml

gradient. The material from the enzyme dissociation procedure is layered onto the preformed gradient and centrifuged at room temperature at $1500 \times g$ for 15 min to separate endothelial cells from debris. Figure 2 illustrates a Percoll gradient separation of microvessel endothelial cells. Fractions from the gradient are collected into a sterile 24-well multiplate and examined under the microscope. All wells containing clumps of refractile cells are diluted in medium 199 and centrifuged to remove Percoll. The cells are generally 80% viable by trypan blue exclusion and usually 50% of these will settle out and attach if provided with an appropriate substrate such as collagen, gelatin, or fibronectin. If single cells are required, the clumps of endothelial cells can be further treated with 0.25% trypsin/0.01% EDTA with a further loss in viability. Regardless of further treatment, the cells are washed twice in medium 199 containing plasma-derived serum and plated or frozen.

In Vitro Procedures

Quantitation

Many of the isolated endothelial cells have lost the tubelike appearance of microvessels, but are clumped together. Electronic particle counting is not possible, so cell numbers are best determined by staining with 0.5% crystal violet in 0.1 *M* citric acid and counting in a hemocytometer [1]. The plating efficiency can be estimated from the number of cells within each unit growth area relative to the number of these cells which have attached during the previous 24 h.

Freezing for Long-Term Storage

The dissociated microvessel endothelial cells can be frozen and stored in liquid nitrogen. If bovine or porcine brain has been used as a source for microvessels, many more are usually available than can be set up for culture. The yield of microvessel endothelial cells obtained from the cortices of three bovine brains, for example, is approximately 10^8 cells. For freezing, twice the number of cells necessary for an experiment are resuspended in complete medium, an equal volume of medium 199 containing 20% fetal calf serum and 20% DMSO is added dropwise, and the cells are dispensed into 1-ml cryovials (Nunc) and frozen according to standard techniques [44]. When needed, a vial is thawed by warming in a 37°C shaking water bath, wiped with 70% ethanol, and suspended in 10 ml of complete medium. The cells are centrifuged at $800 \times g$, resuspended in fresh medium, and dispensed as required. There is usually a 50% decrease in viability after freezing and thawing.

Substrate Requirements

Although both attachment substrates and growth factors for endothelial cells are covered elsewhere in this volume, it should be noted that plating of brain microvessel endothelial cells on uncoated tissue grade plastic usually leads to low plating efficiency and perhaps selection for a subpopulation of cells. Collagen and gelatin work equally well with brain microvessel endothelial cells to increase their plating efficiency, but they may interfere with the determination of protein or cell numbers when biochemical studies are to be undertaken. A coating of fibronectin ($10-100$ µg/cm^2; Sigma) does not appear to contribute significantly to protein measurement or interfere with DNA assays and should be used in this situation. For ultrastructural studies, the cells can be grown on fibronectin-coated gelatin membranes (ICN).

Media, Sera, and Supplement Requirements

As has been reported for large vessel endothelial cells by Wall et al. [63], plasma-derived serum supports the growth of brain microvessel endothelial cells but not smooth muscle cells or pericytes. Equine serum, 10% plasma-derived (Hyclone Laboratories), yields good plating efficiency and growth of brain microvessel endothelial cells and tends to inhibit the occurrence of non-factor VIII immunoreactive cells in the culture. In the best situation, no non-factor VIII antigen-reactive cells are observed at confluence, but this occurs in only one in three isolations. Addition of 100 µg/ml heparin and 50 µg/ml endothelial cell growth supplement [56] helps to achieve confluency faster.

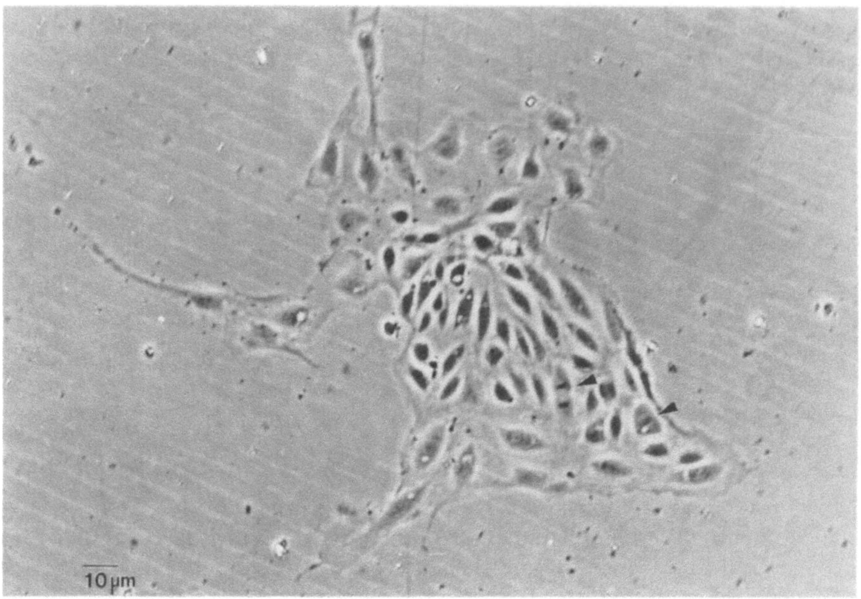

Fig. 3.. Bovine brain microvessel endothelial cells at 3 days in culture. At this early time, a tendency to remain closely associated with the original clump of endothelial cells is apparent as is a more fibroblast-like than epithelial-like morphology. *Arrows* denote mitotic cells and indicate vigorous growth

More recently we have found that the use of MCDB 131 [31] obtained from Clonetics supplemented with 2% fetal bovine serum, 10 ng/ml epidermal growth factor, and 1 µg/ml hydrocortisone eliminates the problem of non-factor VIII immunoreactive cells in the culture. We have had limited experience with the method of brain microvessel endothelial cell cultivation described by Gospodarowicz et al. [26] which utilizes an extracellular matrix synthesized by corneal endothelial cells now obtainable from Accurate Chemical Co. Although it is effective, it is more cumbersome than the use of MCDB 131 plus supplements.

In general, the seeding of 2500–5000 cells per cm² yields confluent cultures containing 100 000 cells per cm² by 1–2 weeks. This represents about four population doublings. Figure 3 is a phase-contrast photomicrograph of bovine brain microvessel endothelial cells after 3 days of culture on fibronectin-coated plastic and in alpha MEM with 10% plasma-derived serum. Under these conditions of culture, they remain closely apposed to one another and in contact with the original colony. They exhibit a more fibroblast-like than epithelial-like morphology.

In Vitro Characterization

Immunoreactive Factor VIII Antigen

The morphology of cultured cells is generally not a reliable characteristic to determine the origin of cell type. There are a variety of cell types, including astrocytes, microglia, smooth muscle cells, pericytes, and fibroblasts, that grow very well in 10%–30% fetal bovine serum and that can readily overgrow a mixed culture. The most reliable marker for the endothelial origin of a cell remains the expression of factor VIII: von Willebrand factor [29]. Synthesized by endothelial cells and megakaryocytes, this large macromolecular complex of blood coagulation factor VIII and von Willebrand factor serves an essential role in the intrinsic coagulation pathway. The presence of this complex in cells or tissues is usually determined with commercially available antibody to factor VIII by immunofluorescence (factor VIII R Ag). All of the authors listed in Tables 1 and 2 who tested for the presence of factor VIII R Ag factor by immunofluorescence or immunoperoxidase utilizing specific antibody reported it to be present, at least initially, in their cells. (Joó and Spatz did not test for it). The Wiebel-Palade body, an organelle often observed in large vessel endothelial cells, has been shown to contain factor VIII: von Willebrand factor [28, 61]. In large vessel endothelial cells [53] there are both a constitutive pathway and a regulated pathway originating from the Weibel-Palade body. Presumably, only a constitutive pathway is present in microvessel endothelial cells from brain as only one author who examined endothelial cells using transmission electron microscopy reported them present [39].

If one has a microscope fitted for epifluorescence, endothelial cells grown on plastic can be used for immunofluorescence. We generally fix just-confluent cultures with a mixture of ethanol and acetone (1 : 1) at –20 °C for 10 min, then air dry. After blocking with normal goat serum (1 : 10 in phosphate buffered saline, PBS), rabbit anti-factor VIII antibody (1 : 10; Behring Diagnostics) absorbed with 100 µg/ml human fibronectin is applied. After 30 min incubation followed by three 10-min washes with PBS, fluorescein- or rhodamine-conjugated goat antirabbit (1 : 50) is applied (30-min incubation) and after three washes the area to be examined is coated with 10% glycerol in PBS. A glass coverslip is applied and a piece of the tissue culture flask big enough to fix on the stage is cut out with a jeweler's saw. The area of interest can then be examined by epifluorescence with the appropriate excitation and filtration. If one does not have access to an epifluorescence microscope, cells can be grown on gelatin-coated coverslips which, after processing as above, can be mounted cell side down on a microscope slide.

Other Endothelial Cell Markers

A variety of other markers have been used with less frequency to establish endothelial origin. Angiotensin converting enzyme participates in blood pressure homeostasis and has been reported to be present in brain microvessel endothelial cells [7, 17, 41]. Freshly isolated microvessels have high activity for alkaline phosphatase and gamma glutamyl transpeptidase, but the activities decrease fairly rapidly in culture. Nonetheless, alkaline phosphatase has been used as a marker for microvessel endothelial cells in vitro [11].

Recently, the highly fluorescent derivative of acetylated low density lipoprotein has been used to characterize endothelial cells [60, 65]. The utility of this marker in vitro is still questionable, however, since there have been conflicting reports of its presence. Carson and Haudenschild [11] and Tontsch and Bauer [57] found it present while Rupnick et al. [41] and Vinters et al. [48] found it to be absent.

Ultrastructural Features

In an extensive cytochemical study of enzymes of the blood-brain barrier, Vorbrodt [59] reported on their distribution in endothelial cells, but it is not yet known whether a similar distribution occurs in vitro. The mitochondrial content of endothelial cells from brain microvessels is about twice that of nonbrain endothelial cells. This may derive from the high energy requirement for active transport [35], but this characteristic has not been used as a marker for endothelial origin. In addition, the existence of unique brain microvessel endothelial cell macromolecules is being investigated [54]. The sine qua non of blood-brain barrier microvessel endothelial cells are the continuous tight junctional complexes and restricted transcellular transport.

Tight Junctions in Thin Sectons: The tight junction, or zonula occludens, was first described in epithelia by Farquhar and Palade [22]. In addition to serving as a permeability barrier for movement of material between cells, it also segregates the apical from the basolateral aspect of the cell. The ultrastructural appearance of this structure in plastic-embedded thin sections consists of a series of punctate contacts in which the outer leaflets of adjacent unit membranes appear to fuse [43]. Reese and Karnovsky [40] showed that in brain endothelium this structure restricted the movement of horseradish peroxidase across the endothelium. Considerable progress has been made in determining the protein portion of the tight junction from epithelial cells. The immunolocalization of a 225 000 dalton protein in the tight junctional complex in arterial endothelial cells has established it as one component of this structure [55]. Whether the same element is present in the microvessel endothelial tight junction is unknown. The presence of tight junctional complexes in a culture distinguishes the endothelial cell from many other potential contaminating cell types, and the junctions would be required to obtain meaningful data on transport. The freeze fracture study by Nagy et al. [34] indicates that microvessel endothelial cells should contain complex anastomosing arrays of tight junctions and no gap junctions; such has been our experience in properly prepared cultures.

The restriction of movement of horseradish peroxidase across a monolayer of cultured brain microvessel endothelial cells is a method for examining the nonpermeability of the cells' junctional complexes [19, 20]. Cells can be grown on plastic or on permeable collagen-coated nylon disks [12]. Horseradish peroxidase (Sigma, Type VI) is made up 1 mg/ml in medium 199. Following removal of growth medium, the cell layer is briefly rinsed in serum-free medium and incubated for 5–10 min at 37 °C with the horseradish peroxidase. The horseradish peroxidase is removed and the cell layer is fixed immediately with 2.5% glutaraldehyde and 2% paraformaldehyde in 0.2 M cacodylate buffer (pH 7.4) for 1 h at 4 °C. At the end of fixation, the cultures are rinsed in 0.2 M cacodylate buffer and incubated in 0.05 M TRIS HCl buffer (pH 7.6) containing 0.5 mg/ml 3,3'-diaminobenzidine and 0.01% hydrogen peroxide for 1 h at 4 °C for histochemical demonstration of horseradish peroxidase. The cells are then washed with buffer and postfixed in 1% buffered osmium tetroxide for 1 h, dehydrated, and embedded in Epon-Araldite epoxy resin (Polysciences). Unstained thin plastic sections cut perpendicular to the growth surface are examined by transmission electron microscopy for exclusion by the tight junction. In intact monolayers, very little or no horseradish peroxidase reaction product is seen beneath the endothelial monolayer, and the intercellular clefts are impermeable to horseradish peroxidase (Fig. 4a). Only occasional small deposits of the tracer decorate the apical surface of the cells, since most of the horseradish peroxidase is washed away during fixation. When 7- to 17-day-old confluent bovine brain microvessel endothelial cells are exposed to hyperosmotic solutions of arabinose or urea, interendothelial tight junctions become

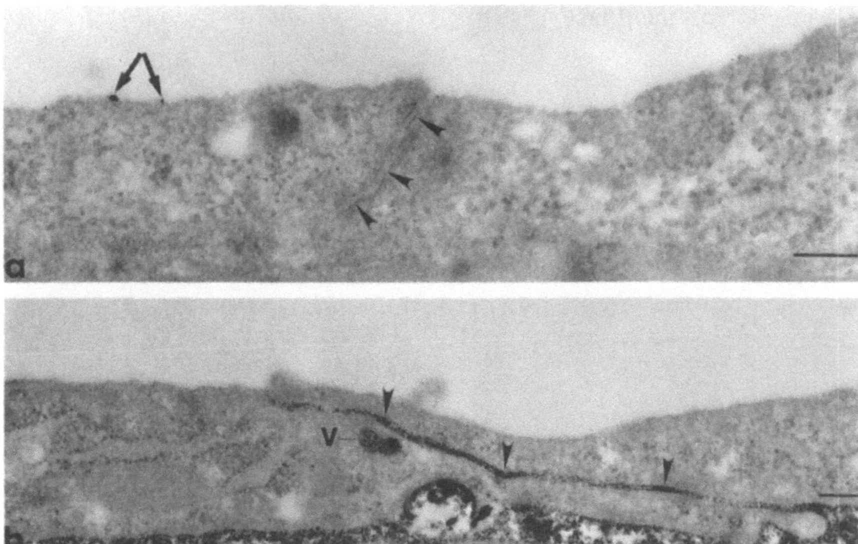

Fig. 4. Brain microvessel endothelial cell culture grown on collagen-coated substrate exposed to horseradish peroxidase for 5 min. *Arrows* denote infrequently seen reaction product on the cell surface of unmanipulated cell layer. Interjunctional clefts (*arrowheads*) and basal surface are free of reaction product (*bar*, 0.2 µm). **b** Bovine brain microvessel endothelial cell culture incubated with 3 M urea and horseradish peroxidase for 3 min at 37 °C. Horseradish peroxidase penetrates the intercellular clefts (*arrowheads*) between successive permeable tight junctions and forms dense deposits on the basal cell surface. Occasional cytoplasmic vesicles (*v*) are labelled with the tracer. *Bar*, 0.2 µm

reversibly permeable to horseradish peroxidase [19, 20]. The tracer penetrates the interjunctional regions between successive tight junctions and accumulates under the basal cell surface (Fig. 4b). The low endocytotic level of these cells is also apparent.

Tight Junctions by Freeze Fracture: Cultured microvessel endothelial cells provide a unique system for studying tight junctions by freeze fracture because the fracture planes often show large uninterrupted expanses of these structures [45, 46]. Brain microvessel endothelial cells are grown on fibronectin-coated glass or plastic coverslips to confluence and prepared for freeze fracture, essentially as described by Pauli et al. [38]. They are fixed for 5 h in 3% glutaraldehyde buffered with 0.2% sodium cacodylate (pH 7.4) and placed in 30% glycerol in 0.2 M cacodylate buffer at 4°C overnight or longer. The following day 3-mm plastic coverslips with attached cells are cut with scissors or glass ones are carefully broken with a diamond stylus to fit into the freeze fracture specimen carrier. The carrier with specimens is frozen in a slurry of Freon 22 cooled with liquid nitrogen, fractured at –115°C, and shadowed in a Balzers BAF 301 Freeze-Etch unit. After thawing, the platinum shadowed carbon replicas are cleaned in sodium hypochlorite solution for 2 h, rinsed in distilled water, and mounted on copper grids. Replicas are examined in a transmission electron microscope at 60 kV. Figure 5 shows a high magnification replica demonstrating the intramembranous particles characteristic of the tight junction on the E face.

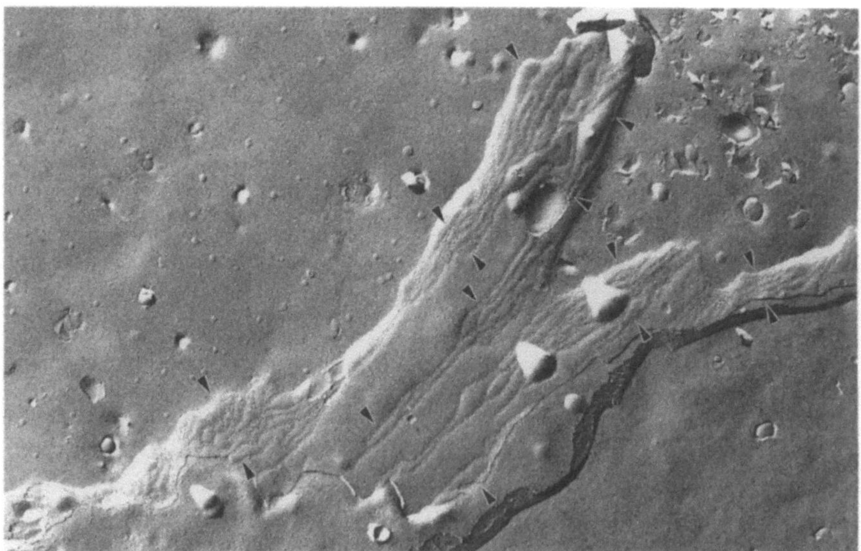

Fig. 5. Electron micrograph of platinum-carbon replicas obtained by freeze-fracture. The regions of intramembranous particles characteristic of the tight junction are denoted (*arrowheads*). Note also the absence of endocytotic vesicles

Discussion

Table 1 summarizes methods reported by investigators for isolating brain microvessel endothelial cells. Basically, all methods begin with mechanical and/or enzymatic destruction of the nervous tissue and isolation of microvessels. Some then utilize enzymatic treatment to disrupt the microvessel basement membrane and free the endothelial cells. In the absence of steps such as sieving on nylon mesh to remove large vessels, contamination by large vessel endothelial cells and smooth muscle cells becomes much more likely. Those methods which do not include enzymatic digestion of the microvessels, or one of short duration, require the endothelial cells to migrate out of the ends of the short segments. Thus, it generally takes much longer to establish cultures, and pericyte contamination becomes a bigger problem. Complete removal of the basement membrane and dissociation of pericytes from endothelial cells is preferable but requires the extra Percoll gradient step. Of course, even cultures

Table 1. Reported methods for isolation of brain microvessel endothelial cells for culture

Reference	Species	Isolation method
Panula et al. [37]	Rat	Gray and white matter; minced; nylon sieving; sucrose density gradient centrifugation
Phillips et al. [39]	Bovine, rat	White matter; minced; 0.5% Trypsin for 5 min
De Bault and Cancilla [15] De Bault et al. [16]	Mouse	Gray and white matter; minced; homogenized; nylon mesh
Bowman et al. [6] Bowman et al. [7]	Rat	Gray matter; minced; homogenized; glass bead filtration; 0.1% collagenase; dextran centrifugation; percoll gradient centrifugation
Spatz et al. [52]	Rat	Gray and white matter; minced; homogenized; 0.01% Trypsin-collagenase 5 min; sucrose density gradient centrifugation
Diglio et al. [17]	Rat	Gray matter; homogenized; 0.05% collagenase 15–20 min; dextran centrifugation; glass bead filtration; sieving on nylon
Bowman et al. [9]	Bovine	Gray matter; minced; dispase isolated; collagenase/dispase dissociated up to 16 h; percoll gradient centrifugation
Goetz et al. [25]	Bovine	Gray matter; minced; 0.5% collagenase/dispase 1 h; nylon mesh; glass bead filtration
Carson and Haudenschild [11]	Bovine	Gray matter; homogenized; nylon sieving; 0.1% collagenase dissociation 16 h; Trypsin-EDTA
Rupnick et al. [41]	Rat	Minced; 0.5% collagenase 30 m; homogenized; dextran gradient centrifugation; percoll gradient centrifugation
Vinters et al. [58]	Human	Gray matter; homogenized; nylon sieving; 0.1% collagenase 2–3 min
Tontsch and Bauer [57]	Porcine	Minced; homogenized; rate velocity sedimentation; 0.075% collagenase 10 min

derived from methods which include all or most of the steps in Fig. 1 may still contain some pericytes. Their presence can be minimized, however, by using selective media, cloning desired colonies, or mechanically destroying the contaminating cells [66].

The large range in the concentration of enzymes and in the dissociation times that have been used for isolating the endothelial cells, as summarized in Table 1, is difficult to explain. Apparently, microvessel endothelial cells are not greatly affected by long incubations in proteases because of their low levels of endocytosis. In contrast, long dissociation times tend to minimize pericyte contamination, suggesting that these cells internalize the enzymes and become damaged by them.

The method employed for isolating and cultivating the cells should be guided by the study for which the cells are to be used. Rupnick et al. [41] have clearly demonstrated that in the absence of any attempt to select for vessel size from white matter of brain, four morphologically distinct, factor VIII immunoreactive cell types can be derived in cultures from rat brain vessels. Some of these may be derived from larger vessels or veins and some may come from those areas of the brain (choroid plexus, area postrema, supraoptic crest, intercolumnar tubercle, neurohypophysis, and median eminence), which contain fenestrated or otherwise leaky microvessels. For the investigator who is prepared to clone or physically weed out contaminating cells, initial purity of the culture is not critical. Because cells are seeded at lower density with this method, establishing cultures requires a longer time during which the processes of selection and adaptation may lead to diminution of endothelial properties. Large quantities of cells with high purity and a high level of expression of differentiated characteristics are best obtained by using brains from large animals and employing a more thorough isolation procedure.

Table 2 summarizes the different media, serum additions, supplements, and substrates employed by other investigators, as well as their experience with subcultivating the cells. Table 2 indicates that the basic medium used is not of critical importance. Medium 199 or a variation by Lewis et al. [32] has been used frequently. In contrast, we have found that the type of serum being used is extremely important. The particular batch of serum affects plating efficiency as well as growth [8, 24]. Some batches will encourage vigorous growth of pericytes, if they are present in the culture, and other batches actually appear to inhibit endothelial cell proliferation. The basis for this variation is not known, but it is a common observation in cell culture. We normally screen several lots, compare them, then purchase a large amount and store it frozen. Most recently, we have found that equine plasma-derived serum (Hyclone Laboratories) reproducibly stimulates endothelial, but not pericyte, growth. Additions of heparin (Sigma) and endothelial cell growth supplement (Biomedical Technologies) are beneficial, but may interfere with certain types of studies. For example, Chung-Welch et al. [13] have reported that heparin inhibits prostaglandin synthesis in lung microvascular endothelial cells. Use of low concentrations of serum or serum-free media should be investigated more thoroughly since it would provide an opportunity to understand the minimal requirements for survival and function. There does not appear to be an absolute requirement for a special substrate, but several authors report a higher

Table 2. Reported conditions for cultivation of brain microvessel endothelial cells

Reference	Medium	Supplements	Substrate	Subculti-vation
Panula et al. [37]	M199; newborn bovine serum	Nerve growth factor	TC plastic	Not reported
Phillips et al. [39]	M199; 20% FBS	None	TC plastic	Yes
De Bault and Cancilla [15] De Bault et al. [16]	Modified M199; 20%–30% FBS	None	TC plastic	Yes
Bowman et al. [7] Bowman et al. [6]	Ham's F12; 20% bovine calf serum	None	Collagen	Not reported
Spatz et al. [52]	Modified M199; 20%–30% FBS	None	TC plastic	Yes
Diglio et al. [17]	Dulbecco's MEM; 20% FBS	150 µg/ml ECGS		Not reported
Bowman et al. [9]	αMEM; 10% PDS	None	Fibronectin	Not reported
Goetz et al. [25]	RPMI 1640; 20% FBS	None	Fibronectin, gelatin, polylysine	Yes
Carson and Haudenschild [11]	Dulbecco's MEM; 15% equine PDS	Heparin, ECGS, retinal extract	Fibronectin, gelatin	Yes
Rupnick et al. [41]	M199; 20% FBS	Heparin, ECGS	Gelatin	Yes
Vinters et al. [58]	Modified M199; 20%–30% FBS	None	None	Yes

M199, Medium 199; TC, tissue culture; FBS, fetal bovine serum; MEM, minimal essential medium; ECGS, Endothelial cell growth supplement; PDS, plasma-derived serum

plating efficiency if fibronectin, gelatin, or collagen is provided. In summary, a variety of media, supplements, and substrates can be used to generate usable cultures of brain microvessel endothelial cells.

The characterization and use of serially subcultivated brain microvessel endothelial cells has not been well documented. In our experience, serially subcultivated cells undergo alterations in morphology (becoming more spindle-shaped), and factor VIII immunoreactivity, which is readily detectable by immunofluorescence in primary culture, now requires more sensitive methods such as immunoperoxidase for detection. De Bault [14] and Tontsch and Bauer [57] found that factor VIII immunoreactivity was rapidly lost from their cerebral microvessel endothelial cell cultures, indicating loss of expression of differentiated function. In an adult organism, the endothelium is a slowly renewing population [21], and it might be expected that stimulation of cell proliferation and adaptation to cell culture would cause alterations in certain properties of the cells. Assuming that the goal of cultivating endothelial cells is to learn how endothelium in general, and brain microvessels in particular, function in vivo, it is probably best to work with primary cultures. Because large animal brains are readily available from slaughterhouses and a large number of microvessel endothelial cells can be obtained from them, primary cultures provide an abundance of cells for experimentation. There are situations, however, in which serial subcultivation can be an advantage. Rutten et al. [42] have been able to achieve high electrical resistance in confluent cultures derived from certain cloned lines of brain microvessel endothelial cells, demonstrating the formation of continuous tight junctions – and a sealed membrane – by the cells.

The principal impediment to successful primary cultures of microvessels from any source is contamination with pericytes. The ratio of pericytes to endothelial cells in vivo is quite high [51]. Pericytes have many of the properties of smooth muscle cells and can inhibit growth of endothelial cells [36]. Recent evidence suggests that activation of TGF-β in the presence of endothelial cells and pericytes may cause this inhibition [3]. Since pericytes from retina [8, 24] and smooth muscle cells [18] grow well in media containing 10% – 30% serum, they should always be viewed as a potential contaminant when high concentrations of serum are used. Cloning can be used to separate endothelial cells from contaminating adventitial cells, but this method requires considerable time to obtain sufficient endothelial cells to work with. Loss of differentiated markers is also much more likely.

The potential to increase our understanding of microvascular function is enhanced by the relative ease with which brain microvessel endothelial cells can be obtained. They are an obvious choice for studying transport functions, for example, although they have been little used for this purpose. Transcellular transport studies require endothelial cell cultivation on a permeable support and the formation by the cells of a continuous, sealed membrane, and there are certain problems associated with obtaining such cultures [2]. Another area warranting further investigation is the differences in properties and functions of microvessel beds from different regions of the brain. Gross et al. [27] presented data showing differences in solute flux in gray and white matter; differences must exist in many other properties as well. In the future, greater refinements in culturing brain microvessel endothelial cells will be required to understand brain microvasculature function.

Acknowledgement. We gratefully acknowledge the expert editorial assistance of Susan Siefert.

References

1. Absher M (1973) Hemocytometer counting. In: Kruse PF, Patterson MKJ (eds) Tissue culture: methods and applications. Academic, San Francisco, pp 395–397
2. Albelda SM, Sampson PM, Haselton FR, McNiff JM, Mueller SN, Williams SK, Fishman AP, Levine EM (1988) Permeability characteristics of cultured endothelial cell monolayers. J Appl Physiol 64:308–322
3. Antonelli-Orlidge A, Saunders KB, Smith SR, d'Amore PA (1989) An activated form of transforming growth factor β is produced by cocultures of endothelial cells and pericytes. Proc Natl Acad Sci USA 86:4544–4548

4. Arthur RE, Shivers RR, Bowman PD (1987) Astrocyte-mediated induction of tight junctions in brain capillary endothelium: an efficient in vitro model. Dev Brain Res 36:155–159
5. Betz AL, Goldstein GW (1986) Specialized properties and solute transport in brain capillaries. Annu Rev Physiol 48:241–250
6. Bowman PD, Betz AL, Goldstein GW (1979) Characteristics of cultured brain capillaries. J Cell Biol 83:95
7. Bowman PD, Betz AL, Ar D, Wolinsky JS, Penney JB, Shivers RR, Goldstein GW (1981) Primary culture of capillary endothelium from rat brain. In Vitro 17:353–362
8. Bowman PD, Betz AL, Goldstein GW (1982) Primary culture of microvascular endothelial cells from bovine retina: selective growth using fibronectin coated substrate and plasma derived serum. In Vitro 18:626–632
9. Bowman PD, Ennis SR, Rarey KE, Betz AL, Goldstein GW (1983) Brain microvessel endothelial cells in tissue culture: a model for study of blood-brain barrier permeability. Ann Neurol 14:396–402
10. Bowman PD, Rarey K, Rogers C, Goldstein GW (1985) Primary culture of capillary endothelial cells from the spiral ligament and stria vascularis of bovine inner ear. Cell Tissue Res 241:479–486
11. Carson MP, Haudenschild CC (1986) Microvascular endothelium and pericytes: high yield, low passage cultures. In Vitro Cell Dev Biol 22:344–354
12. Cereijido M, Dolan WJ, Rotunno CA, Sabatini DD (1978) Polarized monolayers formed by epithelial cells on permeable and translucent support. J Cell Biol 77:853–880
13. Chung-Welch H, Shepro D, Dunham B, Hechtman HB (1988) Prostacyclin and prostaglandin E2 secretions by bovine pulmonary microvessel endothelial cells are altered by changes in culture conditions. J Cell Physiol 135:224–234
14. De Bault LE (1982) Isolation and characterization of the cells of the cerebral microvessels. Adv Cell Neurobiol 3:339–371
15. De Bault LE, Cancilla PA (1979) Gamma-glutamyl transpeptidase in isolated brain endothelial cells: induction by glial cells in vitro. Science 207:653–655
16. De Bault LE, Henriquez E, Hart MN, Cancilla PA (1981) Cerebral microvessels and derived cells in tissue culture. II. Establishment, identification, and preliminary characterization of an endothelial cell. In Vitro 17:480–494
17. Diglio CA, Grammas P, Giacomelli F, Wiener J (1982) Primary culture of rat cerebral microvascular endothelial cells. Isolation, growth and characterization. Lab Invest 46:554–563
18. Diglio CA, Grammas P, Giacomelli F, Wiener J (1986) Rat cerebral microvascular smooth muscle cells in culture. J Cell Physiol 129:131–141
19. Dorovini-Zis K, Bowman PD, Betz AL, Goldstein GW (1984) Hyperosmotic arabinose solutions open the tight junctions between brain capillary endothelial cells in tissue culture. Brain Res 302:383–386
20. Dorovini-Zis K, Bowman PD, Betz AL, Goldstein GW (1987) Hyperosmotic urea reversibly opens the tight junctions between brain capillary endothelial cells in cell culture. Neuropathol Exp Neur 46:130–140
21. Engerman RL, Pfaffenback D, Davis MD (1967) Cell turnover of capillaries. Lab Invest 17:738–743
22. Farquhar MG, Palade GE (1963) Junctional complexes in various epithelia. J Cell Biol 17:375–412
23. Ford DH (1976) Blood-brain barrier: a regulatory mechanism. In: Ehrenpreis S, Kopin IJ (eds) Reviews of neuroscience. Raven, New York, pp 1–42
24. Gitlin JD, d'Amore PA (1983) Culture of retinal capillary cells using selective growth media. Microvasc Res 26:74–80
25. Goetz IE, Warren J, Estrada C, Roberts E, Krause DN (1985) Long-term serial cultivation of arterial and capillary endothelium from adult bovine brain. In Vitro Cell Dev Biol 21:172–180
26. Gospodarowicz D, Massoglia S, Cheng J, Fujii DK (1986) Effect of fibroblast growth

factor and lipoproteins on the proliferation of endothelial cells derived from bovine adrenal cortex, brain cortex, and corpus luteum capillaries. J Cell Physiol 127:121–136
27. Gross PM, Spositi NM, Pettersen SE, Fenstermacher JD (1986) Differences in function and structure of the capillary endothelium in gray matter, white matter and circumventricular organ of rat brain. Blood Vessels 23:261–270
28. Hormia M, Lehto V-P, Virtanen I (1984) Intracellular localization of factor VIII-related antigen and fibronectin in cultured human endothelial cells: evidence for divergent routes of intracellular translocation. Eur J Cell Biol 33:217–228
29. Jaffe EA (1984) Synthesis of factor VIII by endothelial cells. In: Jaffe EA (ed) Biology of endothelial cells. Nijhoff, Boston, pp 209–214
30. Joó F, Karnushina I (1973) A procedure for the isolation of capillaries from rat brain. Cytobios 8:41–48
31. Knedler A, Ham RG (1987) Optimized medium for clonal growth of human microvascular endothelial cells with minimal serum. In Vitro Cell Dev Biol 23:481–491
32. Lewis LJ, Hoak JC, Maca RD, Fry GL (1973) Replication of human endothelial cells in culture. Science 181:453–454
33. Lierse W, Horstmann E (1965) Quantitative anatomy of the cerebral vascular bed with especial emphasis on homogeneity and inhomogeneity in small parts of the gray and white matter. Acta Neurol Scand [Suppl]14:15–19
34. Nagy Z, Peters H, Hüttner I (1984) Fracture faces of cell junctions in cerebral endothelium during hyperosmotic conditions. Lab Invest 50:313–322
35. Oldendorf WH, Cornford ME, Brown WJ (1977) The large apparent work capability of the blood-brain barrier: a study of the mitochondrial content of capillary endothelial cells in brain and other tissues of the rat. Ann Neurol 1:409–417
36. Orlidge A, d'Amore PA (1987) Inhibition of capillary endothelial cell growth by pericytes and smooth muscle cells. J Cell Biol 105:1455–1462
37. Panula P, Joo F, Rechardt L (1978) Evidence for the presence of viable endothelial cells in cultures derived from dissociated rat brain. Experientia 34:95–97
38. Pauli BU, Weinstein RS, Soble LW, Alroy J (1977) Freeze-fracture of monolayer cultures. J Cell Biol 72:763–769
39. Phillips P, Kumar P, Kumar S, Waghe M (1979) Isolation and characterization of endothelial cells from rat and cow brain white matter. J Anat 129:261–272
40. Reese TJ, Karnovsky MJ (1967) Fine structural localization of a blood-brain barrier to exogenous peroxidase. J Cell Biol 34:207–217
41. Rupnick MA, Carey A, Williams SK (1988) Phenotypic diversity in cultured cerebral microvascular endothelial cells. In Vitro Cell Dev Biol 24:435–444
42. Rutten MJ, Hoover RL, Karnovsky MJ (1987) Electrical resistance and macromolecular permeability of brain endothelial monolayer cultures. Brain Res 425:301–310
43. Schneeberger EE, Lynch RD (1984) Tight junctions: their structure, composition, and function. Circ Res 55:723–733
44. Shannon JE, Macy ML (1973) Freezing, storage, and recovery of cell stocks. In: Kruse PF, Patterson MKJ (eds) Tissue culture: methods and applications. Academic, San Francisco, pp 712–718
45. Shivers RR, Bowman PD (1985) A freeze-fracture paradigm of the mechanism for delivery and insertion of gap junction particles into the plasma membrane. J Submicrosc Cytol 17:199–203
46. Shivers RR, Bowman PD, Martin K (1985) A model for de novo synthesis and assembly of tight intercellular junctions. Ultrastructural correlates and experimental verification of the mode revealed by freeze-fracture. Tissue Cell 17:417–440
47. Shivers RR, Arthur FE, Bowman PD (1988) Induction of gap junctions and brain endothelium-like tight junctions in cultured bovine endothelial cells: local control of cell specialization. J Submicrosc Cytol Pathol 21:1–14
48. Shivers RR, Pollock M, Bowman PD, Atkinson BG (1988) The effect of heat shock on primary cultures of brain capillary endothelium: inhibition of assembly of zonulae occludentes and the synthesis of heat-shock proteins. Eur J Cell Biol 46:181–195

49. Siakotos AN (1974) The isolation of endothelial cells from normal human and bovine brain. In: Fleischer S, Packer L (eds) Methods Enzymol. Academic Press, San Francisco, XXXII part B, pp 717–722
50. Siakotos AN, Rouser G, Fleischer S (1969) Isolation of highly purified human and bovine brain endothelial cells and nuclei and their phospholipid composition. Lipids 4:234–239
51. Simionescu M, Ghinea N, Fixman A, Lasser M, Kukes L, Simionescu N, Palade GE (1988) The cerebral microvasculature of the rat: structure and luminal surface properties during early development. J Submicrosc Cytol Pathol 20:243–261
52. Spatz M, Bembry J, Dodson RF, Hervonen H, Murray MR (1980) Endothelial cell cultures derived from isolated cerebral microvessels. Brain Res 191:577–582
53. Sporn LA, Marder VJ, Wagner DD (1987) Von Willebrand factor released from Weibel-Palade bodies binds more avidly to extracellular matrix than that secreted constitutively. Blood 69:1531–1534
54. Sternberger NH, Sternberger LA (1987) Blood-brain barrier protein recognized by monoclonal antibody. Proc Natl Acad Sci USA 84:8169–8173
55. Stevenson BR, Anderson JM, Bullivant S (1988) The epithelial tight junction: structure, function and preliminary biochemical characterization. Mol Cell Biochem 83:129–145
56. Thornton SC, Mueller SN, Levine EM (1983) Human endothelial cells: use of heparin in cloning and long term serial cultivation. Science 222:623–625
57. Tontsch U, Bauer H-C (1989) Isolation, characterization, and long-term cultivation of porcine and murine cerebral capillary endothelial cells. Microvasc Res 37:148–161
58. Vinters HV, Reave S, Costello P, Girvin JP, Moore SA (1987) Isolation and culture of cells derived from human cerebral microvessels. Cell Tissue Res 249:657–667
59. Vorbrodt AW (1988) Ultrastructural cytochemistry of blood-brain barrier endothelia. Prog Histochem Cytochem. Gustav Fischer Verlag, New York, vol 18, pp 1–99
60. Voyta JC, Via DP, Butterfield CE, Zetter BR (1984) Identification and isolation of endothelial cells based on their increased uptake of acetylated low-density lipoprotein. J Cell Biol 99:2034–2040
61. Wagner DD, Olmsted JB, Marder VJ (1982) Immunolocalization of von Willebrand protein in Weibel-Palade bodies of human endothelial cells. J Cell Biol 95:355–360
62. Wagner RC, Matthews MA (1975) The isolation and culture of capillary endothelium from epididymal fat. Microvasc Res 10:286–297
63. Wall RT, Harker LA, Quadracci LJ, Striker GE (1978) Factors influencing endothelial cell proliferation in vitro. J Cell Physiol 96:203–215
64. Williams SK, Gillis JF, Matthews MA, Wagner RC, Bitensky MW (1980) Isolation and characterization of brain endothelial cells: morphology and enzyme activity. J Neurochem 35:374–383
65. Yablonka-Reuveni Z (1989) The emergence of the endothelial cell lineage in the chick embryo can be detected by uptake of acetylated low density lipoprotein and the presence of a von Willebrand-like factor. Dev Biol 132:230–240
66. Zetter BR (1984) Culture of capillary endothelial cells. In: Jaffe EA (ed) Biology of endothelial cells. Nijhof, Boston, pp 14–26

Microvascular Endothelial Cells from Heart*

H.M. Piper, R. Spahr, S. Mertens, A. Krützfeldt, and H. Watanabe

Introduction

Endothelial cells from macrovascular and microvascular vessels differ in many
of their physiological properties [20, 21, 36, 55, 72]. For a detailed analysis of
the metabolism and the physiological function of the microvascular endothelium, cultures of microvascular cells are required. In cardiac tissue the mass of
coronary endothelial cells is much smaller than the mass of cardiomyocytes
(endothelium, 2%–3% of total heart mass) [3, 33, 57]. The endothelial contribution to the metabolic response of the whole heart cannot be accurately
determined unless exclusive metabolic properties of the endothelial cells are
investigated.

A number of approaches have been proposed for the isolation and culture
of microvascular coronary endothelial cells. Coronary endothelial cells have
been isolated by collecting only the perfusate from a heart perfused with
protease solutions [15, 43]. Probably because larger parts of the vascular bed
are rapidly occluded when perfused with proteolytic enzymes, the yield of cells
from this method is very low, and it is uncertain whether it also leads to an
isolation of predominantly microvascular endothelial cells. Others have first
isolated microvessel fragments from the heart tissue [22, 23], i.e., pieces of
small arterioles, venules, and capillary tubes, from which cultures can subsequently be established. Since microvessels are composed of endothelial cells,
pericytes and smooth muscle cells, measures to select the growth of the endothelial cell type in culture have to be used. Endothelial cells may also be
isolated separately from the venular or the arterial side of the coronary microcirculatory bed. For this purpose, large hearts have to be used in which both
the coronary arteries and the coronary sinus can be cannulated. Microspheres
of defined size are infused into the microvascular bed from one side of the
coronary system. When the beads occlude small vessels, endothelial cells adhere to them and can be removed by reversing the flow direction [52]. It has
been claimed that a very pure, topologically selected population of endothelial
cells can thus be obtained [53]. Again, the total number of isolated endothelial
cells is extremely small.

The procedure described in this chapter was designed to isolate and culture
a statistical average of all endothelial cells. For this purpose, the heart is first

---

* This work was supported by the Deutsche Forschungsgemeinschaft (Pi 162/6-1).

completely dissociated in single cells. Because the vast majority of all endothelial cells in heart tissue is contained in small vessels and the capillary bed, the statistical average represents, predominantly, the microvascular endothelium. Selection of endothelial cells from dissociated heart tissue is achieved by the techniques of cell isolation and cell culture.

Culture of Microvascular Coronary Endothelial Cells

Protocols

Cell Isolation

Steps 1 to 6 of the following protocol are identical with the protocol for cardiomyocyte isolation described in Chap. 3 (this volume). Indeed, from a single heart both cardiomyocytes and endothelial cells can be isolated. The only difference in the requirements for the isolation of these two cell types is that the concentration of collagenase and Ca^{2+} must be higher when isolating endothelial cells either alone or with cardiomyocytes.

Materials:
The following materials are needed for cell isolation:

Perfusion system (see Fig. 1, Chap. 3, this volume)
Langendorff system, consisting of:

1. Top reservoir (100 ml), double-walled, temperature-controlled
2. A glass-coil heart exchanger with two cannulas fitted to its outlet (in the shape of an inverted Y, for the simultaneous perfusion of two hearts; replace by a single cannula for the perfusion of one heart), the distance between top reservoir and cannulas is 100 cm
3. Connection by a double-walled, temperature-controlled glass tube between the heat exchanger and top reservoir, containing a flow reducer
4. Funnel, which can be moved below the cannulas to collect the fluid, connected to a tube leading back to the top reservoir
5. Roller pump for pumping the fluid back to the top reservoir
6. Temperature-controlled water circulator, for 37 °C temperature control of the Langendorff system
7. Pasteur pipette, for gassing the top reservoir

Instruments:

- 2 Scissors (coarse and fine)
- 2 Small forceps
- 2 Large Petri dishes (200 mm in diameter)
- 2 Scalpels and a watchglass (or a tissue chopper)
- Nylon mesh (mesh size 200 µm)
- 2 50-ml centrifuge tubes

- 1 50-ml glass beaker
- 1 50-ml Teflon or siliconized glass beaker
- Disposable 5-ml pipette with bore size approximately 2 mm in diameter (e.g., from Falcon or Greiner)
- Centrifuge

Media and Supplements:

- Buffer 1 (in mM): NaCl 110, KCl 2.6, KH_2PO_4 1.2, $MgSO_4$ 1.2, $NaHCO_3$ 25, glucose 11 at 37 °C; continuously gassed with 95% O_2/5% CO_2 (gives pH 7.4). Prepare 1 l, warm up, and gasequilibrate before the experiment.
- Ca stock solution: 100 mM $CaCl_2$ in H_2O.
- Saline: 9 g NaCl/l, ice cold.
- Bovine serum albumin (BSA), fraction V. BSA preparations freed from fatty acids are recommended.
- Collagenase: Crude collagenase, from clostridium histolyticum. An appropriate batch has to be selected. Suitable collagenases can be obtained, e.g., from Worthington, Serva, and Sigma.
- Trypsin (1 : 250) from Biochrom.
- EDTA-buffer (composition in mM): NaCl 125.0, KCl 2.6, KH_2PO_4 1.2, Hepes 10.0, NaEDTA 0.5; pH 7.4.
- Phosphate-buffered saline (PBS); composition in mM: NaCl 137.0, KCl 2.6, KH_2PO_4 1.5, Na_2HPO_4 8.1; pH 7.4.
- Trypsin-EDTA solution: Dissolve 50 mg trypsin (trypsin 1 : 250) and 20 mg EDTA in 100 ml PBS (mixture commercially available; e.g., from Flow Laboratories).

Procedure: To initially establish cultures of coronary endothelial cells, the endothelial material from four rat hearts is required. For this purpose four hearts are perfused simultaneously on two Langendorff systems (2 × 2). Steps 1 to 5 describe the perfusion protocol for the simultaneous perfusion of two hearts on one Langendorff system.

1. Fill the Langendorff system with 120 ml buffer 1, gas upper reservoir with 95% O_2/5% CO_2 through the tip of a Pasteur pipette. Make up collagenase solution immediately before starting the experiment, i.e., dissolve 40 mg in 10 ml buffer 1, with the addition of 25 µl Ca stock. (The exact amount of collagenase has to be determined empirically for a given batch.)

2. Fill the large Petri dishes with ice-cold saline. Kill the animal by deep ether anesthesia. Open chest with coarse scissors. Pour ice-cold saline on the heart. Excise the heart with an intact aortic arch and immerse in the first of the Petri dishes. Free the aorta from mediastinal tissue and transfer to the second Petri dish.

3. Start flow of perfusion system (1 drop/s per cannula), place beaker under cannulas and remove collecting funnel. Mount the hearts on the cannulas as follows: while the heart is being immersed in saline, open the aortic lumen with two forceps, lift the heart to a cannula, slip aorta over cannula (avoid penetration of aortic valve), fix aorta with a crocodile clamp. Repeat the procedure with the second heart. Replace the clamps by threads. For the devitalization of epicardial mesothelial cells, flush the outer surface of the hearts with 70%

ethanol. By adjusting the flow reducer, perfuse each heart with approximately 2 drops/s (10 ml/min per two hearts). Collect the initial 80 ml of effluent in a beaker and discard.

4. By placing the funnel underneath the hearts and switching on the pump, start recirculating the perfusate. Add the 10 ml of collagenase solution to the top reservoir. Keep the flow rate at 2 drops/s per two hearts and continue for 30 min. (The appropriate time depends on concentration and activity of the given batch of collagenase.)

5. Near the end of the 30-min recirculation period, add 400 mg BSA plus 30 ml recirculating medium into a 50-ml Teflon or siliconized glass beaker. Gas with 95% O_2/5% CO_2 through the tip of a Pasteur pipette and maintain at 37°C.

6. Take the four hearts off the cannulas of both Langendorff systems, remove the atria with fine scissors, and cut the ventricular tissue in two to four pieces. Chop these pieces with two scalpels in very small chunks (if available use tissue chopper and cut in 0.7×0.7 mm pieces). Transfer the minced tissue into the 50-ml beaker containing the recirculated medium. Triturate gently with 5 ml disposable pipette twice per min, during a 10- to 15-min incubation period (well-perfused tissue will dissolve during this time).

7. Filter the material through the nylon mesh and divide into two 50-ml centrifuge tubes. Spin the tubes at 25 g for 3 min. (The cardiomyocytes are contained in the pellet, see Chap. 3, this volume). After centrifugation add the supernatant from both tubes plus 200 mg BSA, 10 mg trypsin and 30 μl Ca stock into a 50-ml Teflon or siliconized glass beaker. Gas with 95% O_2/5% CO_2 through the tip of a Pasteur pipette and maintain at 37°C. Stir with a suspended bar at 200 rpm for 30 min.

8. Centrifuge the cell suspension at 250 g for 10 min. Discard the supernatant and resuspend the pellet in 75 ml of culture medium 1.

Note: The material thus isolated still contains a lot of debris. In the culture protocol outlined below the debris is removed after 4 h, when the intact endothelial cells have attached to the dishes. With the debris, all other cell types which do not attach as readily to culture dishes as endothelial cells are removed from the isolated material. We found a previous purification of the isolated cell material on a density gradient (prepared with Percoll from Pharmacia) was not necessary, and even disadvantageous since a exclusive selection of the viable endothelial cells cannot be obtained and part of the viable cells were lost. The presence of debris, however, can hinder the attachment of viable cells if the plating density of the isolated material is too large. The plating density specified should not, therefore, be exceeded.

Cell Culture

Materials: For cell culture the following materials are needed:

Culture Dishes:

- 100-mm plastic culture dishes (Falcon, type 3003)

                                                    H. M. Piper et al.

Media:

- Culture medium 1: Medium 199 with Earle's salts, buffered by 26 mM NaHCO$_3$ – 5% CO$_2$ at pH 7.4, 10% newborn calf serum and 10% fetal calf serum, penicillin 250 IU/ml, streptomycin 250 µg/ml, amphotericin B 12.5 µg/ml, gentamycin 50 µg/ml. Medium 199 can be purchased as dry powder from numerous suppliers (e.g., Boehringer Mannheim; Gibco)
- Culture medium 2: as culture medium 1, but with reduced concentrations of antibiotics, i.e., penicillin 100 IU/ml, streptomycin 100 µg/ml, and amphotericin B 5 µg/ml

Procedure:

1. Add 25 ml of the suspension of newly isolated cell material (75 ml from four rat hearts) to each of three 100-mm Petri dishes and incubate at 37°C with 5% CO$_2$.
2. 4 h after plating, shake the dishes vigorously and decant the medium containing debris and nonattached cells. Add 40 ml of culture medium 1 to each 100-mm dish. Renew the medium 24 h after plating. At that time 1×10^6 attached cells are contained in each 100-mm dish.
3. Change the medium again 2 days after plating and replace by culture medium 2, containing reduced concentrations of antibiotics. Renew this medium every subsequent day.
4. After 5 days, the dishes contain confluent endothelial monolayers with $6-10 \times 10^6$ cells/100-mm dish ($1-2 \times 10^5$ cells/cm^2).
5. Cells in confluent dishes can be further multiplied after trypsinization. For this purpose, wash 100-mm Petri dishes containing confluent monolayers of endothelial cells twice with EDTA-buffer, then add 2 ml trypsin-EDTA solution per dish and incubate at 37°C (without CO$_2$). After 10 min, add 10 ml culture medium per dish. Bring the cells into suspension by pipetting. The volume is adjusted sufficiently to plate the cells on an area three to four times larger than covered before (media volumes: 40 ml per 100-mm dish, 10 ml per 60-mm dish). After 4 days, again confluent cultures are obtained.

Note: To perform the isolation procedure under sterile conditions is not normally required. It is of prime importance, however, to keep the perfusion system clean. In general it is sufficient to flush the system after use with 1 l of distilled water. This has to be done immediately, otherwise proteins contained in the perfusate will dry on the glass. The perfusion system and all glassware should be then flushed with 70% ethanol for 30 min. The perfusion system should be subsequently dried by a stream of clean gas (e.g., filtered compressed air). With these simple means and the use of high concentrations of antibiotics during the first days of culture, cultures free of microbial growth can normally be achieved. The described mixture of antibiotics does not seem to be harmful to the endothelial cells. In case microbial growth cannot be prevented by this mixture, other antibiotics may have to be used.

Cell Yield

The yield of cells establishing in culture was determined in cultures washed vigorously after 4 h and again after 24 h. After 4 h, 10^6 cells per 100-mm dish had attached, the same number of cells had spread after 24 h, indicating the viability of these cells. As documented by staining with acetylated low density lipoprotein (ac-LDL) and anti-α-smooth muscle actin, these cultures are virtually free of contaminating cells (see below). Cell proliferation starts 24 h after cell isolation. Because endothelial cells are plated as single cells and not as cell clusters, they are, during the initial phase of the culture, separated from each other. They exhibit polymorphic shapes when spreading. The yield of viable endothelial cells establishing in culture, as indicated by their spreading within the first 24 h, amounts to approximately 10^6 endothelial cells per gram of rat heart.

Only in some of the publications describing procedures for the isolation of coronary endothelial cells has the yield of viable cells been specified. The total endothelial cell content of heart tissue has been estimated as 10^8 cells per gram [57]. Simionescu and Simionescu [57] and Mistry and Drummond [41] estimated that about 10^7 cells per gram can be isolated through an isolation of microvessels. It was not, however, demonstrated how many of the endothelial cells contained in the microvessel fragments would establish in culture. According to the protocol described by Nees et al. [43] $1-2 \times 10^5$ cells per gram of heart can be established in culture. Diglio et al. [15] isolated about 20 cell clusters per rat heart, which proliferated as colonies. This amounts to a yield of less than or equal to 2×10^2 cells per gram of heart, i.e., only two cells per million endothelial cells in the heart establish in culture. From this initial result, the material for culture propagation is even further selected by growing cells from single colonies. When cells are removed from microvessels in the intact heart by the infusion and subsequent washout of microspheres [53], again, only an extremely small fraction of endothelial cells are selected from the coronary vasculature. Exact figures about the initial yield of the method have not been reported. Comparison of the yield of the cited methods with yield of method described above, i.e., 10^6 cells per gram of heart, shows that this method provides the highest yield of coronary endothelial cells establishing in culture reported so far.

Identification of Culture Purity

Light microscopy of cell cultures is only of limited use in the identification of microvascular endothelial cells. Cells prepared according to the described protocol attach separately and spread in polymorphic shapes. They obtain a polygonal shape only when the culture reaches confluency. Early confluent cultures have a "cobblestone" monolayer morphology (Fig. 1). As known also for macrovascular endothelial cell cultures, the uniformity of this morphology depends on the serum supplement in the medium. When coronary endothelial cells are not subcultivated after reaching confluency, their cultures become very dense, and cell "sprouting" can be observed as reported before for macro-

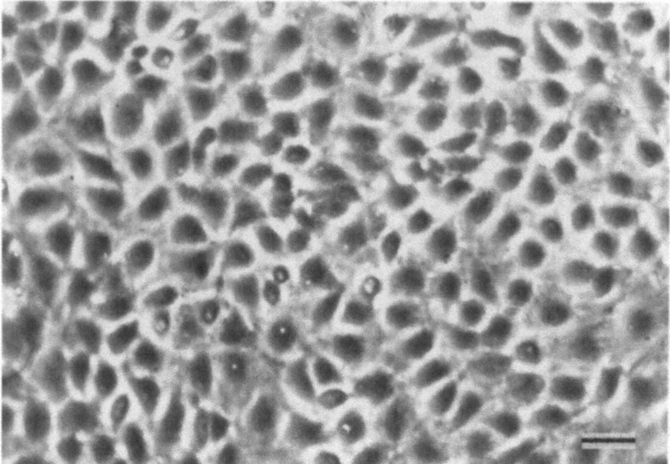

Fig. 1. Phase contrast light microscopic picture of confluent 14-day-old culture of coronary microvascular endothelial cells. (*Bar,* 50 µm)

and microvascular endothelial cell cultures [2, 5, 55, 56, 60]. Sprouting endothelial cells leave the single-cell layer and extend below the main layer in a tubelike form. In confluent primary cultures, older than 5 days, a small percentage (< 10%, see below) of cells can also be detected which stain positively with anti-α-smooth muscle actin. These cells grow predominantly on top of the endothelial monolayer.

The endothelial marker von Willebrand factor [31] is only weakly expressed in coronary microvascular endothelial cells from the rat. Not all cultured endothelial cells show strong staining for von Willebrand factor [34, 49, 64]. Not all reports on positive staining are reliable either [39]. Coronary endothelial cells from the rat also contain only a few Weibel-Pallade bodies (unpublished observation) which represent storage organelles for von Willebrand factor [31]. These structures [68] are present in large number in endothelial cells from large vessels of human origin [27, 35], but they are rare in all capillary endothelial cells, both in vivo and in vitro [4, 14, 15, 17, 34, 43, 48, 62, 66]. In the rat, even in macrovascular endothelial cells, Weibel-Pallade bodies are absent or too few to be easily found by electron microscopy [48].

For these reasons, the endothelial nature of the isolated cells must be identified by other properties. One such indicative property is the uptake of ac-LDL by the living cell [18, 65]. This can be monitored using a fluorescently labeled derivative, i.e., 1,1′-dioctadecyl-1-3,3,3′,3′-tetramethyl-indo-carbocyanine perchlorate (DiI-ac-LDL, Biomedical Technologies). In 4-day-old cultures of coronary microvascular endothelial cells prepared according to the above procedure, 99% of the cells take up DiI-ac-LDL. The percentage of cells taking up the DiI-ac-LDL can be determined when the nuclei of all cells contained in the culture are visualized by staining with bisbenzimide.

Staining the culture with an antibody against α-smooth muscle actin (α-sma) allows determination of the number of pericytes and smooth muscle cells,

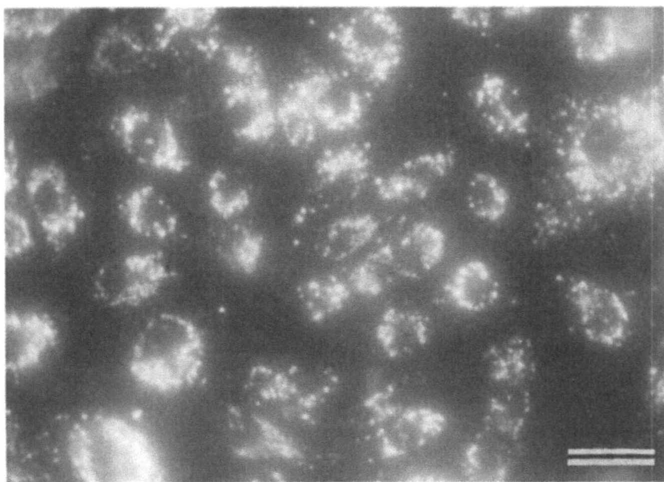

Fig. 2. Accumulation of acetylated LDL by coronary microvascular endothelial cells in 14-day-old non-confluent culture on glass coverslip. The fluorescently labelled acetylated LDL (DiI-ac-LDL) is accumulated around the cell nuclei. (*Bar*, 50 µm)

both of which contain this isoform of α-actin [1, 59]. In cultures prepared according to the above protocol, the number of cells positive for α-sma balances the number of cells positive for ac-LDL. In 14-day-old cultures 5% of all cells stain for α-sma and 95% are intensely stained with DiI-ac-LDL (Fig. 2). Considering the larger number of pericytes (in myocardium, approximately 10% of the number of capillary endothelial cells) [58, 63] compared with smooth muscle cells in heart tissue, the majority of the α-sma-positive cells are likely to be pericytes.

Staining with DiI-ac-LDL:

Materials:

The following solutions and equipment are needed:

- DiI-ac-LDL solution (concentration 200 µg/ml) from Biomedical Technologies
- Fixative solution: 3% formaldehyde in PBS
- Embedding solution: 90:10 glycerol-PBS
- Bisbenzimide solution: 1 µg bisbenzimide per ml PBS (bisbenzimide, Hoechst dye 33258, from Riedel-de Haen)
- Coverslips: 35-mm culture Petri dishes containing glass coverslips (10 × 10 mm)

Procedure:

1. Trypsinize endothelial cells (see above) from culture dishes and prepare a suspension in culture medium 2. Seed them in a dilute suspension on glass

coverslips as follows: Place 100 µl of suspension containing $2-5 \times 10^4$ cells in a drop on the coverslips contained in the Petri dishes. When the cells have attached, 2 h later, fill up the medium in the Petri dishes with 3 ml of culture medium 2. There is little uptake of DiI-ac-LDL during the first 24 h after trypsinization, therefore study cultures 2 days after trypsinization at which time they are still subconfluent.

2. Remove the medium 2 days after seeding and wash the dishes containing the coverslips three times with culture medium 2 (37 °C), then add 1 ml culture medium 2 plus 50 µl of the DiI-ac-LDL solution per dish. Incubate at 37 °C in presence of 5% CO_2.

3. After 4 h, remove the DiI-ac-LDL solution from the dishes and wash the dishes, first three times with culture medium 2 (37 °C), then three times with PBS (37°).

4. Remove PBS and add 3 ml of the fixative solution per dish. After 20 min decant the fixative solution and rinse in distilled water (5 s at room temperature, RT).

5. Remove the water and incubate the dishes with 1 ml of bisbenzimide for 2 min (RT). Wash three times with PBS (RT). With fine forceps remove the coverslips from the culture dishes and drain the liquid on paper tissue.

6. Place a small drop of the embedding solution on a slide. Put the coverslip turned upside down on this drop. Do not use nail polish, since its solvents extract the lipids from the cells!

7. Examine the slides under a fluorescence microscope. For illumination of DiI-ac-LDL staining, use an excitation filter of 530–560 nm (rhodamine combination), for bisbenzimide staining an excitation filter of 340–380 nm.

Staining for α-Smooth Muscle Actin:

Materials:

The following materials are needed for staining:

- AB 1: Monoclonal mouse antibody to α-smooth muscle actin (Sigma, # A 2547): Dilute 1:50 with PBS containing 0.1% BSA.
- Ig solution: Normal rabbit immunoglobulin fraction (Dako, # X903): Dilute 1:20 with PBS containing 0.1% BSA.
- AB 2: Fluorescein-isothiocyanate (FITC) conjugated rabbit antimouse immunoglobulin (Dako, # F261): Dilute 1:20 with PBS containing 0.1% BSA, keep dark.
- Dishes: 100-mm Petri dish.
- Cells: Grown on glass coverslips (10×10 mm) which are contained in 35-mm culture dishes.

Procedure:

1. Wash coverslips with PBS (37°) three times. Remove PBS and fill Petri dishes with 3 ml ice-cold methanol. After 10 min, remove methanol and wash three times with PBS (RT).

2. Cover the bottom of a 100-mm Petri dish with a filter paper soaked with

PBS. Remove the coverslips from the 35-mm dishes and place them on the wet filter paper.

3. Place a drop of 100 µl AB 1 onto each coverslip. Close the lid of the 100-mm Petri dish and incubate for 45 min at 37°C (no CO_2!). Thereafter, wash the coverslips three times with PBS and place them on wet filter paper again.

4. Add a drop of 100 µl of the Ig solution on each cover slip. Incubate for 30 min at 37°C. Then wash the coverslips three times with PBS and place them on wet filter paper again.

5. Add a drop of 100 µl AB 2 on each cover slip. Incubate for 45 min at 37°C. Then wash the coverslips three times with PBS.

6. For nuclear staining with bisbenzimide, continue with step 5 of the protocol in "Staining with DiI-ac-LDL". For embedding only, continue with step 6 of the previous protocol. With these specimens nail polish can be used.

7. Examine the slides under a fluorescence microscope. For illumination of FITC staining use an excitation filter of 450–490 nm, for bisbenzimide staining an excitation filter of 340–380 nm.

Culture Conditions for Selection Against Nonendothelial Cell Growth

The protocol for cell isolation and culture contains three steps for selection against the growth of nonendothelial cells. First, the 30-min incubation of the isolated cell suspension with trypsin represents an important step in selection of endothelial cells. Why this treatment is selective is as yet unknown. Non-muscle cells may be damaged or growth inhibited by a prolonged incubation with trypsin. Alternatively, trypsinization may particularly reduce the ability of nonendothelial cells to attach early after seeding on culture dishes. Second, endothelial cells attach more rapidly than smooth muscle cells, pericytes, and fibrocytes to culture dishes [8, 15, 17, 23, 57]. All cells not firmly attached 4 h after plating are, therefore, removed from the cultures. Third, smooth muscle cells and pericytes grow less rapidly than endothelial cells [29]. For this reason, a major increase in the percentage of nonendothelial cells cannot be expected before confluency of endothelial cells is reached and endothelial growth becomes contact inhibited. Confluent cultures of microvascular coronary endothelial cells can indeed be overgrown by cells staining positive for α-smooth muscle actin. These contaminating cells proliferate predominantly on top of the confluent endothelial monolayer. Once the percentage of smooth muscle cells or pericytes has become large, it can suppress endothelial cell proliferation in a further subculture [46].

The possibility of an overgrowth of confluent endothelial cultures by other cells can be reduced by additional means to select against the proliferation of nonendothelial cells. In cultures prepared according to the protocol outlined in this chapter, overgrowth with fibroblasts (negative for uptake of ac-LDL and negative for α-smooth muscle actin) was never observed. In order to prevent increasing contaminations of confluent endothelial cultures, therefore, the small number of smooth muscle cells and pericytes must be further reduced. The two approaches described in detail below are pancreatin subcultivation and the use of media conditioned by confluent endothelial cultures.

Pancreatin predominantly removes endothelial cells from subconfluent cultures, leaving nonendothelial cells behind [8]. The conditioned media contains factors inhibiting the growth of smooth muscle cells and pericytes [8, 9, 28, 55]. Both treatments had only unsignificant effects when they were applied to cultures prepared according to the protocol described in this chapter and the results were compared after 14 days.

Pancreatin Treatment

For pancreatin treatment pancreatin solution from Gibco (# 043-5720, tenfold concentration) is needed. Dilute before use 1 : 10 with PBS.

1. Use subconfluent cultures (2–3 days after cell isolation on 100-mm Petri dishes). Wash the cultures twice with EDTA buffer, then add 3 ml pancreatin solution per 100-mm dish, and incubate at room temperature (20 °C). After 5 min approximately 50% of the cells detach when dishes are shaken gently. Incline the dish and remove the supernatant containing these cells with a pipette.
2. Centrifuge the supernatant at 250 g for 10 min. Discard the supernatant from this centrifugation and resuspend the pellet in culture medium 2. The cells are then plated on the same number of 100-mm dishes as they were harvested from (40 ml culture medium 2 per 100-mm dish). Change media every day.
3. These cultures become confluent 3–4 days after pancreatin treatment.

Treating Cultures with Conditioned Medium

Treatment of cultures with conditioned medium calls for culture medium 2*: as culture medium 2, with 20% newborn calf serum, no fetal calf serum.

1. Conditioned medium is obtained from confluent porcine aortic endothelial cultures: Harvest endothelial cells from porcine aortas and culture them on 100-mm Petri dishes as described in Chap. 15. These cultures are confluent 5 days after seeding. To obtain conditioned medium, add 50 ml of culture medium 2* per 100-mm dish 5 and 7 days after seeding and remove the Medium two days later.
2. This conditioned medium is used 1 : 1 diluted with culture medium in cultures of microvascular endothelial cells. It can be stored at 0–5 °C for 3 days.

Other Means of Selecting Against Nonendothelial Cell Growth

In the literature a number of other means which enhance the selectivity of the culture conditions against the proliferation of nonendothelial cells have been proposed. We tested the use of the hyaluronic acid as attachment substrate [45] and heparin as culture medium supplement (100 µg/ml) [8, 19, 28, 29, 45, 51, 54] and found no further reduction in the small number of nonendothelial cells present in 14-day-old cultures.

Plasma-derived serum: Smooth muscle cells and pericytes need platelet derived growth stimulants (e.g., platelet-derived growth factor, PDGF) to proliferate, whereas endothelial cells do not. Complete sera used for cell culture contain PDGF. In platelet-poor plasma (plasma-derived serum) smooth muscle cells and pericytes do not proliferate [8, 13, 25, 67, 71].

D-Valine media: In media supplemented with D-valine instead of L-valine, fibroblast and smooth muscle cell growth is inhibited, but that of endothelial cells is not [24, 26, 47]. Such media are commercially available (Gibco).

Thimerosal: Wagner and Matthews [66] reported that treatment of mixed cultures with the mercurial compound thimerosal can selectively intoxicate proliferating fibroblasts. The procedure, however, does not seem to be as selective and can also damage the endothelial cell population [4].

Energy Metabolism of Microvascular Coronary Endothelial Cells

Normoxia

Coronary endothelial cells, in confluent monolayers of 2-week-old cultures, contain high levels of adenine nucleotides: 21.4 nmol ATP/mg protein, 1.8 nmol ADP/mg protein, and 0.5 nmol AMP/mg protein [37]. Calculated from these values, the adenylate energy charge $EC = (ATP + 0.5\ ADP)/(ATP + ADP + AMP)$ amounts to 0.94. This energy charge indicates a well-energized metabolic state. Half of the protein of the cultured cells can be removed by brief treatment with trypsin, so that the adenine nucleotide content per milligram of the remaining cell protein is twice as high as the values given above. On a cell protein base, therefore, the endothelial adenine nucleotide content exceeds that of the myocardial tissue. This finding is consistent with the results previously reported by Nees et al. [42].

In the heart, fatty acids and lactate are the most important energetic substrates under aerobic conditions. Coronary endothelial cells are also able to utilize lactate and palmitate as fuels for oxidative phosphorylation; but they even more readily oxidize amino acids [37, 61]. Among the metabolic substrates tested, glucose is of predominant importance for coronary endothelial cells, in confluent monolayer cultures. But almost all energy obtained by catabolizing glucose is generated glycolytically. This is not due to a lack of oxygen, as demonstrated by the finding that variation of the oxygen tension between 3 and 700 torr did not alter this metabolic behavior [37, 40].

In the presence of 5 mM exogenous glucose alone, the portion of glucose diverted to the hexose monophosphate shunt is small, but it accounts for almost all CO_2 produced from glucose (Fig. 3). Interestingly, at concentrations of glucose below 1 mM, the amount of glucose oxidized in the Krebs cycle is greater than at higher concentrations, and lactate production is conversely decreased [37]. Thus, oxidative metabolism is inhibited at physiological glucose concentrations. The inhibitory effect of glucose on mitochondrial oxidation also affects other substrates degraded in the Krebs cycle, i.e., palmi-

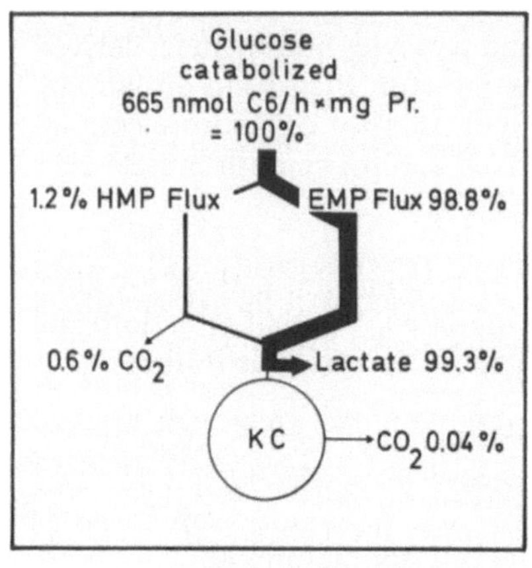

Fig. 3. Scheme of glucose catabolism of coronary endothelial cells in 14-day-old confluent cultures, incubated in Tyrode's solution with 5 mM glucose. *Rates* are given as percentages of total glucose breakdown. *HMP,* hexose monophosphate pathway; *EMP,* Embden-Meyerhof pathway; *KC,* Krebs cycle. [From 61]

tate, lactate, and amino acids. The oxidation of these substrates (300 µM palmitate, 1 mM lactate, 0.5 mM glutamine) is reduced by 50% or more in the presence of 5 mM glucose [37, 61]. The inhibitory effect of glucose on mitochondrial respiration has been termed the "Crabtree effect" [69].

The capacity of the hexose monophosphate pathway greatly exceeds its activity under control conditions. When cellular reduced nicotinamide adenine dinucleotide phosphate (NADPH) is oxidized by the presence of methylene blue (0.4 mM), the flux through the hexose monophosphate pathway becomes greatly stimulated, comprising 81% of total glucose catabolism (Fig. 4). This demonstrates that microvascular coronary endothelial cells possess a large potential capacity of the hexose monophosphate pathway. A similar observation has been made before on microvascular endothelial cells of other origin [16]. The large capacity of this pathway may support endothelial cells in withstanding oxidative stress.

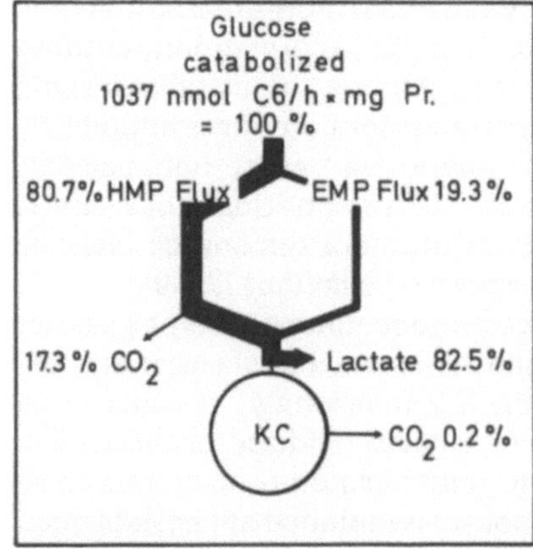

Fig. 4. Scheme of glucose catabolism of coronary endothelial cells in 14-day-old confluent cultures, incubated in Tyrode's solution with 5 mM glucose and 0.4 mM methylene blue. *Rates* are given as percentages of total glucose breakdown. *Pr,* protein; *HMP,* hexose monophosphate pathway; *EMP,* Embden-Meyerhof pathway; *KC,* Krebs cycle. [Data are from 37]

The oxygen uptake of coronary endothelial cells is low when compared with that of the whole heart. In stirred suspensions of 2-week-old cultures, oxygen is taken up at a rate of 8 nmol/min per milligram cell protein when 0.2 mM palmitate is the only exogenous substrate [40]. Under the same supply conditions, the beating myocardium requires 50 to 100 nmol O_2/min per milligram protein at a low to moderate work load. Oxygen consumption by coronary endothelial cells in fact comes close to the demand of the arrested myocardium. When the ATP : O ratio is taken as 2.8, the oxygen consumption mentioned above is equivalent to 45 nmol ATP/min per milligram protein. When glucose is added to the cells, oxygen consumption drops by 50% [40], equivalent to 22 nmol ATP/min per milligram protein, but then lactate is produced at a rate of 25 nmol/min per milligram protein corresponding to 25 nmol ATP/min per milligram protein (1 mol ATP/mol lactate), so that the calulated total energy production amounts to 47 nmol ATP/min per milligram potein, i.e., the same as in the presence of palmitate.

The described results on glucose metabolism in cultures of coronary endothelial cells resemble in several respects those obtained on cultures of microvascular endothelium from penis corpus cavernosum [16] and on isolated brain microvessels [10, 30]. High activity of glutamine-metabolizing enzymes has also been reported for pulmonary microvascular cells [38]. This suggests great similarity of microvascular endothelium of different sources in the investigated aspects of energy metabolism.

It may be of significance for the heart as a whole that coronary microvascular endothelial cells break down glucose only to lactate. This is because lactate is a much better oxidative substrate for the cardiomyocyte than glucose. In producing lactate they are supported by the cellular contents of the vascular space. Leucocytes [50] and lymphocytes [32] can also express a pronounced Crabtree effect, and erythrocytes obligatorily produce lactate, since they lack oxidative metabolism.

Hypoxia

At oxygen tensions above 3 torr, the uptake of oxygen by coronary endothelial cells is constant; below 3 torr oxygen consumption gradually declines (Fig. 5). The rate of oxygen consumption is half-maximal at a pO_2 of 0.8 torr. The relation between the rate of oxygen uptake and oxygen tension apparently reflects Michaelis-Menten kinetics. The sensitivity of oxygen uptake for oxygen tension was found to be the same when exogenous glucose, palmitate, glutamine were supplied, and when only endogenous substrates could be oxidized. In the presence of glucose, however, oxygen uptake rates above 3 torr O_2 are 50% lower than in its absence, due to the contribution of glycolysis to energy metabolism.

In isolated mitochondria, 0.2 torr represents the K_m of oxygen consumption by the cytochrome c oxidase [11]. Whole cells generally have a reduced apparent affinity to oxygen, due to intra- and extracellular gradients of oxygen. For the investigation of the oxygen sensitivity of endothelial cells we used [40] a special incubation system [oxystat, 44] in which the formation of an

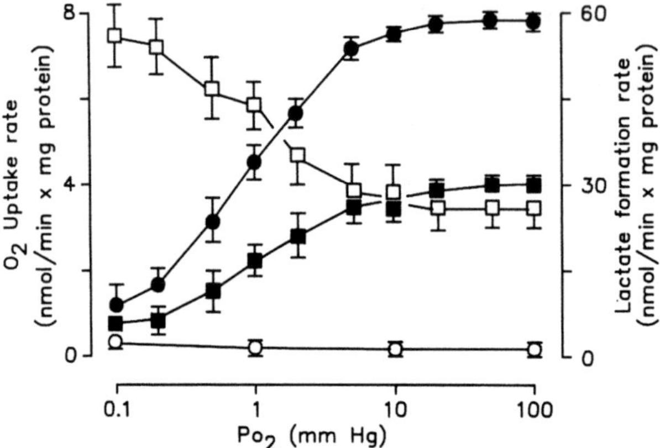

Fig. 5. Coronary endothelial cells from 14-day-old confluent cultures, suspended in Tyrode's solution with 5 mM glucose (squares) or 100 µM palmitate plus 0.5 mM glutamine (circles) under variation of the pO$_2$: Relation of oxygen, uptake (solid symbols) and lactate production (open symbols) to medium pO$_2$ ($\bar{x}\pm$S.D.; n, five cultures). The methods are described in [40]

extracellular oxygen gradient is prevented by rapid mixing of the medium. Therefore, the pO$_2$ of the half-maximal oxygen consumption by coronary endothelial cells, i.e., 0.8 torr, indicates that within the endothelial cell there is a gradient of 0.6 torr towards the inner mitochondrial membrane. This value closely relates with estimates of intracellular oxygen gradients in other cell types [12]. The sensitivity of individual endothelial cells to oxygen is comparable with that of cardiomyocytes [70]. Physiologically, oxygen tensions of 1 torr and below are extremely low, since they are two orders of magnitude below the normal arterial oxygen tension. They will occur locally, however, whenever the heart responds with signs of anaerobic metabolism to a mismatch between oxygen supply and demand.

The aerobic production of lactate in the presence of glucose, 5 mM, is unaffected by a reduction of medium pO$_2$ from 100 to 10 torr. At 1 and 0.1 torr lactate production increases inversely with respect to the decrease in oxygen consumption (Fig. 5). At a pO$_2$ of 0.1 torr lactate production is increased 2.2-fold as compared with well-oxygenated conditions. This means that the Pasteur effect is smaller in coronary endothelial cells than in cardiac muscle cells. This finding is consistent with the pronounced Crabtree effect under aerobic conditions.

In anoxia (pO$_2 \leq 1$ torr) with exogenous supply of glucose, the adenine nucleotide contents of coronary endothelial cells remain unaltered for several hours [40]. The increase in glycolytic energy production is sufficient to fully compensate for the lack of respiratory ATP supply. The calculated rates of ATP production under aerobic conditions (in the presence of 5 mM glucose: 47 nmol ATP/min per milligram protein) and at 0.1 torr O$_2$ (55 nmol ATP/min per milligram protein; see Fig. 5) are nearly identical.

The importance of glucose for hypoxic energy production in coronary endothelial cells was tested in experiments in which glucose was omitted from the incubation medium and replaced by substrates which can only be used oxidatively [40]. Under these conditions the production of lactate is very small and can be fully accounted for by the degradation of glycogen. In the absence of exogenous supply of glucose, cellular contents of ATP decrease at oxygen tensions less than or equal to 2 torr.

It was reported before [7] that in the anoxic perfused heart the endothelium is more resistant to hypoxia-reoxygenation than the cardiomyocyte. The metabolic behavior of isolated microvascular coronary endothelial cells indicates that the coronary endothelium possesses greater energetic stability than cardiomyocytes under hypoxic conditions, both because its energy demand is lower and its glycolytic capacity is larger.

Conclusions

Using the procedure described in this chapter, cultures of coronary endothelial cells can be obtained which are virtually free of nonendothelial cells within the first 2 weeks after isolation. Even though the exact origin of the cells establishing in culture is not known, it seems reasonable to assume that they represent a statistical average of all coronary endothelial cells since the culture is derived from a complete dissociation of heart tissue. Since capillary endothelium comprises more than 90% of all endothelial cells contained in the coronary circulation [6, 33], it seems justified to use the term "microvascular" for this cell preparation [cf. 23, 57].

With this method about 1% of the endothelial cells contained in myocardial tissue establish in culture and proliferate. This yield is the highest value reported in the literature. When cell cultures are derived from an extremely selected population of cardiac endothelial cells with unknown topological origin [e.g., from about 0.0002% as in 15], it remains uncertain what these cells actually represent. If isolated from a heart perfused in a Langendorff system [15], it cannot even be excluded that they are derived from aortic endothelium (since the heart is mounted with the aorta on the perfusing cannula) or from the endocardial cell layer (since in a protease-perfused Langendorff heart the aortic valve usually does not remain completely occluded). An extremely low yield may be justified when the location of the isolated cells in the vascular bed can be determined. Isolation of endothelial cells with the use of microspheres theoretically provides cells from a specified luminal diameter in the vascular bed. Whether this indeed is the case, however, is difficult to prove. For all purposes requiring isolated endothelial cells as a model to study endothelial cell function in a specific localization within the vascular bed, cultures from hyperselected cell populations confer another disadvantage: Unless the experimental needs for the cell material are very low, these cells have to be multiplied several times in culture, and during this proliferation they may lose many of the features characteristic for the part of the vascular bed from which they were isolated. In cultures prepared according to the described protocol, early stages (4 days) could be compared with later ones (14 days) since the initial yield is already sufficient for biochemical determinations [37]. It was found that the basic features of endothelial energy metabolism remain constant within the first 2 weeks in culture. Whether this applies also to other physiological properties remains to be investigated.

References

1. Absher M, Woodcock-Mitchell J, Mitchell J, Baldor L, Low R, Warshaw D (1989) Characterisation of vascular smooth muscle cell phenotype in long-term culture. In Vitro Cell Dev Biol 25:183–192
2. Allikmets EY, Danilov SM (1986) Mitogen-induced disorganization of capillary-like structures formed by human large endothelial cells in vitro. Tissue Cell 18:481–489
3. Anversa P, Levicky V, Beghi C, McDonald SL, Kikkawa Y (1983) Morphometry of exercise-induced right ventricular hypertrophy in the rat. Circ Res 52:57–64
4. Balconi G, Dejana E (1986) Cultivation of endothelial cells: limitations and perspectives. Med Biol 64:231–245
5. Banerjee DK, Ornberg RL, Youdim MBH, Heldman E, Pollard HB (1985) Endothelial cells from bovine adrenal medulla develop capillary-like growth patterns in culture. Proc Natl Acad Sci USA 82:4702–4706
6. Bassingthwaighte JB, Yipintsoi T, Harvey RB (1974) Microvasculature of the dog left ventricular myocardium. Microvasc Res 7:229–249
7. Buderus S, Siegmund B, Spahr R, Krützfeldt A, Piper HM (1989) Resistance of coronary endothelial cells to anoxia-reoxygenation in isolated guinea pig hearts. Am J Physiol 257:H488–H493
8. Carson MP, Haudenschild CC (1986) Microvascular endothelium and pericytes: high yield, low passage cultures. In Vitro Cell Dev Biol 22:344–354
9. Castellot JJ, Addanizio ML, Rosenberg R, Karnovsky ML (1981) Cultured endothelial cells produce a heparinlike inhibitor of smooth muscle cell growth. J Cell Biol 90:372–379
10. Chan CT, Brecher P, Haudenschild C, Chobanian AV (1979) The effect of cholesterol feeding on the metabolism of rabbit cerebral microvessels. Microvasc Res 18:353–369
11. Chance B (1965) Reaction of oxygen with the respiratory chain in cells and tissue. J Gen Physiol 49:163–188
12. Clark A, Clark PAA, Connett RJ, Gayeski TEJ, Honig CR (1987) How large is the drop in PO_2 between cytosol and mitochondrion? Am J Physiol 252:C583–C587
13. Dickinson ES, Slakey LL (1982) Plasma-derived serum as a selective agent to obtain endothelial cultures from swine aorta. In Vitro 18:63–70
14. Diglio CA, Grammas P, Giacomelli F, Wiener J (1982) Primary culture of rat cerebral microvascular endothelial cells. Lab Invest 46:554–563
15. Diglio CA, Grammass P, Giacomelli F, Wiener J (1988) Rat heart derived endothelial and smooth muscle cell cultures: isolation, cloning and characterization. Tissue Cell 20:477–492
16. Dobrina A, Rossi F (1983) Metabolic properties of freshly isolated bovine endothelial cells. Biochim Biophys Acta 762:295–301
17. Folkman J, Haudenschild CC, Zetter BR (1979) Long-term culture of capillary endothelial cells. Proc Natl Acad Sci USA 76:5217–5221
18. Gaffney J, West D, Arnold F, Sattar A, Kumar S (1985) Differences in the uptake of modified low density lipoproteins by tissue cultured endothelial cells. J Cell Sci 79:317–325
19. Gerhart DZ, Broderius MA, Drewes LR (1988) Cultured human and canine endothelial cells from brain microvessels. Brain Res Bull 21:785–793
20. Gerlach E, Nees S, Becker BF (1985) The vascular endothelium: a survey of some newly evolving biochemical and physiological features. Basic Res Cardiol 80:459–474
21. Gerritsen ME (1987) Functional heterogeneity of vascular endothelial cells. Biochem Pharmacol 36:2701–2711

22. Gerritsen ME, Burke T (1985) Insulin binding and effects of insulin on glucose uptake and metabolism in cultured coronary microvessel endothelium. Proc Soc Exp Biol Med 180:17–23
23. Gerritsen ME, Cheli CD (1983) Arachidonic and prostaglandin endoperoxide metabolism in isolated rabbit and coronary microvessels and isolated and cultivated coronary microvessel endothelial cells. J Clin Invest 72:1658–1671
24. Gilbert SF, Migeon BR (1975) D-Valine as a selective agent for normal human and rodent epithelial cells in culture. Cell 5:11–17
25. Gitlin JD, D'Amore PA (1983) Culture of retinal capillary cells using selective growth media. Microvasc Res 26:74–80
26. Gumkowski F, Kaminska G, Kaminski M, Morrissey LW, Auerbach R (1987) Heterogeneity of mouse vascular endothelium. Blood Vessels 24:11–23
27. Haudenschild CC, Cotran RS, Gimbrone MA, Folkman J (1975) Fine structure of vascular endothelium in culture. J Ultrastruct Res 50:22–32
28. Herbert JM, Maffrand JP (1989) Heparin interactions with cultured human vascular endothelial and smooth muscle cells: incidence on vascular smooth muscle cell proliferation. J Cell Physiol 138:424–432
29. Herman IM (1987) Extracellular matrix–cytoskeletal interactions in vascular cells. Tissue Cell 19:1–19
30. Hingorani V, Brecher P (1987) Glucose and fatty acid metabolism in normal and diabetic rabbit cerebral microvessels. Am J Physiol 252:E648–E653
31. Hormia M, Virtanen I (1986) Endothelium – an organized monolayer of highly specialized cells. Med Biol 64:247–266
32. Hume DA, Radik JL, Ferber E, Weidemann MJ (1978) Aerobic glycolysis and lymphocyte transformation. Biochem J 174:703–709
33. Hyde DM, Buss DD (1986) Morphometry of the coronary microvasculature of the canine left ventricle. Am J Anat 177:415–425
34. Irving MG, Roll RJ, Huang S, Bissell DM (1984) Characterization and culture of sinusoidal endothelium from normal rat liver: lipoprotein uptake and collagen phenotype. Gastroenterol 87:1233–1247
35. Jaffe EA, Nachman RL, Becker CG, Minick CR (1973) Culture of human endothelial cells derived from umbilical veins. Identification by morphologic and immunologic criteria. J Clin Invest 52:2745–2756
36. King GL, Buzney SM, Kahn CR, Hetu N (1982) Differential responsiveness to insulin of endothelia and support cells from micro- and macrovessels. J Clin Invest 71:974–979
37. Krützfeldt A, Spahr R, Mertens S, Siegmund B, Piper HM (1990) Metabolism of exogenous subtrates by coronary microvascular endothelial cells in culture. J Mol Cell Cardiol (in press)
38. Leighton B, Curi R, Hussein A, Newsholme EA (1987) Maximum activities of some key enzymes of glycolysis, glutaminolysis, Krebs cycle and fatty acid utilization in bovine pulmonary endothelial cells. FEBS Lett 225:93–96
39. Matsuoka T, Tavassoli M (1988) A modified method for application of indirect immunofluorescent staining for factor VIII/vWF to capillary endothelial endothelia. Am J Med Sci 196:107–110
40. Mertens S, Noll T, Spahr R, Krützfeldt A, Piper HM (1990) The energetic response of coronary endothelial cells to hypoxia. Am J Physiol 258 (in press)
41. Mistry G, Drummond GI (1983) Heart microvessels: presence of adenylate cyclase stimulated by catecholamines, prostaglandines, and adenosine. Microvasc Res 26:157–169
42. Nees S, Gerlach E (1983) Adenine nucleotides and adenosine metabolism in cultured coronary endothelial cells: formation and release of adenine compounds and possible functional implications. In: Berne RM, Rall TW, Rubio R (eds) Regulatory function of adenosine. Nijhoff, The Hague, pp 347–360
43. Nees S, Gerbes AL, Gerlach E (1981) Isolation, identification and continuous culture of coronary endothelial cells from guinea pig heart. Eur J Cell Biol 24:287–297

44. Noll T, DeGroot H, Wissemann P (1986) A computer-supported oxystat system maintaining steady-state O_2 partial pressures and simultaneously monitoring O_2 uptake in biological systems. Biochem J 236:765–769
45. Orlidge A, D'Amore PA (1986) Cell specific effects of glycosaminoglycans on the attachment and proliferation of vascular wall components. Microvasc Res 31:41–53
46. Orlidge A, D'Amore PA (1987) Inhibition of capillary endothelial cell growth by pericytes and smooth muscle cells. J Cell Biol 105:1455–1462
47. Picciano PT, Johnson B, Walenga RW, Donovan M, Borman BJ, Douglas WHJ, Kreutzer DL (1984) Effects of D-valine on pulmonary endothelial cell morphology and function in cell culture. Exp Cell Res 151:134–147
48. Reidy MA, Chopek M, Chao S, McDonald T, Schwartz SM (1989) Injury induces increase of von Willebrand factor in rat endothelial cells. Am J Pathol 134:857–864
49. Rone JD, Goodman AL (1987) Heterogeneity of rabbit aortic endothelial cells in primary culture. Proc Soc Exp Biol Med 184:495–503
50. Rossi F, Zatti M (1966) Effect of phagocytosis on the carbohydrate metabolism of polymorphonuclear leucocytes. Biochim Biophys Acta 121:110–119
51. Rupnick MA, Carey AW, Williams SK (1988) Phenotypic diversity in cultured cerebral microvascular endothelial cells. In Vitro Cell Dev Biol 24:435–444
52. Ryan US, White LA, Lopez M, Ryan JW (1982) Use of microcarriers to isolate and culture pulmonary microvascular endothelium. Tissue Cell 14:597–606
53. Schelling ME, Meininger CJ, Hawker JR Jr, Granger HJ (1988) Venular endothelia cells from bovine heart. Am J Physiol 254:H1211–H1217
54. Schini V, Grant NJ, Miller RC, Takeda K (1988) Morphological characterization of cultured bovine aortic endothelial cells and the effects of atriopeptin II and sodium nitroprusside on cellular and extracellular accumulation of cyclic GMP. Eur J Cell Biol 47:53–61
55. Schor AM, Schor SL (1986) The isolation and culture of endothelial cells and pericytes from the bovine retinal microvasculature: a comparative study with large vessel vascular cells. Microvasc Res 32:21–38
56. Schwartz SM (1978) Selection and characterization of bovine aortic endothelial cells. In Vitro 14:966–980
57. Simionescu M, Simionescu N (1978) Isolation and characterization of endothelial cells from the heart microvasculature. Microvasc Res 16:426–452
58. Sims DE (1986) The pericyte – a review. Tissue Cell 18:153–174
59. Skalli O, Pelte MF, Peclet MC, Gabbiani G, Gugliotta P, Bussolati G, Ravazzola M, Orci L (1989) α-Smooth muscle actin, a differentiation marker of smooth muscle cells, is present in microfilamentous bundles of pericytes. J Histochem Cytochem 37:315–321
60. Smith P (1989) Effect of hypoxia upon growth and sprouting activity of cultured aortic endothelium from the rat. J Cell Sci 92:505–512
61. Spahr R, Krützfeldt A, Mertens S, Siegmund B, Piper HM (1989) Fatty acids are not an important fuel for coronary microvascular endothelial cells. Mol Cell Biochem 88:59–64
62. Spatz M, Bembry J, Dodson RF, Hervonen H, Murray MR (1980) Endothelial cultures derived from isolated cerebral microvessels. Brain Res 191:577–582
63. Tilton RG, Kilo C, Williamson JR (1979) Pericyte–endothelial relationship in cardiac and skeletal muscle capillaries. Microvasc Res 18:325–335
64. Tontsch U, Bauer HC (1989) Isolation, characterization and long-term cultivation of porcine and murine cerebral capillary endothelial cells. Microvasc Res 37:148–161
65. Voyta JC, Via DP, Butterfield CE, Zetter BR (1984) Identification and isolation of endothelial cells based on their increased uptake of acetylated-low density lipoprotein. J Cell Biol 99:2034–2040
66. Wagner RC, Matthews MA (1975) The isolation and culture of capillary endothelium from epididymal fat. Microvasc Res 10:286–297
67. Wall RT, Harher LA, Quadracci LJ, Striker GE (1978) Factors influencing endothelial cell proliferation in vitro. J Cell Physiol 96:203–214

68. Weibel ER, Pallade GE (1964) New cytoplasmatic components in arterial endothelia. J Cell Biol 23:101–112
69. Wenner CE (1979) Pasteur and Crabtree effects – assay in cells. Methods Enzymol 55:289–297
70. Wittenberg BA, Wittenberg JB (1985) Oxygen pressure gradient in isolated cardiac myocytes. J Biol Chem 260:6548–6554
71. Wren FE, Schor AM, Schor SL, Grant ME (1986) Modulation of smooth muscle cell behaviour by platelet-derived factors and the extracellular matrix. J Cell Physiol 127:297–302
72. Zetter BR (1981) The endothelial cells of large and small blood vessels. Diabetes 30 [Suppl 2] : 24–28

Macro- and Microvascular Endothelial Cells from Human Tissues

V. W. M. van Hinsbergh, M. A. Scheffer, and E. G. Langeler

Introduction

Endothelial cells represent the inner lining of all blood vessels. Recent progress in understanding the metabolism of endothelial cells has revealed that they are involved in many (patho)physiological processes. These cells prevent coagulation of the blood and produce factors that are involved in the regulation of fibrinolysis. They metabolize and respond to vasoactive substances and play a role in the regulation of the vascular tone. The endothelium also actively regulates the influx of fluid, macromolecules, and various types of leukocytes into the vessel wall and other tissues. Furthermore, endothelial cells play a central role in neovascularization, a process that is important in reparative processes and that also plays a role in various pathological processes.

Endothelial cells have a large synthetic capacity. It has been estimated that the whole population of endothelial cells in the human represents a mass of about 720 g [92]. Taken together these cells represent a large secretory and regulatory organ, which is maximally exposed to the blood. In addition, endothelial cells often have specialized functions, which depend on the tissue in which they are located.

Progress in the culture of endothelial cells has contributed much to the rapid and still increasing understanding of these functions. Endothelial cells can be cultured from human and animal blood vessels. (The culture of endothelial cells from animal blood vessels is described in other chapters of this book.) Often the use of human endothelial cells will be preferred over that of animal cells. Species differences may have an effect on the metabolic regulation of the cells, on the sensitivity towards specific mediator peptides, and on the applicability of cDNA probes and monoclonal antibodies. In this chapter an overview is given of the culture of human macro- and microvascular endothelial cells.

Description of Methods and Materials

Materials and Preparations for Isolation and Culture of Endothelial Cells

Equipment

Laboratory equipment needed for the culture of human endothelial cells includes a CO_2 incubator, laminar flow hood, inverted phase-contrast microscope, centrifuge, endotoxin-free pipettes, flasks, and tissue culture dishes (e.g., Corning plastics). In addition, for the isolation of endothelial cells from arteries and veins one uses cannulae and/or other tools as depicted in Fig. 1, scalpel, Kocher scissors, surgical silk (e.g., Ethicon Leinenzwirn 3.5, order no. EH 808), scissors, and forceps. For the isolation of foreskin microvascular endothelial cells a fluorescence-activated cell sorter (FACS), rubber plug (about 6 cm in diameter), needles, forceps, and scalpel or razor blade are used. For the culture of human endothelial cells on microcarriers Biosilon (Nunc, Roskilde, Denmark) and cytodex 3 (Pharmacia) are required, and for the culture of human endothelial cells on porous filters Transwell TC-grade poly-carbonate filters (Costar).

Chemicals

Medium 199 (M199) supplemented with 20 mM HEPES was obtained from Flow Laboratories (Irvine, Scotland); sodium bicarbonate and L-glutamine (2 mM) were added separately.

Penicillin/streptomycin was purchased from Boehringer Mannheim, a stock solution (50 000 IU/ml penicillin, 50 mg/ml streptomycin) was prepared in sterile pyrogen-free 0.9% NaCl. This stock solution was diluted 500-fold in M199 supplemented with 20 mM HEPES (final concentration 100 IU/ml peni-cillin, 0.10 mg/ml streptomycin). This was done immediately before use of a bottle of medium because prolonged presence of streptomycin may accelerate oxidation of the medium.

Gelatin was purchased from Merck. No endotoxin contamination was found in our batch by Limulus assay. Gelatin (1%, w/v) was dissolved in sterile pyrogen-free water, sterilized by autoclavation (45 min, 120 °C, 1 bar), and stored until use.

Fibronectin, a cryoprecipitate of human plasma containing predominantly human fibronectin, was obtained from a local blood transfusion service and was used as a crude fibronectin preparation. Purified human fibronectin was

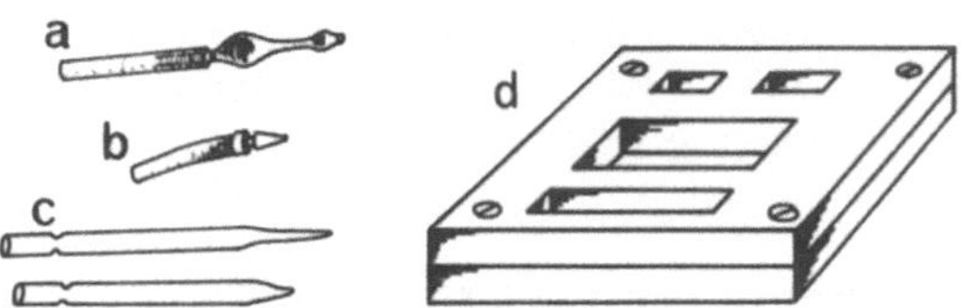

Fig. 1 a–d. Small instruments that are used for the isolation of endothelial cells. **a** Siliconized glass cannula to perfuse an umbilical vein. On one side a sterile tube is connected that can be clamped and to which syringes can be mounted. The other side is pushed into a vein. The small widening prevents separation of the vein from the cannula after fastening the cord to it. **b** Cannula to perfuse an umbilical artery. **C** Two wedges prepared from pasteurian pipettes to stretch the umbilical artery before insertion of a cannula. **d** Pieces of blood vessels can be clamped with the luminal side on top between two perspex plates, the upper one of which contains several slits. The slit is filled with collagenase solution, and after incubation the endothelial cells can be collected

prepared by gelatin-sepharose chromatography according to the method of Vuento and Vaheri [87]. It was finally dialyzed against 10 mM CAPS buffer (pH 11) supplemented with 1 mM CaCl$_2$ and 150 mM NaCl, and subsequently stored in 1 mg portions at –20 °C.

Heparin (5000 IE/ml) was purchased from Leo Pharmaceuticals.

Collagenase (type 1, CLS) was obtained from Worthington; collagenase/dispase from Boehringer Mannheim.

Trypsin (1 - 300) was obtained from ICN (Cleveland, OH, USA) or from Gibco.

Pyrogen-free human serum albumin (20% stock solution) was purchased from the Central Laboratory of the Blood Transfusion Service (Amsterdam, The Netherlands).

Endothelial Cell Growth Factor (See "Preparation of Endothelial Cell Growth Factor").

Recombinant basic fibroblast growth factor (b-FGF) was obtained from Boehringer Mannheim; recombinant human acidic fibroblast growth factor (a-FGF) was a gift from Dr. T. Maciag (Rockville, MD, USA).

3,3'-Dioctadecylindocarbocyanine (DiI) was purchased from Molecular Probes Inc.

Collection and Storage of Blood Vessels

A sterile technique is utilized in all manipulations of the blood vessels. After delivery umbilical cords are put in ice-cold buffer (140 mM NaCl, 4 mM KCl, 11 mM D-glucose, 10 mM HEPES, pH 7.3, 100 IU/ml penicillin, 0.10 mg/ml streptomycin). The cords can be stored in a refrigerator until use (good results are obtained at least up to 48 h). Isolation of endothelial cells proceeds more easily with cords that have been stored in the cold for 6 h or more.

Arteries and Veins. At autopsy or surgery the vessels or pieces thereof are immediately put into, stored in (up to 6 h tested), and rinsed with ice-cold buffer (see above). Poor results are obtained with vessels obtained when autopsy is performed more than 6 h after death.

Pieces of foreskin obtained at circumcision are stored in the same buffer.

Subcutaneous and omental fat tissue is stored in the same buffer supplemented with 0.5% pyrogen-free human serum albumin. Albumin is added to prevent possible deleterious effects of free fatty acids.

Coating of Tissue Culture Wells and Coverslips

All dishes and flasks used to culture endothelial cells are coated with fibronectin or gelatin. Both types of coating give similar results. For coating the dishes or flasks are overlayed with a thin layer of 1% gelatin or with 10 µg/cm^2 fibronectin in M199 and incubated at room temperature for at least 30 min. This solution is removed by aspiration immediately before seeding the cells.

For immunofluorescence and immunospectrophotometric studies human endothelial cells are cultured on coverslips. The coverslips used are round,

14 mm in diameter, and fit into 24-multiwell dishes, or rectangular (24 mm × 10 mm, which can be mounted in a fluorospectrophotometer cuvette); they are defatted and sterilized by washing in 70% ethanol and 96% ethanol and subsequently air dried in a laminar flow hood. The coverslips are incubated for 45 min at room temperature in 1% gelatin solution. After removal of the gelatin solution the attached gelatin is cross-linked by incubation in 0.5% glutaraldehyde in phosphate buffered saline (PBS). The glutaraldehyde solution is removed and the coverslips are washed vigorously twice with M199, incubated in M199 for 15 min, and subsequently washed again vigorously several times with M199. The coverslips are stored dry until use.

Sera, Culture Media, and Other Solutions

Human serum is prepared from freshly collected blood obtained from healthy donors. The sera of 15–25 subjects are pooled and stored at 4 °C (for up to 3 months) or at –80 °C. Before use the sera are centrifuged and filtered through 0.45 μm Acrodisc filters at room temperature. The quality of every batch of human serum is tested before use. Normally human serum is not heat-inactivated. Only in a few cases has heat inactivation improved the quality of the serum, particularly with respect to foreskin microvascular endothelial cells.

Newborn calf serum was purchased from Gibco. It was stored at –20 °C. Newborn calf serum was heat inactivated by incubation for 30 min at 56 °C.

The standard culture medium that we use for the culture of endothelial cells from umbilical cord and other arteries and veins is M199 supplemented with 20 mM HEPES (pH 7.4), 20% human serum, 50 μg/ml crude endothelial cell growth factor (ECGF), 5 U/ml heparin, 100 IU/ml penicillin, and 0.1 mg/ml streptomycin. Instead of 20% human serum, 10% human serum and 10% heat-inactivated newborn calf serum can also be used. For human foreskin microvascular endothelial cells the medium is supplemented with an additional 10% serum (final concentration 20% human serum and 10% newborn calf serum).

A stock solution of 0.1 g/ml *collagenase* (Worthington type 1, CLS) was made in sterile PBS, sterilized by filtration through 0.45 μm Acrodisc filters, and stored in small portions at –20 °C. For the isolation of endothelial cells, the stock solution was thawed and diluted 100-fold with M199 supplemented with 20 mM HEPES and penicillin/streptomycin (final concentration 0.10%) and warmed to 37 °C.

In contrast to the collagenase of Worthington, *collagenase/dispase* (Boehringer Mannheim) had to be dissolved immediately before use. Storage at –20 °C resulted in a considerable loss of activity. Collagenase/dispase was dissolved, sterilized (0.22 μm filter), and diluted in the same way as described for collagenase. It was used at a final concentration of 0.05%.

Trypsin/EDTA solution, 0.05% (w/v) trypsin, 137 mM NaCl, 5.4 mM KCl, 4.2 mM NaHCO$_3$, 5 mM D-glucose, 0.67 mM EDTA, pH 7.3 and wash-buffer Sol-A made up of 137 mM NaCl, 5.4 mM KCl, 4.2 mM NaHCO$_3$, 5 mM D-glucose, pH 7.3 were used to wash the cells before detachment by trypsin/EDTA solution.

Preparation of Endothelial Cell Growth Factor

A crude preparation of ECGF can be prepared according to the method described by Maciag et al. [56]. A bovine brain (about 600 g) is obtained aseptically from a local slaughterhouse. After removal of blood-containing regions the brain is cut into pieces that are homogenized for 3 min in 0.1 M NaCl in an ice-cooled blender (total 1050 ml NaCl solution). The pH is kept at 7.0 during this procedure. The homogenate (pH 7.0) is stirred for 2 h at 4 °C, and subsequently centrifuged for 40 min at $13\,800 \times g$ at 4 °C. To the obtained supernatant 0.5% (w/v, final concentration) streptomycin sulfate is added and the mixture is incubated for at least 1 h to extract lipid material. The pH should be checked and kept at 7.0. The mixture is centrifuged for 40 min at $13\,800\,g$. The obtained supernatant is lyophilized. The lyophilized preparation can be stored at 4 °C for 6 months. It is dissolved in, e.g., M199 supplemented with 20 mM HEPES and 5 U/ml heparin (optional, but it improves the stability of the growth factor). The crude ECGF preparation can be further purified by ammonium sulfate precipitation and heparin-sepharose affinity chromatography as described by Burgess et al. [14].

Preparation of DiI-Acetylated LDL

Preparation of LDL: LDL is prepared from freshly obtained serum prepared from the blood of a healthy donor by gradient ultracentrifugation according to the method of Redgrave et al. [70]. The LDL fraction is sliced from the tube, the protein content is determined, and the LDL is dialyzed overnight against PBS at 4 °C.

Acetylation of LDL: The procedure is performed at 0 °C. To a solution of 2 ml LDL (1 – 1.5 mg/ml) 2 ml saturated Na-acetate is added and, subsequently, by addition of small amounts and under continuous stirring, 3 – 4.5 µl acetic anhydride. This is stirred for 30 min, and the LDL solution is subsequently dialyzed against PBS, and stabilized with 1% albumin or 20% lipoprotein-depleted human serum [8].

Preparation of DiI-Acetylated LDL: In 1 ml DMSO, 3 mg DiI (3,3′-dioctadecylindocarbocyanine) is dissolved. With 2 ml lipoprotein-depleted serum 1 mg acetylated-LDL (1 ml, sterilized by filtration) is mixed and 50 µl 3 mg/ml DiI in DMSO is added. The mixture is incubated for 18 h at 37 °C. Unbound DiI is removed from the DiI-acetylated-LDL preparation by separation over Sephadex G-50 and the DiI-acetylated-LDL is sterilized by filtration [67].

Precautions

The use of human material requires safety precautions. Human endothelial cells may be, or become, infected with viruses or mycoplasmas. This not only results in altered properties of endothelial cells, but also offers the possibility that pathological material may become transferred. Contamination with endotoxin will activate the endothelial cells. Therefore, all materials and solutions should be pyrogen free.

The use of fungizone (Flow Laboratories) is often advocated to prevent contamination with yeasts or molds. Although its use cannot always be circumvented, e.g., in primary culture of foreskin microvascular endothelial cells, its use in experiments should be strongly rejected. Fungizone (amphotericin B) forms complexes with cholesterol and in this way it forms pores in the membranes. Hence, the cell properties will become altered. It is difficult to remove it from the cells once it has been used.

Culture media should not be stored for long periods. The antioxidants in M199 become consumed after several weeks. Human endothelial cells are rather sensitive to oxidative damage.

Isolation and Culture of Human Endothelial Cells

Umbilical Vein and Artery Endothelial Cells

Endothelial cells are isolated from umbilical arteries and veins as described by Jaffe et al. [45] and Gimbrone et al. [30]. The isolation and culture of umbilical vein endothelial cells have been described in detail by other authors [31, 44, 59]. Therefore, described below is our routine procedure to isolate endothelial cells both from an artery and from a vein from an umbilical cord.

The umbilical cords are collected and stored at 4 °C as described in "Collection and Storage of Blood Vessels". The umbilical cord is inspected for clamped or otherwise damaged areas. These areas are discarded. Next, a small piece of one end of the cord is cut and discarded. One of the arteries is cannulated and rinsed with cord buffer. To facilitate cannulation of an artery, the freshly cut artery is distended by one or two wedge-shaped glass bars prepared from a pasteurian pipette (Fig. 1 c). The cannula (Fig. 1 b) is quickly inserted and tightly mounted with surgical silk. The artery is slowly rinsed with cord buffer, draining into a sterile waste beaker. No air should be introduced into the lumen of the vessel. During this procedure a marked distension of the blood vessel usually occurs. Subsequently, a small piece of the other end of the vessel is cut and another cannula is inserted. The artery is filled with 0.1% collagenase solution [1] or 0.05% collagenase/dispase solution via a small tube that is mounted to the cannula. Both tubes are clamped with Kocher scissors and the filled distended vessel is incubated for 20 min at 37 °C in a beaker containing 0.9% NaCl. After the incubation the clamps are removed. The contents of the vessel are removed by flushing the vessel with 25 ml M199 and collected in a 50-ml centrifuge tube (Falcon plastics). Subsequently the umbilical vein is cannulated with two cannulas as depicted in Fig. 1a, rinsed with cord buffer, filled with 0.1% collagenase or 0.05% collagenase/dispase, and incubated for 15 min at 37 °C in the 0.9% NaCl containing beaker and processed in a manner similar to that indicated for the umbilical artery. The cells

---

[1] Solutions mentioned are those described in "Sera, Culture Media, and Other Solutions".

in the collected washings of the umbilical artery and vein are each centrifuged (5 min, 200 g) and suspended in standard culture medium. They are seeded in fibronectin-coated or gelatin-coated (see "Coating of Tissue Culture Wells and Coverslips") multiwell dishes (total area $20-25$ cm^2) or in a similarly coated T25 flask and incubated at 37 °C under 5% CO_2/95% air atmosphere.

The recovery of endothelial cells from the umbilical vein is less by this procedure than when only the umbilical vein is used. When only the umbilical vein is used the procedure is identical to that described for the umbilical artery, except for the other type of cannulas. The cells are washed the next day and propagated at 37 °C under 5% CO_2/95% air atmosphere in standard culture medium, which is renewed every $2-3$ days. When the cells have become confluent, they are usually detached by treatment with trypsin/EDTA and are passaged with a split ratio 1:3.

Serial propagation of human umbilical vein and artery endothelial cells can be performed for up to $30-70$ population doublings [33, 59, 80].

Endothelial Cells from Adult Arteries and Veins

Segments of human aorta, carotid, iliac, and renal artery, and from vena cava are obtained from autopsy; segments of saphenous vein from surgery. They are collected and stored as described above. Intercostal arteries and other small blood vessel branches are closed by ligation with surgical silk. The vessels are cannulated with siliconized glass cannulas and rinsed with cord buffer. Endothelial cells are detached from these vessels by a 20-min incubation at 37 °C in 0.1% collagenase in M199, and collected in one or two 50-ml centrifuge tubes by flushing the vessel with M199. After addition of 10% heat-inactivated serum the cells are centrifuged for 5 min at 200 g and resuspended in standard culture medium. The cells are seeded in $6-12$ wells (16 mm in diameter) that have been coated with 10 µg/cm^2 human fibronectin, and incubated at 37 °C under 5% CO_2/95% air atmosphere. Usually 4 h after seeding, endothelial cell spreading is clearly visible. Cell isolates prepared from arteries from elderly people are then washed several times rather vigorously to remove debris and smooth muscle cells, while endothelial cells remain attached to the dish. In many preparations of adult artery and vein endothelial cells we have observed that a part of the isolated endothelial cells do not adhere and spread. The attached cells are further propagated in standard culture medium at 37 °C under 5% CO_2/95% air atmosphere. The medium is renewed every 2 to 3 days (3 days only during the weekend; some cell detachment is observed after 3 days). When the cells become confluent they are detached with trypsin/EDTA solution and passaged with a split ratio 1:4. Serial propagation can be performed for $40-70$ population doublings (Fig. 2) [41, 81]. Whereas human vein endothelial cells could be kept in a normal morphology for $12-14$ passages (split ratio 1:5), endothelial cells isolated from the arteries of adult people could only be propagated for $5-7$ passages in a normal small diameter morphology [81]. Thereafter the diameter of the cells increased and hence the cell density decreased.

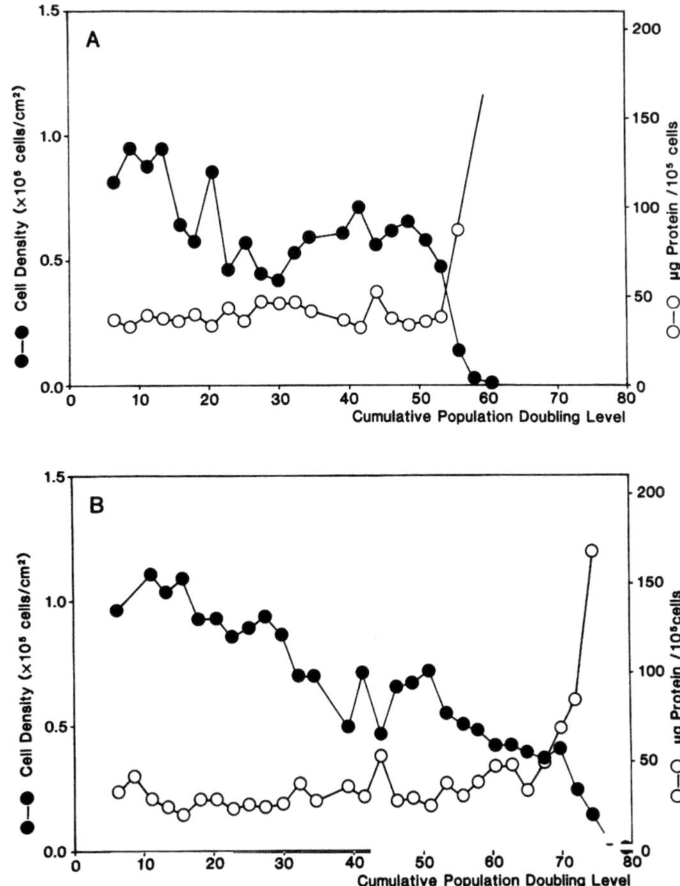

Fig. 2a, b. Serial propagation of endothelial cells from **a** human iliac artery and **b** iliac vein of a 33-year-old donor. Cells have been subcultured weekly with a split ratio 1:5

Small Pieces of Blood Vessels

When small pieces of blood vessels are obtained from surgery the vessel can be stretched on a sterile perspex plate, its luminal side on top (prevent drying!). A second sterile perspex plate with various holes is placed upon it in such a way that only endothelial cell-covered surface is exposed in the hole (see Fig. 1 d). Next, the hole is filled with 0.1% collagenase solution for 20 min at 37 °C. After incubation the isolated endothelial cells are obtained by aspiration and gentle washing of the exposed area. The cells are spun down by centrifugation for 5 min at 200 g, resuspended in standard culture medium, and seeded on fibronectin-coated dishes and cultured as described above.

Foreskin Microvascular Endothelial Cells

Endothelial cells from human foreskin microvessels are isolated using the method described by Davison et al. [21] and separated from contaminating cells by FACS as described by Voyta et al. [86].

Isolation Procedure: A piece of foreskin is washed and stretched on a rubber plug by needles. Very thin slices are cut with a scalpel; if possible the epidermis is discarded. The pieces are incubated in trypsin solution (0.3% trypsin, 1% EDTA, 137 mM NaCl, 5.4 mM KCl, 5 mM D-glucose, pH 7.3) for 30 min at 37°C and washed with M199. The incubated pieces are pressed in serum-containing medium to push cells out of them. Pieces of skin and epithelial layer (floats) are removed. The isolated cells are centrifuged for 2 min at 1500 rpm.

The cells are resuspended in M199 supplemented with 20 mM HEPES, 20% human serum, 10% newborn calf serum (heat inactivated) 2 mM L-glutamine, 2.5 µg/ml fungizone (only in primary culture), and penicillin/streptomycin. (Note: the use of crude ECGF, hypothalamus extract, stimulates the growth of melanocytes [90]; therefore, purified or recombinant b-FGF or a-FGF should be preferred.) The cells are then seeded in fibronectin-coated dishes. Two or three days after isolation, the cells are detached by trypsin/EDTA and passaged in order to remove epithelial cells and part of the melanocytes.

Separation of Endothelial Cells: Prior to cell sorting, the cells are incubated for 4 h at 37°C in M199 supplemented with 20 mM HEPES, 20% (lipoprotein-depleted) serum, 10 µg/ml DiI-acetylated LDL, and penicillin/streptomycin. During this period the cells become loaded with DiI-acetylated LDL. This can be checked by fluorescence microscopy (see "DiI-Acetylated LDL Uptake"). Endothelial cells and macrophages are labelled by DiI-acetylated LDL. However, only endothelial cells are recovered during subculturing. The labeled cells are washed once with PBS and then detached with trypsin/EDTA solution to obtain a single cell suspension. The trypsin is neutralized by washing the cells with M199 supplemented with 10% serum. Immediately before cell sorting, the cells are resuspended in serum-free M199. Cells and collection tubes are cooled on ice during the whole procedure.

DiI-acetylated LDL labeled cells are sorted from other cell types using a Beckton-Dickson FACS IV cell sorter. The 514 nm wavelength of an argon laser is used for excitation. The fluorescence emission above 550 nm is collected. Sampling gates are set by use of positive and negative cells (a small sample of the cell suspension or, if a limited number of cells is available, by the use of umbilical vein endothelial cells and fibroblasts). Cells are collected into tubes containing 10% human serum. After completion of cell sorting, the cells are spun down, washed once, and resuspended in standard culture medium (with 20% human serum and 10% heat-inactivated newborn calf serum), and seeded on fibronectin-coated dishes [82]. In several cases a second separation by FACS has been necessary to obtain a purified population of endothelial cells. Although the cells could be propagated for 6–7 passages (split ratio 1:3) without the addition of ECGF, the addition of the growth factor improves the propagation rate and markedly increases the life span of the cultures (over 18 passages).

Additional Culture Techniques

Storage of Endothelial Cells in Liquid Nitrogen

Human endothelial cells from various types of arteris and veins and from foreskin microvessels can be stored in liquid nitrogen and propagated later. To that end, confluent early passage endothelial cells ($25-75$ cm^2) are detached from the dishes by treatment with trypsin/EDTA solution. Trypsin activity is stopped by the addition of a fivefold volume of 20% serum in M199. The cells are spun down by centrifugation for 5 min at 200 g, and subsequently resuspended in 0.5 ml M199 supplemented with 20 mM HEPES, 20% human serum, and penicillin/streptomycin. The suspension is cooled to $0\,^{\circ}C$, and subsequently slowly diluted with an equal volume of 20% DMSO in the same medium (final DMSO concentration 10%). The cell suspension is transferred to plastic tubes (round-bottomed tubes with a screwing cap for freezing cells, cat no. 121 261, Greiner, Alphen a/d Rijn, The Netherlands) and closed with a screwing cap. The tubes are slowly cooled ($30\,^{\circ}C$/h) to $-70\,^{\circ}C$ and subsequently transferred into liquid nitrogen.

For further propagation of the cells the tubes are rapidly thawed in a waterbath at $37\,^{\circ}C$, and diluted in standard culture medium. The cells are spun down by centrifugation for 5 min at 200 g, resuspended in standard culture medium, seeded on fibronectin-coated dishes, and further propagated under standard conditions.

Human Endothelial Cells on Porous Filters

Confluent primary cultures are seeded on Transwell polycarbonate filters (pore size 0.4 or 3 µm; exposed area 0.33 cm^2), which have been coated with 10 µg/ml human fibronectin (30 min at room temperature). To obtain high density cultures confluent cells are detached from the culture dishes with trypsin/EDTA solution and seeded on a filter surface that is twofold smaller than the original dish surface. Nonattached cells are removed 4 h after seeding. Filters are incubated at $37\,^{\circ}C$ under 5% CO_2/95% air atmosphere in the standard culture medium, and the culture medium replaced at least every other day. Longer intervals between the medium renewals interfere with the formation and maintenance of an intact monolayer. Experiments can be done $4-7$ days after seeding the cells. The characteristics of these monolayers have been documented by Langeler et al. [54, 55].

Human Endothelial Cells on Microcarriers

Human endothelial cells can be cultured on Biosilon (Nunc) or Cytodex 3 (Pharmacia) microcarriers. To that end, the microcarriers are coated with human fibronectin. Endothelial cells obtained from dishes by trypsin/EDTA treatment are suspended in standard medium and mixed with the microcarriers in an appropriate concentration. The mixture is incubated in a tube at 37°C under 5% CO_2/95% air atmosphere and gently moved at 10-min intervals. After 2 h the microcarriers are poured into a bacterial plastic petric dish and maintained there under standard culture conditions until they are used in further experiments. The microcarriers can also be transferred into a Techne 500 ml microcarrier stirrer flask (no. 7609, Techne). The microcarriers are gently suspended by intermittent stirring (1 min stirring, 3 min off) at 25 rpm (Techne stirrer MCS-104L and Techne waterbath MWB 10L). The whole system is kept under 5% CO_2/95% air atmosphere (inflated from a gas cylinder via a water flask, a 0.22 µm filter, and sterile tubes). One-half of the medium is renewed daily with standard culture medium. Stirring must be as gentle as possible; human endothelial cells are easily damaged by continuous stirring.

Identification of Human Endothelial Cells

Morphological, ultrastructural, immunological, and biochemical criteria have been used to identify endothelial cells in culture. We describe here several generally used criteria and methods to establish the nature of cultured human endothelial cells. The identification and characterization of human endothelial cells is further discussed in another section of this chapter.

Morphology

The morphology of human endothelial cells can be checked using a Leitz Diavert phase contrast microscope with large distance objectives (10 × , Leitz). Human endothelial cells show a cobblestone morphology at confluency (Fig. 3a, b). Upon prolonged maintenance in the confluent state the monolayer starts to disorganize. First a few "sprouting" cells are observed (Fig. 3c); subsequently the whole culture becomes disorganized. Disorganization of the endothelial cell monolayer is delayed by the presence of ECGF in the culture medium. The disorganization is reversed when the cells are detached by trypsin/EDTA solution and placed on new fibronectin-coated dishes.

Ultrastructure

Evaluation of human endothelial cells by transmission electron microscopy can be performed by washing the surface of the culture dish with 1.5% glutaraldehyde in 100 mM cacodylate buffer, pH 7.4, for 10 min at room tempera-

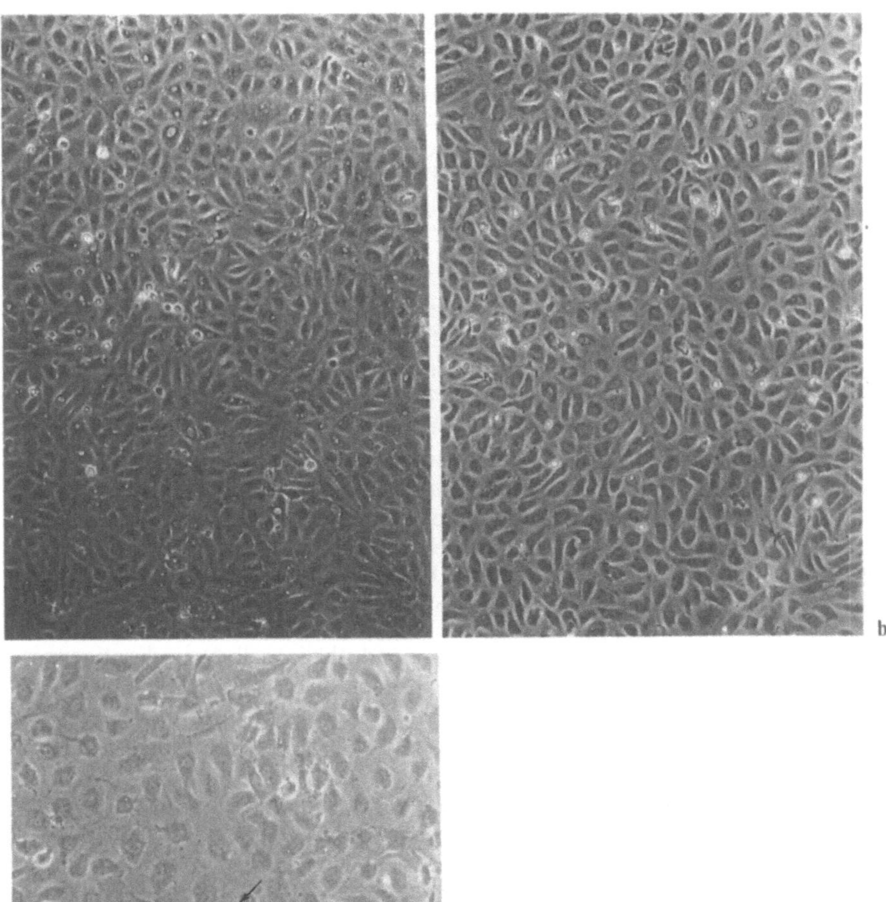

Fig. 3 a–c. Phase-contrast photomicrographs of human endothelial cells from **a** aorta and **b** umbilical artery. In **c** a culture of umbilical artery endothelial cells with "sprouting" cells (*arrows*) is shown

ture. The cells are postfixed with 1% OsO_4 in phosphate buffer with 0.05 M potassium hexacyanoferrate (II) [60]. Following dehydration in a graded 70% (v/v) to 100% (v/v) ethanol series, the cells are embedded in situ in Epon. Ultrathin sections are stained and visualized with a Philips EM 300 electron microscope at 80 kV. Weibel-Palade bodies [89] are characteristic for endothelial cells and are found relatively frequently in umbilical vein endothelial cells.

Immunological Criteria

Materials: Rabbit antihuman von Willebrand factor (vWF) antiserum was purchased from the Central Red Cross Blood Transfusion Laboratory (Amsterdam, The Netherlands). Fluoresceine-conjugated swine antirabbit immunoglobulins (FITC swine antirabbit Ig) were obtained from DAKO Immunoglobulins (Denmark). Rhodamine-labeled *Ulex europaeus* lectin I (1 mg/ml) was purchased from E.Y. Laboratories. The endothelial cell-specific monoclonal antibodies PAL-E and EN-4 can be purchased from SANBIO.

To prepare PPD mountant, 100 mg of *p*-phenylenediamine (Sigma) is dissolved in 10 ml PBS (pH 7.4) and added to 90 ml glycerol. The final pH is adjusted to approximately 8.0 with 0.5 M carbonate-bicarbonate buffer (pH 9.0). The mixture is stored in the dark at $-20\,°C$.

Procedures: Immunofluorescent staining of endothelial cell-derived antigens is performed on cell monolayers that have been cultured on gelatin-coated coverslips (see "Coating of Tissue Culture Wells and Coverslips").

The coverslips are washed with PBS, fixed in 80% (v/v) acetone for 10 min at 4°C, and rinsed again with PBS.

Von Willebrand factor is demonstrated by indirect immunofluorescent staining. After washing the fixed cells on coverslips with PBS, they are incubated for 30 min at room temperature with rabbit antihuman vWF antiserum (1:80 dilution in 10% pig serum), followed by two washes in PBS and a 30-min incubation with FITC-labeled swine antirabbit Ig (1:50 in PBS). After two additional washes with PBS, the cells are mounted with PPD mountant under glass coverslips and investigated using a Leitz epi-illumination fluorescence microscope.

Ulex europaeus lectin I is demonstrated by immunofluorescent staining. After washing the coverslips with PBS, the fixed cells are incubated for 30 min at room temperature in rhodamine-labeled *Ulex europaeus* lectin I (1 mg/ml, diluted 1:10 in PBS supplemented with 1 mM $CaCl_2$ and 1 mM $MgCl_2$). After two washes with PBS, the cells are mounted with PPD mountant under a glass coverslip and investigated using fluorescence microscopy.

The presence of the endothelial antigens that are recognized by the monoclonal antibodies PAL-E and EN-4 can be demonstrated by indirect fluorescence microscopy. Acetone (80%)-fixed cells are washed twice with PBS and are incubated with the various monoclonal antibodies for 30 min at room temperature. These are then washed again twice and are subsequently incubated for another 30 min with FITC-goat antimouse Ig (1:50 in PBS). The cells are mounted in PPD mountant under glass coverslips and examined using a Leitz epi-illumination fluorescence microscope.

DiI-Acetylated LDL Uptake

Uptake of DiI-acetylated LDL is demonstrated as described by Pitas et al. [67] with endothelial cells cultured on gelatin-coated coverslips (see "Coating of Tissue Culture Wells and Coverslips"). The cells are washed twice with M199 supplemented with 1% albumin and incubated for 4 h with DiI-acetylated LDL (20 µg/ml) in M199 supplemented with 20 mM HEPES, 10% (lipo-protein-depleted) serum, and penicillin/streptomycin. Subsequently the cells are washed five times with M199 supplemented with 1% albumin and fixed by incubation for 20 min at room temperature in 4% paraformaldehyde in PBS. The coverslips are washed again three times with Hank's salt solution (Flow Laboratories) and investigated using fluorescence microscopy with a Leitz Laborlux D fluorescence microscope (excitation 580 nm).

Angiotensin-Converting Enzyme

Angiotensin-converting enzyme (ACE) can be demonstrated by indirect im-munofluorescence microscopy or its enzymatic activity can be measured in the conditioned medium or cell extracts of human endothelial cells [47, 79].

Discussion and Perspective

Other Types of Human Endothelial Cells

In addition to the types of endothelial cells described in "Isolation and Culture of Human Endothelial Cells", the isolation and culture of microvascular en-dothelial cells from human fat, brain, adult dermis, kidney, and liver have been described.

Subcutaneous Fat Tissue Microvascular Endothelial Cells and Omental Fat Tissue Mesothelial Cells

It has been reported that omental fat tissue is a good source of human mi-crovascular endothelial cells. However, the epitheloid cells obtained from hu-man omental tissue using the method of Kern et al. [49] have recently been characterized as mesothelial cells [84]. These cells contain an abundant pres-ence of cytokeratins 8 and 18, show only a diffuse and faint staining of vWF, and do not present the endothelial cell antigens that are recognized by the monoclonal antibodies Pal-E and EN-4.

From subcutaneous fat a small number of capillaries have been isolated using the same procedure. These endothelial cells in culture contain vWF in discrete granular structures [84].

Brain Microvascular Endothelial Cells

Brain microvascular endothelial cells can be isolated from human cortex [28]. To that end, human brain tissue obtained from autopsy 6 h after death is rinsed and homogenized. The homogenate is sieved through nylon or polypropylene filters (350 and 110 μm mesh). The filtrate is washed and the microvessels are collected on a 20 μm mesh nylon filter. The cells can be grown under standard conditions. Primary cultures contain vWF.

Microvascular Endothelial Cells from Adult Dermis

Davison et al. [22] describe the isolation and culture for microvascular endothelial cells obtained from human adult dermis. The cells are cultured up to six passages on fibronectin-coated dishes in a complex medium that contains a high concentration of human serum (50%), additional growth factors, and agents that increase the cellular cAMP level. The cells contain vWF and Weibel-Palade bodies.

Renal Microvascular Endothelial Cells

Renal microvascular endothelial cells have been isolated and cultured from unused donor kidneys [29, 91]. Following dissection of cortex from medulla, cortical tissue is minced, and incubated in Dulbecco's modified essential medium supplemented with 0.7 mg/ml collagenase and 0.25 mg/ml DNase for 2 h at 25 °C. Microvascular segments are recovered at the interface of a step Percoll gradient (30% – 50% in M199). Collected microvascular segments are seeded on Primaria dishes (Falcon) and cultured under standard culture conditions. The cells contain vWF and ACE.

Sinusoidal Endothelial Cells from Human Liver

Sinusoidal endothelial cells from human liver have been isolated by collagenase perfusion and centrifugal elutriation [12, 50–52, 77]. These cells can be kept in maintenance culture for short periods of time. The cells have been characterized on the basis of their morphology and ultrastructure. They can also be identified by their uptake of FITC-labeled formaldehyde-treated human albumin [6].

Endothelial Cells from Human Arteries and Veins

Endothelial cells from human pulmonary artery and vein [47], varicose vein [72], and saphenous vein [88] have been isolated in a way similar to that described in "Isolation and Culture of Human Endothelial Cells". Instead of isolation of endothelial cells by incubation of blood vessels with collagenase, endothelial cells from arteries and veins are also isolated by gently removing the cells with a cotton swab or by carefully scraping the cells from the vessel.

Culture Conditions

Advantages and Disadvantages of Subculturing

Inoculates of human endothelial cells are often contaminated by blood cells. It is therefore very difficult to exclude the presence of a few monocyte-derived macrophages in the primary culture. The macrophages have the potency to produce the cytokines interleukin-1 (IL-1) and tumor necrosis factor (TNF), which activate endothelial cells and dramatically change their properties (see "Modulation of Endothelial Cell Phenotype"). The macrophages can not be subcultured and hence are lost after passage of the endothelial cells. Therefore, for many studies, cells that have been passaged two or three times are preferred. However, this is not a general rule. The production of prostacyclin by cultured endothelial cells decreases rapidly during subculturing of animal [32] and human endothelial cells. Prostacyclin in its turn elevates the cAMP level [1, 39]. Processes, which result directly or indirectly in the release of arachidonic acid from phospholipids in the cell, have therefore to be evaluated in primary or early passage cells. Special care has to be taken to avoid activation by monocyte-derived macrophages. For studies on gene structure and gene regulation, the cells can be propagated in bulk quantities, provided that ECGF and serum are continuously present. Umbilical vein and adult vein (vena cava, saphenous vein) and foreskin microvascular endothelial cells can be kept in a small diameter morphology for more than ten passages, but endothelial cells from adult artery often become enlarged after four to seven passages. The metabolism of these enlarged cells is probably altered. They start to spontaneously release urokinase-type plasminogen activator [81].

Endothelial Cell Growth Factors

In newborn calf serum or fetal bovine serum the growth of human endothelial cells depends on the presence of ECGF, which is identical to a-FGF and a member of the heparin-binding growth factor family [13]. It is also called heparin-binding growth factor-1 (HBGF-1). It can be replaced by b-FGF (HBGF-2), another member of the heparin-binding growth factor family [34] (see Chap. 16 of this book). These growth factors act on the endothelial cell via binding to a specific receptor. Recently a separate type of growth factor, platelet-derived endothelial cell growth factor (PD-ECGF) has been isolated and cloned [43, 62]. This acts as a real mitogen on human endothelial cells. Clemmons et al. [17] have also identified a growth factor for endothelial cells from human platelets. These platelet-derived growth factor(s) may underlie the previous observation that subcultured human endothelial cells proliferate for several passages in human serum without ECGF, whereas they do not grow in fetal bovine serum in the absence of ECGF [58, 80].

Serum-Containing and Semisynthetic Media

The selection of a good type and batch of serum is very critical in culturing human endothelial cells. Many authors report the culture of human endothelial cells in fetal bovine serum, but our own experience with fetal bovine serum (heat-inactivated and in the presence of ECGF) is unsatisfactory. We have obtained the best results with pooled human serum. It can be replaced by human plasma anticoagulated with heparin. If no human plasma components should be present in the experiments, replacement of human serum by newborn calf serum (heat-inactivated and supplemented with ECGF) gives better results than fetal bovine serum. This has also been found by Thornton et al. [76].

The presence of serum and the crude preparation of ECGF in the culture introduces an undesired experimental variability. Today, the crude preparation of ECGF can be replaced by purified or recombinant ECGF (a-FGF, 10 ng/ml) or by b-FGF (5–10 ng/ml), combined with heparin. Several attempts have been made in the past to formulate a semisynthetic culture medium for human endothelial cells [9, 24, 57, 78]. These media contain ECGF, epidermal growth factor (EGF), transferrin, insulin, and albumin (pyrogen-free human serum albumin). Addition of human lipoproteins (HDL or low concentrations of nonoxidized LDL) to such a medium improves the growth of endothelial cells [15, 75, 78]. The cells can be propagated for a short period in these media. Hoshi and McKeehan [40] described a serum-free medium for the growth of human umbilical vein endothelial cells. This medium consists of MCDB 107 [35] as nutrient medium supplemented with ECGF, EGF, high-density lipoproteins, and the conditioned medium of human hepatoma HepG2 cells. These authors have identified in a later report [61] two proteinase inhibitors in the hepatoma HepG2 conditioned medium with growth-stimulating properties. Hoshi and McKeehan [41] reported the serial propagation of human endothelial cells from a variety of adult human vessels in a low-serum culture medium. The nutrient medium MCDB 107 was supplemented with 5 µg/ml partially purified brain-derived growth factor (ECGF), 10 ng/ml EGF, and 2% fetal bovine serum. With this culture medium these authors [42] have evaluated the growth requirements of endothelial cells from adult human arteries and veins. The growth of these cells depends on the presence of heparin-binding growth factor-1 (ECGF) and EGF. Lipoproteins, fibronectin, and two liver-derived proteinase inhibitors (see above) probably play largely "supportive" roles in the proliferation of these endothelial cells.

Knedler and Ham [53] have described a defined medium, in which omental fat-derived epitheloid cells, can be propagated. These cells appear to be mesothelial cells (see "Other Types of Human Endothelial Cells"). Dermal microvascular endothelial cells have been propagated in this medium, but they lose their epitheloid morphology during subculturing [7].

Identification and Characterization of Human Endothelial Cells

Identification of Various Types of Human Endothelial Cells

Human endothelial cells in confluent culture display a "cobble-stoned" pattern, which is characteristic for epithelial cells, mesothelial cells, and endothelial cells, but which can occasionally also be found with other types of cells, e.g., rat smooth muscle cells. On the other hand, growing endothelial cells may display a more irregular pattern. Therefore, additional criteria are necessary to establish the endothelial nature of presumed cells. The demonstration of endothelial-specific antigens by monoclonal antibodies or by biochemical assays can provide certainty about the nature of the cells. In particular cases, e.g.. discrimination between mesothelial cells and endothelial cells, both positive and negative criteria have to be used to identify the nature of the cells (see below).

vWF has appeared to be a good marker for human endothelial cells (Fig. 4a). vWF has been demonstrated in a granular pattern in cultured endothelial cells from neonatal and adult arteries and veins, in skin and foreskin microvascular endothelial cells, and in capillary endothelial cells from subcutaneous fat. vWF has been demonstrated perinuclearly in brain capillary endothelial cells. Besides endothelial cells, platelets, megakaryocytes, and mast cells also contain vWF. Cultured mesothelial cells may produce small amounts of vWF, but only a faint and diffuse presence and no granular localization of vWF can be demonstrated. The granular pattern of vWF is related to its presence in storage organelles, the so-called Weibel-Palade bodies, which can be demonstrated by electron microscopy. Recently, it has been demonstrated by immunofluorescence microscopy that the glycoprotein GP 140 is colocalized with vWF with these structures [11, 37].

Rhodamine- or FITC-labeled *Ulex europaeus* lectin-1 (Fig. 4b) has been proposed as a marker for endothelial cells [38]. It recognizes fucose-containing glycoproteins. Besides endothelial cells, epithelial and mesothelial cells are also recognized by *Ulex europaeus* lectin-1. In endothelial cell populations of arteries and veins, it is a convenient marker in addition to vWF. In many of these cultures a few percent of the endothelial cells do not contain the expected large amounts of vWF.

Many monoclonal antibodies have been raised against various endothelial cell antigens. Several of them recognize specifically endothelial cells. They include anti-ACE [4], EN-4 (is similar to EN-3) [20], PAL-E [74], BMA-120 (is BW-200) [2], C-IV [48], HEC-1 [66] and 14E5, 10B9 and 12C6 [36]. Their specificity and use as vascular markers in various tissues has been reviewed recently by Ruiter et al. [71]. Anti-ACE recognizes ACE and reacts with endothelial cells from arteries, veins, and capillaries. The monoclonal antibody PAL-E reacts with cultured endothelial cells from capillaries (foreskin microvessels) and from umbilical vein (Fig. 4d). It does not react with cultured endothelial cells from umbilical arteries and from adult arteries. The monoclonal andibody EN-4 (Fig. 4c), which recognizes endothelial cells of capillaries, but not those of arteries and veins in tissue slices, recognizes also umbilical vein endothelial cells. The antibody BMA-120 reacts with endothe-

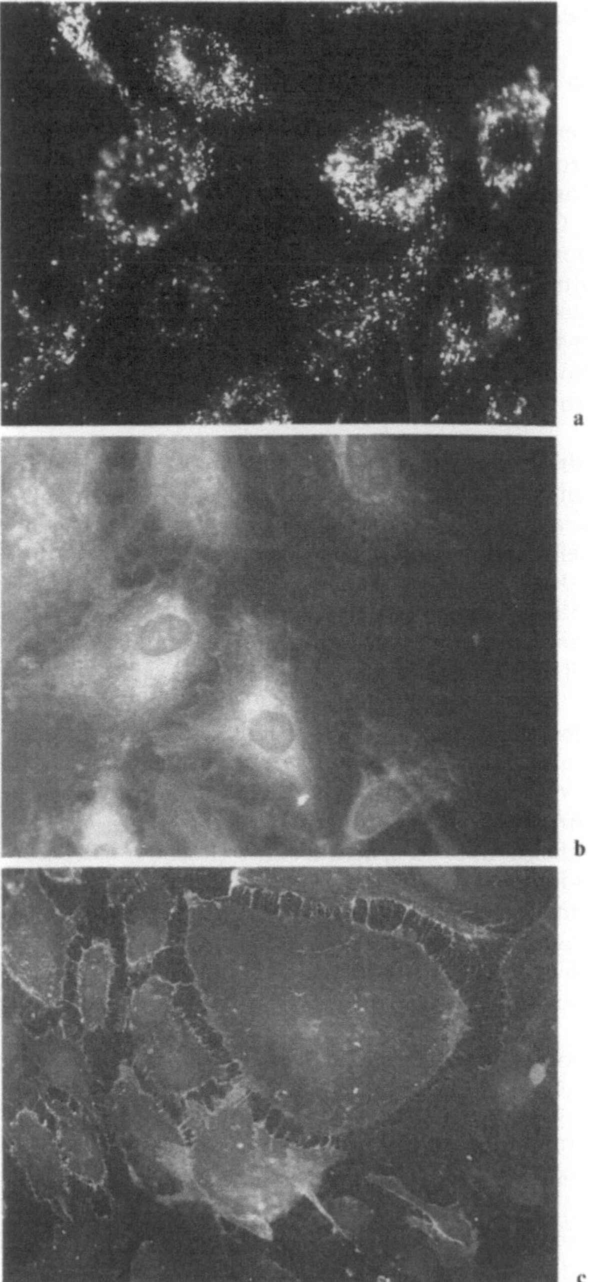

Fig. 4a–g. Characterization of human endothelial cells by fluorescence microscopy. **a** Presence of von Willebrand factor. **b** Interaction of *Ulex europaeus* lectin-1. **c** Binding of monoclonal antibody EN-4. **d** Binding of monoclonal antibody PAL-E. **e** Uptake of DiI-acetylated LDL. **f, inset** Absence of cytokeratin-18. The endothelial cells were cultured from umbilical artery (**a, e**) and vein (**c, f, inset**) and from foreskin microvessels (**b, d**). **f** Presence of cytokeratin-18 in omental tissue-derived mesothelial cells

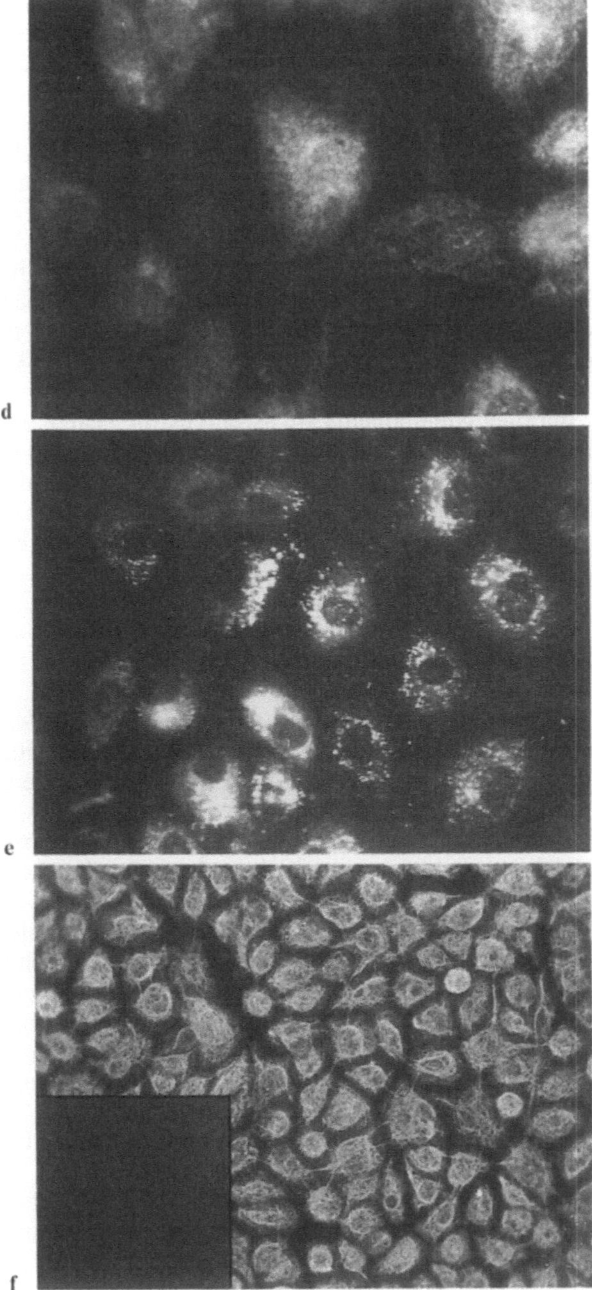

lial cells from arteries, veins, and capillaries in tissue slices. Besides endothelial cells, it also recognizes omental tissue-derived mesothelial cells in culture.

It is anticipated that many other antibodies for endothelial cell proteins will become available in the near future, such as antibodies against thrombomodulin, glycoprotein IIb/IIIa [25], and glycoprotein IIa [85]. In addition, the recognition of endothelial cell determinants, that are specific for endothelial cells in various tissues – as recently developed for murine endothelial cells by Auerbach et al. [5] – will become a helpful tool in the isolation and identification of various types of human endothelial cells.

Several biochemical markers have been used to further identify cultured cells as endothelial cells. They include the uptake of acetylated LDL (Fig. 4e), the production of prostacyclin and prostaglandin E2, the production of tissue-type plasminogen activator, the activation of protein C, and the nonthrombogenicity of the cell surface.

Intermediate filaments are used as cytological markers for various types of cells. Human umbilical vein and artery endothelial cells in culture contain vimentin, but no cytokeratins and desmin (Fig. 4f, inset) [26]. Cytokeratins 8 and 18 are abundantly present in mesothelial cells cultured from omental tissue, and hence are an appropriate discriminator between endothelial and mesothelial cells (Fig. 4f).

Modulation of Endothelial Cell Phenotype

In addition to markers identifying endothelial cells from various types of human blood vessels, markers reflecting the culture state and/or activation of endothelial cells are also desired to fully characterize the cells used in experiments.

Slight contamination of an endothelial cell population with monocyte-derived macrophages may result in the exposure of the cells to the cytokines IL-1 and TNF, which alter many properties of the endothelial cells. These alterations comprise many hemostatic and immunological properties of the endothelial cells and the interaction of leukocytes with them [see for review 18, 68] and changes in the production of matrix components by them [63]. Similar alterations are induced when the cells become exposed to the lymphokine lymphotoxin or to bacterial lipopolysaccharide (endotoxin). Strict endotoxin-free conditions are thus required during the isolation and culture of endothelial cells. As indicators for endothelial cell activation, the leukocyte adhesion molecule ELAM-1 and the coagulation factor tissue factor have been suggested.

Besides these inflammatory mediators, many other components alter the properties of human endothelial cells. Interferon gamma, for example, increases HMC class II antigens and down-regulates ECGF receptors on endothelial cells [27, 69]. The interaction of several vasoactive substances, including thrombin and histamine, with receptors on the cell results in the generation of inositol triphosphate and diacylglycerol by phospholipase C and in a raise in the cytoplasmatic calcium concentration of the cells. They induce a series of metabolic alterations (release of prostacyclin and vWF, stimulation of the production of tissue-type plasminogen activator and endothelium-derived relaxing factor, synthesis of PAF acether, and cellular contraction). Oxidizing conditions also result in an elevation of the cytoplasmatic calcium concentration and may easily damage human endothelial cells.

Latent viral or mycoplasma infections can alter the properties of endothelial cells. The expression of Fc- and C3b-receptors can be induced by such infections [16, 73].

Differences have been observed in the characteristics of subconfluent and confluent endothelial cells. Furthermore, subculturing of the cells under routine conditions results in a loss of the ability to produce prostacyclin and in a change of other endothelial cell properties. At present there is insufficient information available about these alterations to identify specific markers for the "culture state" of human endothelial cells. The spontaneous secretion of urokinase-type plasminogen activator by endothelial cells from human arteries and veins may be a potential marker for the altered behavior of these cells. Whereas endothelial cells from human arteries and veins always produce tissue-type plasminogen activator, they secrete urokinase-type plasminogen activator only when they start to become senescent [81] or when they have been activated by the inflammatory mediators endotoxin, IL-1, and TNF [83].

Changes in gene expression have also been reported when monolayers of human endothelial cells cultured were altered in tube-like structures [59]. Upon organization of the cells in a three-dimensional tubular structure, the level of the c-*sis* (PDGF-B) mRNA transcript decreased, whereas the mRNA transcript for fibronectin increased [46]. Specific mRNA species may become determinants of cell characterization in subcultured cells.

Comparison with Endothelial Cells In Vivo

The ultimate goal of using human endothelial cells in culture is to unravel complicated (patho)physiological processes occurring in the vessel wall in vivo. Although the isolation and culture of endothelial cells has the advantage of investigating endothelial cell-specific processes without interfering with other cells, while increasing the amount of cells for biochemical studies, it is obvious that the culture process per se may also introduce artifacts. Culturing endothelial cells may result in the loss of specific functions and may introduce new metabolic characteristics, which are not present in endothelial cells in vivo. Presently, the growth rate of endothelial cells in vitro largely exceeds that of their in vitro counterparts. Manipulation of the proliferation rate of endothelial cells by growth factors and growth inhibitors may result in culture procedures that can convert rapidly growing cells, necessary for propagation, into quiescent cell populations, desired for experiments. Notwithstanding their limitations, the present procedures of culturing endothelial cells have resulted in many interesting and (patho)physiologically important observations. For example, cultured endothelial cells have been used to identify the endothelium-derived relaxant factor nitric oxide [65], the potent vasoconstrictor endothelin [93], and leukocyte adhesion molecules. The leukocyte adhesion molecule ELAM-1 [10], which was identified in vitro, has recently also been demonstrated in activated postcapillary venules in vivo [19]. Interestingly, ELAM-1 can be induced by inflammatory mediators not only on microvascular endothelial cells in vitro, but also on cultured endothelial cells from neonatal and adult arteries and veins. In vivo, ELAM-1 has not been found in these large vessels. This suggests that the regulation of ELAM-1 in artery and vein endothelial cells has become modified. It is interesting to evaluate whether this modification is limited to specific functions of these cells or whether a general tendency exists, by which various types of endothelial cells (slowly) shift into a common phenotype when they are isolated and propagated in culture. The localization of the endothelial cell in the body makes it half a blood cell, half a tissue cell. Tissue-specific characteristics of endothelial cells are likely induced by paracrine stimulatory and inhibitory signals from their environment. It is conceivable that specific functions of endothelial cells are only conserved when the cells are exposed to these signals. Identification of these presumed factors will increase our insight into the development and regulation of region-specific functions of the endothelium. Establishing and evaluating cocultures of endothelial cells with pericytes [3], smooth muscle cells [64], astrocytes [23], hepatocytes, and other cell types are important steps in achieving this goal.

Acknowledgement: E. G. Langeler is a recipient of Grant 85.097 from the Netherlands Heart Foundation.

References

1. Adams-Brotherton AF, Hoak JC (1982) Role of Ca^{2+} and cyclic AMP in the regulation of the production of prostacyclin by the vascular endothelium. Proc Natl Acad Sci USA 79:495–499
2. Alles JU, Bosslet K (1986) Immunohistochemical and immunochemical characterization of a new endothelial cell-specific antigen. J Histochem Cytochem 34:209–214
3. Antonelli-Orlidge A, Saunders KB, Smith SR, d'Amore PA (1989) An activated form of transforming growth factor β is produced by cocultures of endothelial cells and pericytes. Proc Natl Acad Sci USA 86:4544–4548
4. Auerbach R, Alby L, Grieves J, Joseph J, Lindgren C, Morrisey LW, Sidkey YA, Tu M, Watt SL (1982) Monoclonal antibody against angiotensin-converting and human endothelial cells. Proc Natl Acad Sci USA 79:7891–7895
5. Auerbach R, Alby L, Morrissey LW, Tu M, Joseph J (1985) Expression of organ-specific antigens on capillary endothelial cells. Microvasc Res 29:401
6. Babaev VR, Kosykh VA, Tsibulsky VP, Ivanov VO, Repin VS, Smirnov VN (1989) Binding and uptake of native and modified low-density lipoproteins by human hepatocytes in primary culture. Hepatology 10:56–60
7. Baskin JB, Ham RG (1985) Improved culture conditions for human dermal microvascular endothelial cells. In Vitro 21:18A
8. Basu SK, Goldstein JL, Anderson RGW, Brown MS (1976) Degradation of cationized LDL and regulation of cholesterol metabolism in homozygous familial hypercholesterolemia fibroblasts. Proc Natl Acad Sci USA 73:3178–3182
9. Berliner JA (1981) Regulation of endothelial cell DNA synthesis and adherence. In Vitro 17:985–992
10. Bevilacqua MP, Stengelin S, Gimbrone MA, Seed B (1989) Endothelial leukocyte adhesion molecule I. An inducible receptor for neutrophils related to complement regulatory proteins and lectins. Science 243:1160–1165

11. Bonfanti R, Furie BC, Furie B, Wagner DD (1989) PADGEM (GMP 140) is a component of Weibel-Palade bodies of human endothelial cells. Blood 73:1109–1112

12. Brouwer A, Barelds RJ, de Leeuw AM, Blauw E, Plas A, Yap SH, van den Broek AMWC, Knook DL (1988) Isolation and culture of Kupffer cells from human liver. Ultrastructure, endocytosis and prostaglandin synthesis. J Hepatol 6:36–49

13. Burgess WH, Maciag T (1989) The heparin-binding (fibroblast) growth factor family of proteins. Annu Rev Biochem 58:575–606

14. Burgess WH, Mehlman T, Friesel R, Johnson WV, Maciag T (1985) Multiple forms of endothelial cell growth factor. Rapid isolation and biological and chemical characterization. J Biol Chem 260:11389–11392

15. Chen J-K, Hoshi H, McClure DB (1986) Role of lipoproteins in growth of human adult arterial endothelial and smooth muscle cells in low lipoprotein-deficient serum. J Cell Physiol 129:207–214

16. Cines DB, Lyss AP, Bina M, Corkey R, Kefalides NA, Friedman HM (1982) Fc and C3 receptors induced by herpes simplex virus on cultured human endothelial cells. J Clin Invest 69:123–128

17. Clemmons DR, Isley WL, Brown MT (1983) Dialyzable factor in human serum of platelet origin stimulates endothelial cell replication and growth. Proc Natl Acad Sci USA 80:1641–1645

18. Cotran RS, Pober JS (1989) Effects of cytokines on vascular endothelium: their role in vascular and immune injury. Kidney Int 35:969–975

19. Cotran RS, Gimbrone MA, Bevilacqua MP, Mendrick DL, Pober JS (1986) Induction and detection of a human endothelial activation antigen in vivo. J Exp Med 164:661–666

20. Cui YC, Tai P-C, Gatter KC, Mason DY, Spry CJF (1983) A vascular endothelial cell antigen with restricted distribution in human foetal, adult and malignant tissues. Immunology 49:183–189

21. Davison PM, Bensch K, Karasek MA (1980) Isolation and growth of endothelial cells from the microvessels of the newborn human foreskin in cell culture. J Invest Dermatol 75:316–321

22. Davison PM, Bensch K, Karasek MA (1983) Isolation and long-term serial cultivation of endothelial cells from the microvessels of the adult human dermis. In Vitro 19:937–945

23. DeBalut LE, Cancilla PA (1980) Gamma-glutamyl transpeptidase in isolated brain endothelial cells: induction by glial cells in vitro. Science 207:653–655

24. De Groot PG, Willems C, Gonsalves MD, van Aken WG, van Mourik JA (1983) The proliferation of human umbilical vein endothelial cells in serum-free medium. Thromb Res 31:623–634

25. Fitzgerald LA, Charo IF, Phillips DR (1985) Human endothelial cells synthesize membrane glycoproteins similar to platelet membrane GPIIb and GPIIIa. J Biol Chem 260:10893–10896

26. Franke WW, Schmid E, Osborn M, Weber K (1979) Intermediate-sized filaments of human endothelial cells. J Cell Biol 81:570–580

27. Friesel R, Komoriya A, Maciag T (1987) Inhibition of endothelial cell proliferation by gamma-interferon. J Cell Biol 104:689–696

28. Gerhart DZ, Broderius MA, Drewes LR (1988) Cultured human and canine endothelial cells from brain microvessels. Brain Res Bull 21:785–793

29. Gibbs VC, Wood DM, Garovoy MR (1985) The response of cultured human kidney capillary endothelium to immunologic stimuli. Hum Immunol 14:259–269

30. Gimbrone MA Jr, Cotran RS, Folkman J (1974) Human vascular endothelial cells in culture. Growth and DNA synthesis. J Cell Biol 60:673–684

31. Gimbrone MA Jr, Shefton EJ, Cruise SA (1979) Isolation and primary culture of endothelial cells from human umbilical vessels. Tissue Culture Assoc 4:813–817

32. Goldsmith JC, Jafvert CT, Lollar P, Owen WG, Hoak JC (1981) Prostacyclin release from cultured and ex vivo bovine vascular endothelium. Lab Invest 45:191–197

33. Gordon PB, Sussman II, Hatcher VB (1983) Long-term culture of human endothelial cells. In Vitro 19:661–671

34. Gospodarowicz D, Cheng J, Lirette M (1983) Bovine brain and pituitary fibroblast growth factors: comparison of their abilities to support the proliferation of human and bovine vascular endothelial cells. J Cell Biol 97:1677–1685
35. Ham RG, McKeehan WL (1979) Media and growth requirements. Methods Enzymol 59:44–93
36. Hamburger AW, Reid YA, Pelle BA, Breth LA, Beg N, Ryan U, Cines DB (1985) Isolation and characterization of monoclonal antibodies reactive with endothelial cells. Tissue Cell 17:451
37. Hattori R, Hamilton KK, Fugate RD, McEver RP, Sims PJ (1989) Stimulated secretion of endothelial von Willebrand factor is accompanied by rapid redistribution to the cell surface of the intracellular granule membrane protein GMP-140. J Biol Chem 264:7768–7771
38. Holthöfer H, Virtanen I, Kariniemi A-L, Hormia M, Linder E, Miettinen A (1982) *Ulex europaeus* I lectin as a marker for vascular endothelium in human tissues. Lab Invest 47:60–66
39. Hopkins NK, Gorman RR (1981) Regulation of endothelial cell cyclic nucleotide metabolism by prostacyclin. J Clin Invest 67:540–546
40. Hoshi H, McKeehan WL (1984) Brain- and liver cell-derived factors are required for growth of human endothelial cells in serum-free culture. Proc Natl Acad Sci USA 81:6413–6417
41. Hoshi H, McKeehan WL (1986) Isolation, growth requirements, cloning, prostacyclin production and life-span of human adult endothelial cells in low serum culture medium. In Vitro Cell Dev Biol 22:51–56
42. Hoshi H, Kan M, Chen J-K, McKeehan WL (1988) Comparative endocrinology-paracrinology-autocrinology of human adult large vessel endothelial and smooth muscle cells. In Vitro Cell Dev Biol 24:309–320
43. Ishikawa F, Miyazono K, Hellman U, Drexler H, Wernstedt C, Hagiwara K, Usuki K, Takaku F, Risau W, Heldin C-H (1989) Identification of angiogenic activity and the cloning and expression of platelet-derived endothelial cell growth factor. Nature 338:557–562
44. Jaffe EA (1980) Culture of human endothelial cells. Transplant Proc 12:49–53
45. Jaffe EA, Nachman RL, Becker CG, Minick CR (1973) Culture of human endothelial cells derived from umbilical veins. Identification by morphologic and immunologic criteria. J Clin Invest 52:2745–2756
46. Jaye M, McConathy E, Drohan W, Tong B, Deuel T, Maciag T (1985) Modulation of the *sis* gene transcript during endothelial differentiation in vitro. Science 228:882–885
47. Johnson AR (1980) Human pulmonary endothelial cells in culture. Activities of cells from arteries and cells from veins. J Clin Invest 65:841–850
48. Kaplan KL, Weber D, Cook P, Dalecki M, Rogozinski L, Sepe O, Knowles D, Butler VP (1983) Monoclonal antibodies to E 92, an endothelial cell surface antigen. Arteriosclerosis 3:403–412
49. Kern PA, Knedler A, Eckel RH (1983) Isolation and culture of microvascular endothelium from human adipose tissue. J Clin Invest 71:1822–1829
50. Kirn A, Steffan AM, Bingen A (1980) Isolement et culture de cellules de Kupffer humaines. CR Acad Sci (Paris) 291:249–251
51. Kirn A, Bingen A, Steffan AM, Wild MT, Keller F, Cinqualbre J (1982) Endocytic capacities of Kupffer cells isolated from the human adult liver. Hepatology 2:216–222
52. Kirn A, Gendrault JL, Gut JP, Steffan AM, Bingen A (1982) Isolement et culture de cellules sinusoïdales de foies humain et murin: une nouvelle approche pour l'étude des infections virales du foie. Gastroenterol Clin Biol 6:283–293
53. Knedler A, Ham RG (1987) Optimized medium for clonal growth of human microvascular endothelial cells with minimal serum. In Vitro 23:481–491
54. Langeler EG, van Hinsbergh VWM (1988) Characterization of an in vitro model to study the permeability of human arterial endothelial cell monolayers. Thromb Haemost 60:240–246

55. Langeler EG, Snelting-Havinga I, van Hinsbergh VWM (1989) Passage of low density lipoproteins through monolayers of human arterial endothelial cells. Effects of vasoactive substances in an in vitro model. Arteriosclerosis 9:550–559
56. Maciag T, Cerundolo J, Ilsley S, Kelley PR, Forand R (1979) An endothelial cell growth factor from bovine hypothalamus: identification and partial characterization. Proc Natl Acad Sci USA 76:5674–5678
57. Maciag T, Weinstein R, Stemerman MB, Gilchrest BA (1980) Selective growth of human endothelial cells from both papillary and reticular dermis. J Invest Dermatol 74:256
58. Maciag T, Hoover GA, Stemerman MB, Weinstein R (1981) Serial propagation of human endothelial cells in vitro. J Cell Biol 91:420–426
59. Maciag T, Kadish J, Wilkins L, Stemerman MB, Weinstein R (1982) Organizational behavior of human umbilical vein endothelial cells. J Cell Biol 94:511–520
60. McGee-Russel SM, de Bruijn WC (1971) Image and artifact. Comments and experiments on the meaning of the image in the electron microscope. In: McGee-Russel SM, de Bruijn WC (eds) Cell structure and its interpretation. Arnold, London, pp 115–133
61. McKeehan WL, Sakagami Y, Hoshi H, McKeehan KA (1986) Two apparent human endothelial cell growth factors from human hepatoma cells are tumor-associated proteinase inhibitors. J Biol Chem 261:5378–5383
62. Miyazono K, Okabe T, Urabe A, Takaku F, Heldin C-H (1987) Purification and properties of an endothelial cell growth factor from human platelets. J Biol Chem 262:4098–4103
63. Montesano R, Mossaz A, Ryser J-E, Orci L, Vassalli P (1984) Leukocyte interleukins induce cultured endothelial cells to produce a highly organized, glycosaminoglycan-rich pericellular matrix. J Cell Biol 99:1706–1715
64. Navab M, Hough GP, Stevenson LW, Drinkwater DC, Laks H, Fogelman AM (1988) Monocyte migration into the subendothelial space of a coculture of adult human aortic endothelial and smooth muscle cells. J Clin Invest 82:1853–1863
65. Palmer RMJ, Ferrige AG, Moncada S (1987) Nitric oxide release accounts for the biological activity of endothelium-derived relaxing factor. Nature 327:524–526
66. Parks WM, Gingrich RD, Dahle CE, Hoak JC (1985) Identification and characterization of an endothelial, cell-specific antigen with a monoclonal antibody. Blood 66:816–823
67. Pitas RE, Innerarity TL, Weinstein JN, Mahley RW (1981) Acetoacetylated lipoproteins used to distinguish fibroblasts from macrophages in vitro by fluorescence microscopy. Arteriosclerosis 1:177–185
68. Pober JS (1988) Cytokine-mediated activation of vascular endothelium. Am J Pathol 133:426–433
69. Pober JS, Gimbrone MA (1982) Expression of Ia-like antigens by human vascular endothelial cells is inducible in vitro: demonstration by monoclonal antibody binding and immunoprecipitation. Proc Natl Acad Sci USA 79:6641–6645
70. Redgrave TG, Roberts DCK, West CE (1974) Separation of plasma lipoproteins by density-gradient ultracentrifugation. Anal Biochem 65:42–49
71. Ruiter DJ, Schlingemann RO, Rietveld FJR, de Waal RMW (1989) Monoclonal antibody defined human endothelial antigens as vascular markers. J Invest Dermatol 93:25S–32S
72. Ryan US, White LA (1985) Varicose veins as a source of adult human endothelial cells. Tissue Cell 17:171–176
73. Ryan US, Schultz DR, Ryan JW (1981) Fc and C3b receptors on pulmonary endothelial cells: induction by injury. Science 214:557–558
74. Schlingemann RO, Dingjan GM, Emeis JJ, Blok J, Warnaar SO, Ruiter DJ (1985) Monoclonal antibody PAL-E specific for endothelium. Lab Invest 52:71–76
75. Tauber J-P, Cheng J, Massoglia S, Gospodarowicz D (1981) High density lipoproteins and the growth of vascular endothelial cells in serum-free medium. In Vitro 17:519–530
76. Thornton SC, Mueller SN, Levine EM (1983) Human endothelial cells: use of heparin in cloning and long-term serial cultivation. Science 222:623–625

77. Van Bossuyt H, Reekmans M, van der Spek P, Wisse E (1986) Culture of sinusoidal liver cells from rat and human liver needle biopsies. In: Kirn A, Knook DL, Wisse E (eds) Cells of the hepatic sinusoid, vol 1. Kupffer Cell Foundation, Rijswijk, pp 475–476

78. Van Hinsbergh VWM, Emeis JJ, Havekes L (1983) Interaction of lipoproteins with cultured endothelial cells. In: Thilo-Körner DGS and Freshey RJ (Eds) The endothelial cell – a pluripotent control cell of the vessel wall. Karger, Basel, pp 99–112

79. Van Hinsbergh VWM, Havekes L, Emeis JJ, van Corven E, Scheffer M (1983) Low density lipoprotein metabolism by endothelial cells from human umbilical cord arteries and veins. Arteriosclerosis 3:547–559

80. Van Hinsbergh VWM, Mommaas-Kienhuis AM, Weinstein R, Maciag T (1986) Propagation and morphologic phenotypes of human umbilical cord artery endothelial cells. Eur J Cell Biol 42:101–110

81. Van Hinsbergh VWM, Binnema D, Scheffer MA, Sprengers ED, Kooistra T, Rijken DC (1987) Production of plasminogen activators and inhibitor by serially propagated endothelial cells from adult human blood vessels. Arteriosclerosis 7:389–400

82. Van Hinsbergh VWM, Sprengers ED, Kooistra T (1987) Effect of thrombin on the production of plasminogen activators and PA inhibitor-1 by human foreskin microvascular endothelial cells. Thromb Haemost 57:148–153

83. Van Hinsbergh VWM, van den Berg EA, Turion PNC, Fiers W, Dooijewaard G (1988) Induction of u-PA antigen and mRNA in human endothelial cells by inflammatory mediators (Abstr 48). Fibrinolysis [Suppl 2] 2:23

84. Van Hinsbergh VWM, Kooistra T, Scheffer MA, van Bockel JH, van Muijen GNP (1990) Characterization and fibrinolytic properties of human omental tissue mesothelial cells. Comparison with endothelial cells. Blood 75 (in press)

85. Van Mourik JA, Leeksma OC, Reinders JH, de Groot PG, Zandbergen-Spaargaren J (1985) Vascular endothelial cells synthesize a plasma membrane protein indistinguishable from the platelet membrane glycoprotein IIa. J Biol Chem 260:11 300–11 306

86. Voyta JC, Via DP, Butterfield CE, Zetter BR (1984) Identification and isolation of endothelial cells based on their increased uptake of acetylated-low density lipoprotein. J Cell Biol 99:2034–2040

87. Vuento M, Vaheri A (1979) Purification of fibronectin from human plasma by affinity chromatography under non-denaturing conditions. Biochem J 183:331–337

88. Watkins MT, Sharefkin JB, Zajtchuk R, Maciag TM, d'Amore PA, Ryan US, van Wart H, Rich NM (1984) Adult human saphenous vein endothelial cells: assessment of their reproductive capacity for use in endothelial seeding of vascular prostheses. J Surg Res 36:588–596

89. Weibel ER, Palade GE (1964) New cytoplasmic components in arterial endothelia. J Cell Biol 23:101–112

90. Wilkins L, Gilchrest BA, Szabo G, Weinstein R, Maciag T (1985) The stimulation of normal human melanocyte proliferation in vitro by melanocyte growth factor from bovine brain. J Cell Physiol 122:350–361

91. Wojta J, Hoover RL, Daniel TO (1989) Vascular origin determines plasminogen activator expression in human endothelial cells. Renal endothelial cells produce large amounts of single chain urokinase type plasminogen activator. J Biol Chem 264:2846–2852

92. Wolinsky H (1980) A proposal linking clearance of circulating lipoproteins to tissue metabolic activity as a basis for understanding atherogenesis. Circ Res 47:301–311

93. Yanagisawa M, Kurihara H, Kimura S, Tomobe Y, Kobayashi M, Mitsui Y, Yazaki Y, Goto K, Masaki T (1988) A novel potent vasoconstrictor peptide produced by vascular endothelial cells. Nature 332:411–415

Vascular Endothelial Cell-Synthesized Extracellular Matrices as Attachment Substrates In Vitro

I. M. Herman

Preparation of Endothelial-Synthesized Matrices for Attachment Assays

It has become widely held that the extracellular matrix exerts significant effects on the behavior of vascular and nonvascular cells during embryogenesis, wound healing, and in association with tumor growth [1, 2, 7–11, 17, 22]. In particular, soluble as well as insoluble matrix components alter the shape of living cells, their migratory and reproductive capacitance, as well as their adhesive properties [3–6, 12, 16, 19, 21]. To address the role the molecules of the extracellular compartment play in these events, model systems have been developed. A considerable effort has been focused on revealing the domain structure and function of the matrix molecules and their respective membrane receptors, which are held responsible for signal transduction [5, 13–15, 18, 20]. It is equally important to understand matrix effects on cell behavior to reveal the matrix' linkage(s) with the mechanochemical effector molecules [11]. For the most part, these aspects of the extracellular-cytoplasmic signal transduction pathway remain largely underexplored and unresolved.

In considering the development of models to accurately assess the role the matrix plays in directing vascular cell behavior, one must take into account both its synthesis as well as organization. In developing an in vitro analogue of the basal lamina, which reflects the diversity of blood vessels known to constitute the developing vasculature or those associated with occlusive vascular disease, one is left with an almost inordinate task. With this in mind, however, we have been developing methods that yield defined extracellular matrices of known composition. Most importantly, these matrices have been synthesized and organized by living endothelial cells, one of the major cellular contributors of the functional basal lamina in vivo. While this chapter is not meant to be an inclusive accounting of the effects of the matrix on vascular cell behavior, it should serve as a methodological guide for those workers who are interested in testing the effects of a biosynthesized matrix on vascular endothelial attachment and spreading in vitro.

Vascular Endothelial Cell Culture for Extracellular Matrix Production

Whereas extracellular matrices derived from large and small blood vessel endothelial cells can be prepared and analyzed for their respective effects on endothelial attachment, for the purpose of this chapter, only large blood vessel (aortic) endothelial cells will be considered. In the author's laboratory, bovine aortae are either irrigated with dilute solutions of collagenase or the intima scraped with a sterile scalpel blade to yield endothelial cell primary cultures, which are then subcloned and characterized [12, 23].

For matrix production, aortic endothelial cells are plated at a confluent or "saturating" density. Practically speaking, cells are plated into 6-well, 35-mm diameter dishes at 10^6 cells/well. Routinely, the cells are plated at 110% of this number and then the cultures incubated for 60 min at 37 °C prior to washing and refeeding with Dulbecco's modified Eagle's medium (DMEM) supplemented with 5% calf serum, 2 mM glutamine, 100 U/ml streptomycin, and 100 U/ml penicillin; 50 µg/ml Mefoxin (cephoxatin sulfate) can also be used as an antibiotic. Cells are fed every third day by total media replacement.

Preparation of Endothelial Cell-Synthesized Extracellular Matrices Using Sodium Deoxycholate

1. Plate (as described above) aortic endothelial cells at a confluent density and allow the cells to synthesize and organize the subendothelial matrix for 5–7 days. Feed the cells with DMEM containing 5% calf serum every third day.
2. On the 7th day, postplating, wash the monolayers with phosphate buffered saline (PBS, 0.015 M sodium phosphate, pH 7.5, 0.15 M NaCl) three times.
3. Lyse the confluent monolayer in matrix preparing buffer (MPB, 0.5% sodium deoxycholate, (Sigma, Lot 57C-0456), 0.015 M NaCl, 0.001 M EGTA, 0.001 M phenylmethyl sulfonylfluoride) for 15 min at room temperature. Aspirate the MPB and repeat this step.
4. Wash the DOC matrix five times with PBS. To insure that all the detergent is removed from the matrix prior to the subsequent cell platings, use twice the volume for each of the washes compared with the initial cell lysis steps. For example, 1 ml MPB is used to lyse a confluent endothelial monolayer cultured in 35-mm dishes and 2 ml PBS is used for each 5 min wash.
5. Following the final PBS wash of the DOC matrix, the population of cells to be tested can then be plated. Alternatively, the DOC matrices can be stored at 4 °C in a sterile, moist environment for several weeks. However, in the author's laboratory matrices are prepared and used on the day of preparation.

Preparation of Endothelial Cell-Synthesized Extracellular Matrices Using Ethylene Glycol-bis(B-Aminoethyl Ether)-N,N,N′,N′-Tetraacetic Acid (EGTA)

1. Prepare 7-day postconfluent monolayers of aortic endothelial cells as described above.
2. Wash the confluent endothelial cell monolayer three times with PBS, making certain to keep the monolayer moistened with the buffer.

3. At room temperature, incubate the cells with a sterile solution of 0.02 M EGTA in PBS without Ca^{2+} or Mg^{2+} for 15 min on a swirling platform (300 rpm). During this first EGTA buffer incubation, the monolayer will begin to dislodge from the substrate as an intact sheet of cells. Routinely, cells are grown for attachment assays to be performed in Costar 3524-24 well plates (16-mm diameter). Aspiration of the monolayer in the EGTA-PBS should be done with extreme care so that the underlying matrix is not damaged and the residual cells are not left adherent to the well bottom. To make certain that all cells are removed by the procedure, this step is repeated.
4. Following the second EGTA-PBS wash, the monolayer is washed five times with PBS, again taking precautions to keep the resultant matrix moist. Following the five, 5-min washes this matrix is ready for subsequent cell plating. As with the case with the DOC matrix, the 24-well plates with the bound EGTA matrix can be stored at 4°C for several weeks. However, plates with bound matrix are generally used immediately.

Preparation of Attachment Substrates Using Purified Extracellular Matrix Molecules

Over the past 15 years, there has been a wealth of information obtained regarding the effects that purified matrix molecules exert on living cells during cell attachment, in association with cell migration and accompanying proliferation (see "References" and Fig. 1). Specific domains of extracellular macromolecules have been identified, which are important recognition motifs for other matrix molecules or for their high affinity receptors embedded within the

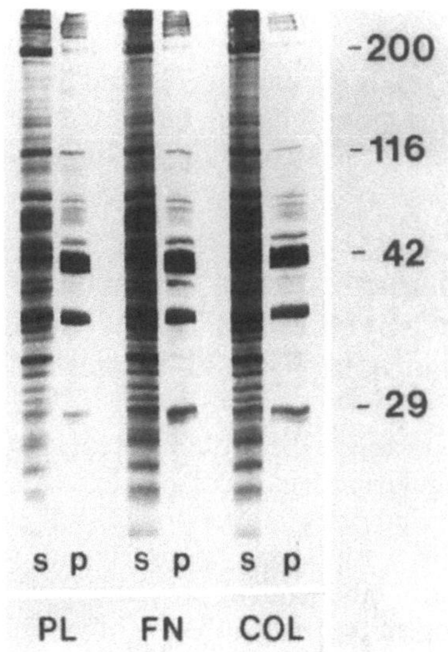

Fig. 1. Analysis of soluble and cytoskeletal-associated proteins present within endothelial cells plated and attached to various matrices. Equal cell numbers were loaded on this 14% SDS polyacrylamide gel. Proteins were stained with Coomassie Brilliant Blue R250. Molecular weights (in kilodaltons) are indicated to the right of the *COL* lanes. *PL*, plastic; *FN*, fibronectin; *COL*, types I and III collagens; *s*, supernatant; *p*, pellet

plasma membranes of living cells. As such, many workers have resorted to adsorption of intact matrix molecules or the active domains therefrom onto the attachment substrate. Many of these molecules have been purified by classical biochemical means and the active peptides have been synthesized. Unfortunately, there is little or no information available regarding: (a) whether the native conformation of these matrix molecules is retained upon spreading onto a planar substrate or (b) what the local concentration of a given matrix molecule actually is in vivo. However, functional assays (cell attachment or cell spreading) are performed with the appropriate controls (adsorption of control molecules or adsorption of serum components alone do not yield comparable results). This, however, still does not address the issue of nativity. With conformationally dependent antibodies directed against specific matrix molecules, one might approach this question. For those workers who wish to perform attachment assays using purified extracellular matrix molecules as substrates, the matrix molecules of choice (or its fragment) can be dissolved in sterile saline prior to dispersion on the attachment surface. For example, purified fibrils of type I and III collagen from rat tail tendon can be dissolved in dilute acetic acid (0.5 N) prior to pipetting onto the surface of the well. In the case of the fibrillar collagens, a "three-dimensional" matrix can be formed by exposure of the acid-soluble protein to vapors of ammonium hydroxide. Washing with sterile PBS prior to cell plating can then be performed in a fashion analogous to that described for the endothelial-synthesized matrices. In addition, purified glycoproteins of the matrix (e.g., fibronectin, laminin, thrombospondin) can be dissolved in sterile PBS and delivered to the well bottom. Adhesion of the matrix molecules can proceed at room temperature or at 37°C for a period up to several hours, although control experiments indicate that binding is maximal within minutes (performing assays to determine the concentration of proteins bound to the surface of the well is appropriate). There are a number of colorimetric assays, which depend on antibody recognition (ELISA, Western blotting), assays that employ radiolabeled ligands (e.g., I^{125}-labeled proteins), or assays that are colorimetric, which are not antibody dependent (Bio-rad protein assay). All of these assays are acceptable, some more sensitive than others. Nevertheless, one should determine the rate and extent of binding for the particular molecule(s) of interest.

Characterization of Endothelial Cells and the Endothelial Cell-Synthesized Extracellular Matrix Using Gel Electrophoresis, Fluorography, and Western Blot Analysis

To examine the proteins present within the endothelial cytoplasm or the extracellular matrix, SDS gel electrophoresis and fluorography of biosynthetically labeled populations or SDS gel electrophoresis followed by Western blotting with cytoplasmic or matrix antibodies can be performed (these are a few of many biochemical procedures that can be used for matrix characterization). For biosynthetic labeling of vascular endothelial cells, it is advisable to include radiolabeled amino acids that will be incorporated into collagenous (proline-

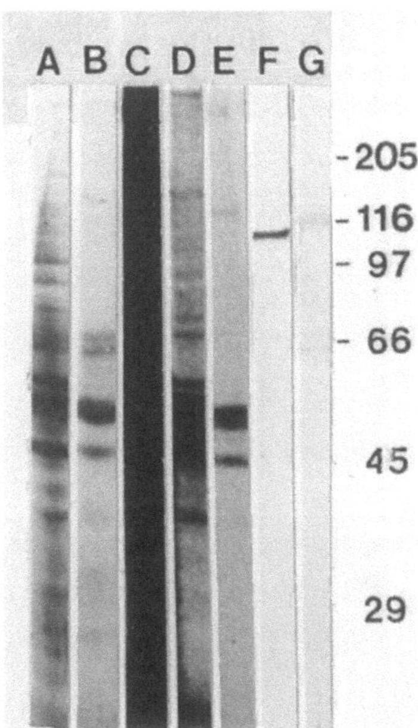

Fig. 2. Biochemical characterization of endothelial-derived extracellular matrices by SDS gel electrophoresis, fluorography, and Western blotting. *Lane A*, EGTA matrix: total protein, silver stain. *Lane B*, DOC matrix: total protein, silver stain. *Lane C*, EGTA matrix: ^{35}S-methionine labeling. *Lane D*, DOC matrix: ^{35}S-methionine labeling. *Lane E*, DOC matrix: ^{14}C-proline labeling. *Lane F*, DOC matrix: anti-type IV collagen Western blot. *Lane G*, EGTA matrix: anti-heparan sulfate proteoglycan Western blot. Molecular weights (in kilodaltons) are indicated to the right of *Lane G*. (Electrophoresis and Western Blotting by Dr. Jeffrey C. Yost; matrix antibodies prepared and characterized by Dr. Steven Ledbetter.)

rich) as well as noncollagenous components. As such dual labeling using [^{14}C]proline and [^{35}S]methionine is preferred. For proline and methionine labeling, 5 µCi/ml media (>250 mCi/mmol) and 20–100 µCi/ml media (>800 Ci/mmol), respectively, are routinely used. For labeling matrix proteins, cells can be cultured in media containing 1% the normal complement of L-methionine for up to 2 days. The cells will not proliferate in this media, but will sustain their initial population plating density. Radiolabeled matrices can then be prepared as described above and scraped into SDS gel sample buffer for polyacrylamide gel electrophoresis (PAGE) and fluorography or TCA precipitation and scintillation counting. Alternatively, proteins present within the cellular or extracellular fractions can be dissolved in SDS gel sample buffer, electrophoresed, and visualized. Additionally, proteins of interest can be transferred to nitrocellulose paper for Western blot analysis (Fig. 2).

Summary

This chapter should serve as a guide for the preparation of extracellular matrices which have been previously synthesized by living vascular endothelial cells. It has been established that these matrices not only influence the attachment, migration (cytoskeletal organization, Fig. 3), and proliferation of endothelial cells, but also the behavior of vascular smooth muscle cells. In an effort to develop a meaningful in vitro analogue of the intact basal lamina, the endothelial-synthesized extracellular matrix represents a reasonable first approximation.

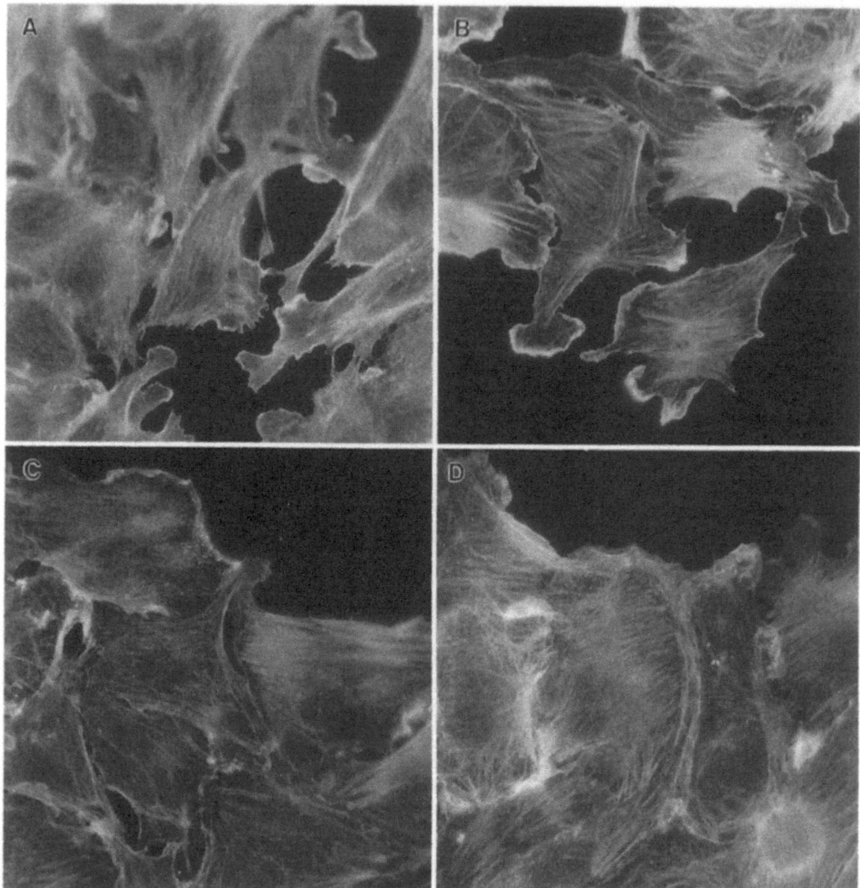

Fig. 3 A–D. Immunofluorescent localization of actin in endothelial cells recovering from injury on defined extracellular matrices. **A** Plastic; **B** fibronectin; **C** type IV collagen; **D** DOC matrix. Note that the organization of the actin cytoskeleton (and the resultant motility following injury) is significantly altered when cells recover on extracellular matrices, i.e., aortic endothelial cells migrate several times more slowly and rearrange the actin cytoskeleton into abundant stress fibers when migrating on matrix. **A–D**, × 780

Acknowledgements. This work was supported in part by NIH HL 33901 and 35570. The author held an American Heart Association Established Investigatorship Award during the course of these studies.

References

1. Baird A, Durkin T (1986) Inhibition of endothelial cell proliferation by type B transforming growth factor: interactions with acidic and basic fibroblast growth factor. BBRC 138:476–482

2. Bissel MJ, Hall HG, Parry G (1982) How does the extracellular matrix direct gene expression? J Theor Biol 99:31–68
3. Burridge K, Fath K, Kelly G, Nuckolls G, Turner C (1988) Focal adhesions: transmembrane junctions between the extracellular matrix and the cytoskeleton. Annu Rev Cell Biol 4:487–526
4. Cheifitz S, Weatherbee JA, Tsang MLS, Anderson JK, Mole JE, Lucas R, Massague J (1987) Transforming growth factor system, a complex pattern of cross-reactive ligands and receptors. Cell 48:409–415
5. Chen WT, Hasegawa E, Hasegawa T, Weinstock C, Yamada K (1985) Development of cell surface linkage complexes in cultured fibroblasts. J Cell Biol 100:1103–1114
6. Dejana E, Colella S, Languino LR, Balconi G, Corboscio GC, Marchisio PC (1988) Fibrinogen induces adhesion, spreading and microfilament organization of human endothelial cells in vitro. J Cell Biol 104:1403–1411
7. Dike LE, Farmer SR (1988) Cell adhesion induces expression of growth-associated genes in suspension-arrested fibroblasts. Proc Natl Acad Sci USA 85:6792–6796
8. Folkman J, Moscona A (1978) The role of cell shape and growth control. Nature 27:345–349
9. Gospodarowicz D, Ill C (1980) Extracellular matrix and control of proliferation of vascular endothelial cells. J Clin Invest 665:1351–1364
10. Hay ED (1982) Interaction of embryonic cell surface and cytoskeleton with extracellular matrix. Am J Anat 165:1–12
11. Herman IM (1987) Extracellular matrix-cytoskeletal interaction in vascular cells. Tissue Cell 19:1–19
12. Herman IM, Castellot JJC (1987) Regulation of vascular smooth muscle cell growth by endothelial-synthesized extracellular matrices. Arteriosclerosis 7:463–469
13. Kleinman H, Kebe RJ, Martin G (1981) Role of collagenous matrices in the adhesion and growth of cells. J Cell Biol 88:473–485
14. Laterra J, Silbert JE, Culp LA (1983) Cell surface heparan sulfate mediates some adhesive responses to glycosaminoglycan-binding matrices, including fibronectin. J Cell Biol 96:112–123
15. Lawler J, Weinstein R, Hynes R (1988) Cell attachment to thrombospondin: the role of arg-gly-asp, calcium and integrin receptors. J Cell Biol 107:2351–2361
16. Madri JM, Pratt BM, Tucker AM (1988) Phenotypic modulation of endothelial cells by TGF-B depends on the composition and organization of the extracellular matrix. J Cell Biol 106:1375–1384
17. McCarthy JB, Basara ML, Palm SL, Sas DF, Furcht LT (1985) The role of cell adhesion proteins, laminin and fibronectin, in the movement of malignant and metastatic cells. Cancer Metastasis Rev 4:125–152
18. Piershbacher MD, Ruoslahti E (1984) Cell attachment activity of fibronectin can be duplicated by small synthetic fragments of the molecule. Nature 309:30–33
19. Roberts AB, Flanders KC, Kondaiah K, Thompson NL, van Obberghen-Schilling E, Wakefield L, Rossi P, de Crombrugghe B, Heine U, Sporn M (1988) TGF-B: biochemistry and roles in embryogenesis, tissue repair and remodeling and carcinogenesis. Recent Prog Horm Res 44:157–197
20. Singer II, Kawka DW, Scott S, Mumford RA, Lark MW (1987) The fibronectin cell attachment sequence arg-gly-asp-ser promotes focal contact formation during early fibroblast attachment and spreading. J Cell Biol 1004:573–584
21. Vlodavsky I, Folkman J, Sullivan R, Fridman R, Ishai-Michaeli R, Sasse J, Klagsbrun M (1987) Endothelial cell-derived basic fibroblast growth factor: synthesis and deposition into subendothelial extracellular matrix. Proc Natl Acad Sci USA 84:2292–2296
22. Yamada K (1983) Cell surface interactions with extracellular materials. Annu Rev Biochem 52:761–799
23. Young WC, Herman IM (1985) Extracellular matrix modulation of endothelial cell shape and motility following injury in vitro. J Cell Sci 73:19–32

Endothelial Cells Grown on Filter Membranes *

D. M. Shasby and S. S. Shasby

Introduction

The endothelium is a single layer of epithelioid cells which lines the blood vessels. Because of this unique geography, the endothelium is potentially important in the control of many of the events of vascular biology. The endothelium could be important in determining the flux of molecules into and out of the vascular space. The endothelium could also control the movement of cells from the blood to the tissues and potentially the flow of tissue lymphocytes back into the circulation. Because of its unique location and own active metabolism the endothelium could also act as a sensor of intravascular events and at the same time an initiator of messages to parts of the vessel wall outside the vessel lumen. All of these potential roles have stimulated investigators for many years. However, the geography of the endothelium has also made it very difficult to precisely investigate these potential functions and ascribe them solely to the endothelium. Other components of the vessel wall may contribute to limiting the flux of molecules from the blood to the tissues, and it is very difficult to sample the contents of the immediate extravascular space. Other components of the vessel wall may also restrict the movement of cells from the blood to the tissues and vice versa. Similarly, the endothelium is not the only cell type in the vessel wall capable of secreting important messenger molecules. Even when these messenger molecules are identified in cultures of endothelium, the potential precision of the message that might reside in the polarity of release of the molecules remains difficult to determine.

During the 1970s there were several lines of epithelial cells available that made it possible to investigate aspects of these cells in culture. In 1976 Misfeldt and colleagues described the culture of Madin Darby canine kidney (MDCK) cells on Millipore filters. The cells formed a polarized monolayer with tight junctions demonstrable on freeze-fracture preparations. The monolayer developed a spontaneous potential difference and had an electrical resistance of 84 ohms $\times$ cm^2 [1]. In 1978 Cereijido et al. [2] reported a similar model in which MDCK cells had been grown on nylon mesh precoated with a collagen solution. The epithelial cells also formed an intact monolayer on this matrix with a transepithelial electrical resistance of approximately 100 ohms $\times$ cm^2. In 1981 Taylor et al. [3] combined the experience that Steve Schwartz had accu-

---

* Supported in part by grants from the National Institutes of Health, HL 33540, HL 36605, and HL14230

mulated in endothelial cell culture with the techniques used for the epithelial cells and reported that they were able to culture bovine aortic endothelial cells on gelatin-treated polycarbonate filters. The cells formed an intact monolayer by transmission electron microscopy, and the monolayer itself restricted the diffusion of albumin and the diapedesis of polymorphonuclear leukocytes. At the same time, McCall et al. [4] described monolayers of endothelial cells grown on polytetrafluoroethylene membranes. In their original description, the endothelial cell covered membranes limited water flow, but did not exhibit sieving of albumin in the aqueous medium. Our laboratory adapted the techniques of Taylor et al. [3] and began to investigate what were some of the determinants of the integrity of the endothelial monolayers and how they might be altered by inflammatory pathways [5–7]. Many laboratories have subsequently modified or adapted the technology to investigate aspects of endothelial biology. This chapter will describe our approach, that of some other investigators, point out some of the current problems, and give some hints as to what may be future directions.

Culture of the Cells

In general, the techniques we and others have used for culture of the endothelial cells are the same as those used for culture of the cells on tissue culture plastic. These are dealt with in detail in other chapters in this series and will not be considered in detail here. Briefly, we have used primarily medium 199 supplemented with MEM vitamins and amino acids, 2 mM glutamine, and 10% fetal bovine serum (FBS) as culture medium. The cells have been isolated by scraping the vessel lumen and passaged with trypsin (0.25%, 15 s) and EDTA (1%) using 3:1 split ratios every 4–5 days. Cells from passes 4–12 have been used. It has been our experience that cells from higher passage numbers do not form as tight a barrier, although other investigators have used higher passaged cells.

We have also used human umbilical vein cells in some studies [8, 9]. We have used only primary cultures of these cells which have been directly plated onto the filters after harvesting the cells from the vessel wall using standard collagenase digestion. These cells have developed good barriers (albumin permeability of 1×10^{-6} cm $\times$ s^{-1}) and have retained a brisk response to hormonal stimulation, a characteristic the other cells lose quickly in passage [9].

Other investigators have used endothelial cells from other sources. Several groups have used microvascular endothelial cells from the brain. Bowman et al. described bovine cerebromicrovascular endothelial cells grown on collagen-coated nylon mesh. They observed tight junctions between the cells and restricted diffusion of sucrose across the monolayers. However, the electrical resistance of these monolayers (<100 ohms $\times$ cm^2) was much less than that expected for the cerebral circulation [10]. More recently Rutten et al. [11] reported that one of five clones of brain microvessel endothelial cells from bovine brain developed electrical resistances of approximately 780 ohms $\times$ cm^2. These cells were grown on type 1 collagen gels. If this work can be repeated it will represent an important quantitative improvement in the barrier that can be produced in vitro.

Preparation of Filters

Our original preparation of the polycarbonate filters was an adaptation of a technique described by Postlethwaite et al. [12] to investigate fibroblast chemotaxis. Briefly the polycarbonate filters (0.45-µm pore diameter, Nuclepore) were individually placed in tissue cassettes (Tissue-Tek, Miles Laboratories), heated to $50\,°C$ in 0.5% acetic acid $\times 20$ min, rinsed $\times 2$, and heated again in 0.05% gelatin (porcine skin, Sigma) at $100\,°C \times 60$ min, air dried, and ethylene oxide sterilized. The filters were then air dried for 48 h. In some instances the filters were first glued (Millipore CF cement) to the bottom of polycarbonate cylinders (ADAPS) which could be fitted into the well of standard 24-well tissue culture plates. The cylinder and attached filter were then given the same treatment as listed above and air dried to remove the ethylene oxide. Just prior to adding the cells to the filter, the filters were covered with a bead of fibronectin (30 $\mu g \times cm^{-3}$) in balanced salt solution for 30 min. The excess fibronectin was aspirated from the filter and the cells were then applied to the filter surface. For filters attached to polycarbonate cylinders, the cells were added to the surface of the filter within the lumen of the cylinder.

More recently, commercial preparations of polycarbonate (Transwell, Costar) and nitrocellulose (Millicel, Millipore) filters attached to plastic rings have become available. The polycarbonate filters have been treated for tissue culture while the nitrocellulose have not. We have utilized both preparations with good success. We have pretreated each with fibronectin prior to plating the cells, although the manufacturers do not recommend any additional treatment to the polycarbonate filters. Gelatin or other matrix should also probably be applied to the nitrocellulose filters.

Not all authors have advocated treatment of the filters with gelatin or another collagenous matrix. We have found that this improves the quality of the preparations and others have confirmed this (J. Shepard, personal communication). Albelda et al. [13] used collagen to coat the surfaces of filters used for epithelial cell cultures, but did not collagen treat filters used for endothelial cell cultures. They observed defects in the endothelial monolayers but not in the epithelial monolayers. As more information has become available about the important role of the matrix in determining cell phenotypes, it has become increasingly evident that careful attention to the matrix may be one of the most important improvements that can be made in this model [14].

Alternative substrates have been used by several investigators. Collagentreated polytetrafluoroethylene (PTFE) supported by a Gore-Tex scrim and preincubated in medium with FBS was used by Baetscher and Brune [15]. This support is very porous (10–15 µm mesh) and has proven to be very useful for the investigation of the chemotaxis of inflammatory cells across the endothelium [16]. Furie et al. [17] used the stromal surface of amniotic membrane as a matrix and support for bovine adrenal microvascular cells. These monolayers restricted the passage of macromolecules and have also proven useful for the investigation of inflammatory cell chemotaxis. While the original publication reported a transendothelial electrical resistance of > 60 ohms $\times$ cm^2 that value has not been repeated from the same laboratory [17, 18].

Plating Cells on the Filters and Culture of the Monolayers

We and most authors have plated the endothelial cells onto the filters at a confluent density of $2-5 \times 10^5$ cells $\times$ cm^2 in a suspension of approximately 1×10^6 cells $\times$ cm^{-3} in tissue culture medium with FBS. The cells are allowed to settle and adhere to the filter surface for 4 h. After 4 h the medium containing the nonadherent cells is aspirated from the surface and replaced with fresh tissue culture medium containing 10% or 20% FBS. We have found that removing the nonadherent cells is an important step and have assumed that lysis of the nonadherent cells over the rest of the culture period disrupts the integrity of the adherent cells.

The culture medium is replaced either daily or every other day depending on the area of the filter. Monolayers have been studied 2–8 days after plating. While we have found little change in the permeability characteristics of the monolayers over this time, the cell-cell junctions viewed on freeze-fracture preparations change from predominantly gap junctions on days 1–5 to predominantly tight junctions by day 7 [19].

Methods of Studying Properties of the Endothelial Monolayers

Taylor's and our initial reports used modified chemotaxis chambers to study the flux of macromolecules and inflammatory cells across the endothelial monolayers [3, 5, 6]. Our subsequent reports have used either filters glued to the polycarbonate cylinders (ADAPS; Fig. 1), filters mounted in modified Ussing chambers (Jim's Instrument Manufacturing) or the commercial preparations discussed above. At present we use either the ADAPS cylinders or the commercial rings for studies in which we want to investigate the effect of an intervention (such as the addition of histamine) on the barrier function of the monolayers. This technique allows for each monolayer to act as its own con-

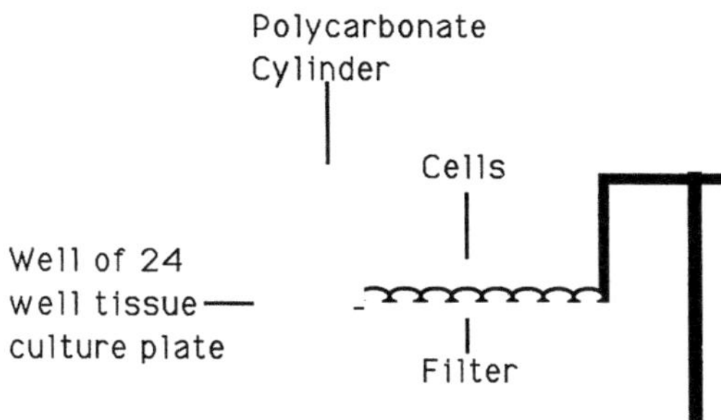

Fig. 1. Polycarbonate cylinder for studying properties of endothelial monolayers

trol, reducing the variance caused by differences between individual preparations. It is also easy to examine a large number of samples with this technique, making it possible to examine all the variables and controls with the same preparation of cells, again limiting the variance of one passage of cells to the next.

Typically, the filter preparation is placed into the well of a 24-well tissue culture plate which has been filled with tissue culture medium containing a known concentration of the macromolecules of interest. The filter and attached cylinder are then suspended in the well and the same medium added to the lumen of the cylinder to balance the fluid levels. A tracer molecule such as radiolabelled albumin is then added to the lumen of the cylinder and the diffusion over time is determined by sampling both chambers. The monolayer is then washed, placed into a new well under the same conditions, the appropriate intervention made, and the flux of the macromolecule again measured over the same time interval. Repeat measurements on the same monolayers have demonstrated an average reproducibility of within 5% of the original measurement.

For transport studies we have chosen to mount the monolayers in modified Ussing chambers which provide better mixing than the above approach [20]. These chambers have now been adapted to accept the Millicel chambers (Jim's Instrument Manufacturing) which facilitates preparation. In the Ussing chambers it is also possible to measure the transendothelial electrical resistance using KCl agar bridges connected to a current voltage clamp (Biomedical Engineering).

Malik's group has altered the ADAPS cylinders in two significant ways. They have added a flotation collar at the top of the cylinder to allow the cylinder and monolayer to float in a larger well, thereby preserving a balance of fluid levels inside and outside the cylinder. They have also placed the cylinder in a larger well that can be stirred, providing better mixing on the abluminal surface and permitting multiple sampling of the abluminal chamber [21]. We still prefer the modified Ussing chambers for transport measurements.

Baetscher and Brune [15] were the first to report measurements of the hydraulic conductivity of endothelial monolayers in vitro, and they made the important observation that placing the monolayers under a hydrostatic pressure caused a gradual decrease in the hydraulic conductivity over several hours. The final hydraulic conductivity measured in this way was quantitatively very similar to measurements obtained from single capillaries in vivo in frog mesentery. These observations have been repeated by Suttorp et al. [22] who have added the additional information that the "pressure induced sealing" of the monolayer also increased the reflection coefficient of the monolayers to albumin from 0.23 to 0.73. We measured hydraulic conductivities for monolayers of porcine pulmonary artery endothelial cells grown on polycarbonate filters. Our measurements were made in the absence of albumin and serum, "without a sealing period," and also produced values similar to those obtained in frog mesenteric capillaries in the absence of albumin, which are higher than those obtained in the presence of albumin [20]. If albumin contributes to a fiber matrix that restricts water flow, as suggested by Michel and Curry, then Baetscher's technique may be a mechanism for establishing a normal amount of albumin in the matrix and an important adjunct to the basic model. The model for hydraulic conductivity measurements described by Suttorp is superior to ours and would be our current choice [22].

Problems with the Current Model and Future Directions

While the hydraulic conductivity measurements and the electrical resistance measurements obtained with models of endothelial cells grown on fibrous or filter supports approximate single capillary measurements in frog mesentery, the permeabilities to macromolecules such as albumin have been 10–100 times greater than those measured in whole organs in vivo. Albelda et al. [13] suggested the increased albumin permeability was due to defects in the monolayers. However, other authors have not made similar observations. We carefully evaluated cell-cell junctions with transmission electron microscopy and did not observe the defects Albelda et al. described [13]. Langeler et al. [23] have recently described a model similar to the one used by Baetscher and Suttorp, and they also did not find defects microscopically nor did they observe transfer of very low density lipoprotein (VLDL) through the monolayers. Had there been defects similar to those described by Albelda et al. [13], VLDL transfer should have occurred. Asymmetric transfer of molecules such as insulin, eicosanoids, and hydroxy fatty acids has also been observed with such monolayers, and defects in the monolayer would seem to preclude such asymmetry [24–26]. The reflection coefficient for albumin of 0.73 reported by Suttorp et al. [22] is also not consistent with the defects Albelda et al. [13] described. We have measured an albumin reflection coefficient of 0.68 to 0.83 using our setup for hydraulic conductivity measurements [20] (Peterson and Shasby, unpublished observations). We have also observed that treatment of endothelial monolayers with agents to increase endothelial cell cAMP, decreases the albumin permeability by approximately 50% (to approximately $5-6 \times 10^{-7}$ cm $\times$ s^{-1}) [9]. This decrease in permeability is not consistent with the bulk of the albumin flux occurring through defects in the monolayer, and represents an albumin permeability very close to values estimated for the coronary microcirculation $(3.7 \times 10^{-7}$ cm $\times$ s$^{-1})$ and for skeletal muscle capillaries $(2-6 \times 10^{-7}$ cm $\times$ s$^{-1})$ [27].

Monolayers used for investigating the cerebromicrovascular have generally been much less restrictive a barrier than the in situ cerebromicrovasculature. The exception to this is the one clone described by Rutten et al. [11]. Recent observations that cocultivation of cerebromicrovascular endothelial cells with astroglia restores the normal tight junction morphology to the endothelial cells may be an important observation for this field [28]. Similar cocultivation of cerebromicrovascular endothelial cells with glia imparted polarity to amino acid transport across the endothelium [29]. To date, most attempts to grow noncerebral microvascular cells on membrane supports have failed to produce monolayers with any junctional complexes [30]. Furie et al.'s report [17] of bovine adrenal microvascular cells grown on serosal amnion may be an exception to this. The observations with the astroglia may be relevant to endothelia from other vascular beds, and suggest that important directions for the future will involve careful consideration of matrix and the influence of other relevant parenchymal cells on endothelial cell phenotype.

Haselton et al. [31] have recently applied multiple tracer analysis to the chromatographic separation of tracer molecules by endothelial cell covered polypropylene beads (Cytodex-3, Kontes). They measured permeabilities for mannitol and sucrose that were ten fold in vivo estimates, and this may indicate a more restrictive barrier than what has been previously measured on filter supports. In addition, this technique exposes much more surface area for transport. It will be of interest to see how use of this approach evolves.

References

1. Misfeldt DS, Hamamoto ST, Pitelka DR (1976) Transepithelial transport in cell culture. Proc Natl Acad Sci USA 73:1212–1216
2. Cereijido M, Robbins ES, Dolan WJ, Rotunno CA, Sabatini DD (1978) Polarized monolayers formed by epithelial cells on a permeable and translucent support. J Cell Biol 77:853–880
3. Taylor RF, Price TH, Schwartz SF, Dale DC (1981) Neutrophil-endothelial cell interactions on endothelial monolayers grown on micropore filters. J Clin Invest 67:584–587
4. McCall E, Povey J, Dumonde DC (1981) Fluid flow studies on vascular endothelium cultured on micropore membranes. Agents Actions 11:654–657
5. Shasby DM, Shasby SS, Sullivan JM, Peach MJ (1982) Role of endothelial cell cytoskeleton in control of endothelial permeability. Circ Res 51:657–661
6. Shasby DM, Shasby SS, Peach MJ (1983) Granulocytes and phorbol myristate acetate increase permeability to albumin of cultured endothelial monolayers and isolated perfused lungs: role of oxygen radicals and granulocyte adherence. Am Rev Respir Dis 127:72–76
7. Shasby DM, Shasby SS (1986) Effects of calcium on transendothelial albumin transfer and electrical resistance. J Appl Physiol 60:71–79
8. Shasby DM, Shasby SS, Peach MJ (1985) Polymorphonuclear leukocyte-arachidonate edema in isolated perfused lungs and monolayers of cultured endothelial cells. J Appl Physiol 59:47–55
9. Carson MM, Shasby SS, Shasby DM (1989) Histamine and inositol phosphate accumulation in endothelium: cAMP and a G protein. Am J Physiol 1 (in press)
10. Bowman PD, Ennis SR, Rarey KE, Betz AL, Goldstein GW (1983) Brain microvessel endothelial cells in tissue culture: a model for study of blood-brain barrier permeability. Ann Neurol 14:396–402
11. Rutten MJ, Hoover RL, Karnovsky MJ (1987) Electrical resistance and macromolecular permeability of brain endothelial monolayer cultures. Brain Res 425:301–310
12. Postlethwaite AE, Snyderman R, Kang AH (1976) The chemotactic attraction of human fibroblasts to a lymphocyte-derived factor. J Exp Med 144:1188–1203
13. Albelda SM, Sampson PM, Haselton FR, McNiff JM, Mueller SN, Williams SK, Fishman AP, Levine EM (1988) Permeability characteristic of cultured endothelial cell monolayers. J Appl Physiol 64:308–322
14. Madri JA, Williams SK (1983) Capillary endothelial cell cultures: phenotypic modulation by matrix components. J Cell Biol 97:153–165
15. Baetscher M, Brune K (1983) An in vitro system for measuring endothelial permeability under hydrostatic pressure. Exp Cell Res 148:541–547
16. Darby H, Brown KA, Anderson RA, Williams BT, Dumonde DC (1988) Transendothelial chemotaxis in vitro of human monocytes. J Immunol Methods 113:157–163
17. Furie MB, Cramer EB, Naprstek BL, Silverstein SC (1984) Cultured endothelial cell monolayers that restrict the transendothelial passage of macromolecules and electrical current. J Cell Biol 98:1033–1041

18. Huang AJ, Furie MB, Nicholson SC, Fischbarg J, Liebovitch LS, Silverstein SC (1988) Effects of human neutrophil chemotaxis across human endothelial cell monolayers on the permeability of these monolayers to ions and macromolecules. J Cell Physiol 135:355–366
19. Shasby DM, Roberts RL (1987) Transendothelial transfer of macromolecules in vitro. Fed Proc 46:2506–2510
20. Shasby DM, Peterson MW (1987) Effects of albumin concentration on endothelial albumin transport in vitro. Am J Physiol 253:H654–H661
21. Cooper JA, DelVecchio PJ, Minnear FL, Burhop KE, Selig WM, Garcia JG, Malik AB (1987) Measurement of albumin permeability across endothelial monolayers in vitro. J Appl Physiol 62:1076–1083
22. Suttorp N, Hessz T, Seeger W, Wilke A, Koob R, Lutz F, Drenckhahn D (1988) Bacterial exotoxins and endothelial permeability for water and albumin in vitro. Am J Physiol 255:C368–C376
23. Langeler EG, Snelting-Havinga I, van Hinsbergh VWM (1989) Passage of low density lipoproteins through monolayers of human arterial endothelial cells. Arteriosclerosis 9 (in press)
24. King G, Johnson SM (1985) Receptor-mediated transport of insulin across endothelial cells. Science 227:1583–1586
25. Shasby DM, Stoll JL, Spector AA (1987) Polarity of arachidonic acid metabolism by bovine aortic endothelial cell monolayers. Am J Physiol 253:H1177–H1183
26. Kaduce TL, Figard PH, Leifur R, Spector AA (1989) Formation of 9-hydroxyoctade-canoic acid from linoleic acid in endothelial cells. J Biol Chem 264:6823–6830
27. Svendsen JH, Paaske WP, Sejrsen P, Haunso S (1989) Capillary permeability of 131l-albumin in canine myocardium as determined by bolus injection, residue detection. Microvasc Res 37:352–356
28. Arthur FE, Shivers RR, Bowman PD (1987) Astrocyte-mediated induction of tight junctions in brain capillary endothelium: an efficient in vitro model. Brain Res 36:155–159
29. Beck DW, Vinters HV, Hart MN, Cancilla PA (1984) Glial cells influence polarity of the blood-brain barrier. J Neuropathol Exp Neurol 43:219–224
30. Meyrick B, Hoover R, Jones MR, Berry LC, Brigham KL (1989) In vitro effects of endotoxin on bovine sheep lung microvascular and pulmonary artery endothelial cells. J Cell Physiol 138:165–174
31. Haselton FR, Mueller SN, Howell RE, Levine EM, Fishman AP (1989) Chromatographic demonstration of reversible changes in endothelial permeability. J Appl Physiol (in press)

Microcarrier Cultures of Endothelial Cells*

R. Spahr and H. M. Piper

Introduction

Microcarrier cultures of endothelial cells have a number of interesting applications. With the use of microcarriers the surface for growing anchorage-dependent cells can be enlarged in a given volume of culture medium. Therefore, microcarrier cultures are an ideal system to produce large quantities of cultured cells. Cells cultured on microcarriers can also be easily transferred from one experimental system to another without trypsinization, e.g., from two separate cultures of endothelial and smooth muscle cells into one defined coculture of both cell types [2]. Furthermore, packed columns of endothelial cells grown on microcarriers represent a useful model for studies on endothelial metabolism requiring a large amount of endothelial material. Such a column can be perfused in the same manner as an isolated organ.

There are a great number of possibilities to vary the conditions for endothelial cell microcarrier cultures. Cell growth and development depend both on the material of the microcarriers and the composition of the culture media. In this chapter some basal alternatives for culturing porcine aortic endothelial cells on microcarriers are described.

Preparing Endothelial Cells for Microcarrier Cultures

Isolation of Endothelial Cells from Porcine Aortas

Materials

The instruments used are a pinboard, pins, scissors, two forceps, a scalpel with a curved blade, 2 50-ml centrifuge tubes, and a centrifuge.

The media needed are:

- Phosphate buffered saline (PBS; composition in mM: NaCl 137.0, KCl 2.6, KH$_2$PO$_4$ 1.5, Na$_2$HPO$_4$ 8.1; pH 7.4). For PBS with aclocillin, dissolve 0.2 g aclocillin (Securopen, Bayer) in 1 l PBS.

---

* This work has been supported by a grant from the Minister für Forschung und Technologie des Landes Nordrhein-Westfalen

– Culture medium 1 (for dish cultures), comprised of medium 199 with Earle's salts, buffered by 26 mM NaHCO$_3$ in 5% CO$_2$ at pH 7.4, 20% newborn calf serum (NCS), penicillin 250 IU/ml, streptomycin 250 µg/ml, amphotericin B 12.5 µg/ml, and gentamycin 50 µg/ml.

The culture dishes used are 100-mm plastic culture dishes (Falcon, type 3003).

Procedure

1. Obtain 25 segments of porcine thoracic aortas, 15–25 cm in length, from a local slaughterhouse. After excision from the animal, immerse them in a flask filled with PBS with aclocillin and bring them rapidly to the laboratory.
2. Cut open the aortic segments in the longitudinal direction and fix them with pins on the board, as flat sheets with the endothelial surface facing upwards. Rinse the surface five times with PBS.
3. Gently scrape off endothelial cells from the aortic intimal surface using the blade of a scalpel. The blade must be inclined at an angle of approximately 60° to the intimal surface. Scrape each aortic segment three to four times along longitudinal, nonoverlapping lines. Strictly avoid scraping a site twice. Suspend the material accumulating on the blade in 50 ml of culture medium 1.
4. Centrifuge the cell suspension at 250 g for 2 min. Discard the supernatant and resuspend the pellet in 250 ml of culture medium 1.
5. Using the method described, an average of 1.5×10^6 viable cells, which can establish in culture, can be obtained per aortic segment.

Note: Slaughterhouse material is not sterile. In our experience it makes no difference whether cell harvesting is performed in sterile air under the laminar flow hood or on the bench. With the use of high concentrations of antibiotics during the isolation procedure and during the first days of culture, cultures free of microbial growth can be achieved. The described mixture of antibiotics does not seem to be harmful to the endothelial cells. In case microbial growth cannot be prevented by this treatment, other antibiotics have to be used. This initial period of culture, which is required to remove microbial contaminants and cell debris, is best conducted on Petri dishes from which media can be easily changed.

Dish Cell Culture

1. Distribute the material from 25 aortas, suspended in 250 ml of culture medium 1, in 50-ml aliquots onto five 100-mm culture dishes and incubate at 37 °C with 5% CO$_2$. Renew the medium after 24 h. At this time, the culture is already subconfluent.
2. Change the medium again 2 days after plating, reduce the concentration of the antibiotics to penicillin 100 IU/ml, streptomycin 100 µg/ml, and amphotericin B 5 µg/ml. Renew this medium again 4 days after plating.

3. After 6 days, the dishes contain confluent endothelial monolayers with approximately $1.5-2.0 \times 10^7$ cells/100-mm dish ($2.5-3.5 \times 10^5$ cells/cm^2).

Note: Six-day-old cultures contain approximately 3% of cells with a positive stain for smooth-muscle α-actin (for the staining procedure see Chap. 11). The number of these cells is distinctly increased when the aortas are scraped very forcefully. Therefore, gentle scraping is required. Of the cells 95%–97% take up acetylated low density lipoprotein labeled with 1,1'-dioctadecyl-1-3,3,3',3'-tetramethyl-indo-carbocyanine perchlorate (DiI-ac-LDL, Biomedical Technologies), which is a functional marker for their endothelial nature [8] (for the procedure see Chap. 11).

Microcarrier Cultures

Microcarrier Culture on Polystyrene Beads

Materials

The glassware used are spinner vessels with a suspended stirrer rod (gross volume 500 ml; Techne, no. UK-F7689), and a stirrer with an interval timer (Techne, model MCS 104 L). The vessels must be siliconized (see below). Microcarrier beads, type Biosilon from Nunc are used, stored in 1-g aliquots.
The media needed are:

- Culture medium 2 (for microcarrier culture), comprised of medium 199 supplemented with Earle's salts, buffered with 25 mM HEPES at pH 7.4, 20% NCS, penicillin 100 IU/ml, and streptomycin 100 μg/ml.
- EDTA buffer (composition in mM): NaCl 125, KCl 2.6, KH$_2$PO$_4$ 1.2, HEPES 10, NaEDTA 0.5; pH 7.4.
- Trypsin-EDTA solution: 50 mg trypsin (trypsin 1:250) and 20 mg EDTA dissolved in 100 ml PBS (mixture commercially available; e.g., from Flow Laboratories).

Procedure

1. Prepare 1-g aliquots of sterile microcarriers. The microcarriers may be weighed under nonsterile conditions and autoclaved. Before use incubate the microcarriers overnight in culture medium 1. In our experience this incubation step is already sufficient to achieve sterility since culture medium 1 contains high concentrations of antibiotics.
2. Wash 100-mm Petri dishes containing confluent monolayers of endothelial cells twice with EDTA buffer, then add 2 ml trypsin-EDTA solution per dish, and incubate at 37°C. After 10 min, add 10 ml culture medium 2 per dish. Bring the cells into suspension by pipetting. Collect the material of up to five dishes in one 50-ml tube in a final volume of approximately 50 ml of culture medium 2. Count the number of cells per ml in a Neubauer chamber.

3. Mix a volume of the suspension containing 5×10^6 cells with 1 g of preincubated microcarrier beads in one spinner flask. Fill up the volume to 100 ml with culture medium 2.
4. As next step transfer the spinner flask to a temperature-controlled incubator (37 °C) containing the magnet stirrer. For the first 24 h, stir the cultures with intermissions (3 min stirring at 35 rpm followed by a 10-min interval). For porcine endothelial cells, microcarrier cultures can also be directly started with continuous stirring instead of the described interval stirring (see below).
5. After the first 24 h of interval stirring, the medium is increased to a total volume of 250 ml culture medium 2, and the cultures are stirred continuously at 35 rpm.
6. Every 24 h, replace ⅔ of the medium by fresh culture medium 2.

Note: Siliconizing of Glassware: Only clean glassware is used. Surfaces which are to be in contact with the cell suspension are rinsed with the silicone solution (e.g., Sigmacote, Sigma). The silicone solution is allowed to run off. The coated surface is dried in air, then washed six times with distilled water. The glassware is then heated to 100 °C for 1 h. Subsequently it has to be autoclaved for tissue culture use. The silicone film is transparent. Culture vessels should be resiliconized when cell-covered microcarriers start to stick to their walls. For this purpose, the previous coating must first be removed: glassware is immersed at 100 °C in 0.5 N NaOH, thereafter washed five times in distilled water and dried.

Properties of the Microcarrier Culture

According to the above protocol cells are seeded at a density of 5×10^6 cells/g Biosilon, corresponding to about 30 cells per bead (see physical data in Table 1). In exponential growth, they grow with a doubling time of approximately

Table 1. Properties of microcarrier beads (according to the manufacturers' specifications)

Product name	Supplier	Material	Diameter (hyd)	Beads / g (dry)	Area (hyd) / g (dry)
			μm	$\times 10^6$/g	cm²/g
Biosilon	Nunc	Solid polystyrene	160–300	0.15	255
Glass microcarrier beads	Sigma	Solid glass	150–210	0.13	325
Ventregel II	Ventrex	Polystyrene, coated with gelatin	150–210	1.7	325
Cytodex 3	Pharmacia	Dextran, coated with collagen I	133–215	4.0	4600
CultiSpher-G	Percell Biolytica	Gelatin	170–270	2–3	17700

hyd, hydrated material; dry, dry material

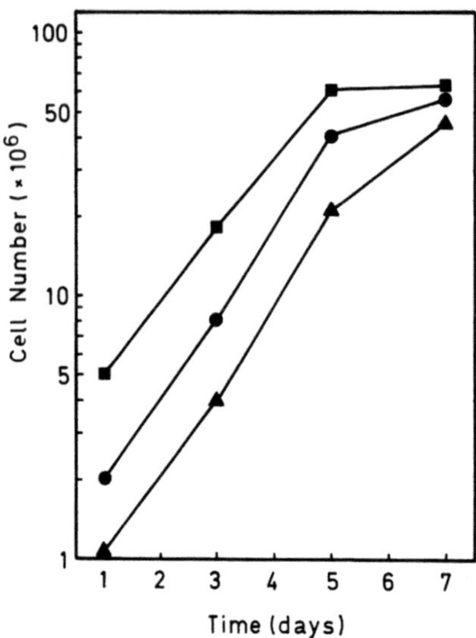

Fig. 1. Effect of the initial seeding density on the growth of porcine aortic endothelial cells on Biosilon microcarriers. The time scale on the *abscissa* is given in days after seeding of cells on microcarriers. Seeding densities were 1×10^6 (*triangles*), 2×10^6 (*circles*), and 5×10^6 cells (*squares*) per gram of Biosilon. Cultures were performed as described in the procedure protocol in the presence of 20% NCS. Cells grown on Petri dishes for 7 days were used to start microcarrier cultures. Mean values from four experiments

20 h (Fig. 1). This rate closely corresponds to the growth rate of porcine aortic endothelial cells cultured on flat tissue-grade polystyrene surfaces [3]. Confluency of cells on the microcarrier beads is reached 5 days after seeding. In confluent cultures clusters of 3–10 microcarriers form spontaneously. The cell density in confluent cultures amounts to 6×10^7 cells/g Biosilon beads, i.e., more then ten times the number of cells seeded. This figure corresponds to approximately 400 cells per microcarrier bead and a surface density of approximately 2.5×10^5 cells/cm^2. The latter value is slightly lower than the density of primary confluent endothelial cell cultures on polystyrene tissue culture Petri dishes (see above).

Variations of the Protocol

Stirring Mode

In the Techne spinner flasks a stirring speed greater than 30 rpm is needed to avoid sedimentation of the polystyrene microcarriers. Stirring speeds up to 60 rpm are not harmful to porcine aortic endothelial cells, but a stirring speed greater than 35 rpm is not required.

For porcine aortic endothelial cells grown on Biosilon beads the initial period of interval stirring is not required. A direct start of continuous stirring at 35 rpm results in identical growth rates. Cell growth starts 24 h after seeding, irrespective of the stirring mode.

Seeding Density

The doubling time of endothelial cells in the phase of exponential growth remains the same when the seeding density is varied between 1 and 5×10^6 cells/g Biosilon beads. When confluency is reached the number of cells per gram of microcarrier material is approximately equal (Fig. 1) indicating that all microcarrier beads are covered with endothelial cell monolayers. This can indeed be demonstrated by light microscopy. With seeding densities of 1×10^6 cells per gram of microcarriers and more, confluency is reached within 1 week.

Serum Type

With the batches of fetal calf serum (Gibco) and newborn calf serum (Biochrom) tested in our laboratory, the growth of porcine aortic endothelial cells was the same (Table 2). The major advantage in using NCS resides in its much lower price. The half-synthetic serum replacement Nu-Serum (Collaborative Research), containing various growth factors and a small percentage of serum, does not support endothelial cell growth to the same extent as fetal calf serum (FCS) or NCS do, as can be seen from a prolonged doubling time (Table 2).

Serum Content

The initial onset of cell growth in microcarrier cultures is distinctly influenced by the addition of serum. Endothelial cells from porcine aorta grow well with a supplement of NCS. Serum batches commercially available, however, vary greatly in their suitability for cell culture. The selection of a suitable batch is best performed in dish cultures. General selection criteria are the growth rate

Table 2. Growth of porcine aortic endothelial cells on microcarriers in presence of 20% serum. Data are given as means $\pm$ S.D., $n = 4$ cultures

Name	Age (days)	Serum type	Doubling time (h)
Biosilon (Nunc)	0	NCS	30.8 ± 3.2
	2	NCS	21.5 ± 1.6
	7	NCS	20.9 ± 3.8
	7	FCS	23.3 ± 1.2
	7	NuS	27.4 ± 2.2
Glass beads (Sigma)	7	NCS	19.9 ± 2.0
Ventregel II (Ventrex)	7	NCS	21.6 ± 1.7
Cytodex 3 (Pharmacia)	7	NCS	21.0 ± 2.0
CultiSpher-G (Percell Biolytica)	7	NCS	No growth

Age, age of cells used for microcarrier cultures (in days after cell isolation); NCS, newborn calf serum (Biochrom); FCS, fetal calf serum (Gibco); NuS, Nu-Serum (Collaborative Research)

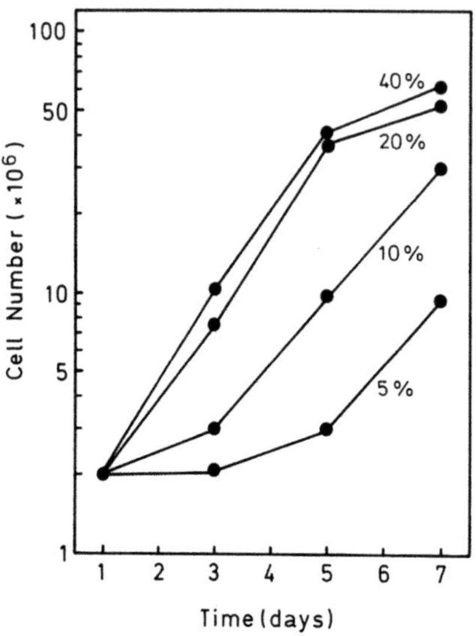

Fig. 2. Effect of the concentration of newborn calf serum (% v/v) on the growth of porcine aortic endothelial cells on Biosilon microcarriers. Cultures were performed as described in the procedure protocol. Cells grown on Petri dishes for 7 days were used to start microcarrier cultures, with a seeding density of 2×10^6 cells/g Biosilon beads. Mean values from four experiments

and development of the typical cobblestone morphology in confluent cultures. For special purposes other criteria might also be relevant (e.g., certain enzyme activities). After selection of a suitable batch, it is advisable to purchase a large quantity.

With the batch of NCS used in our laboratory, a 20% supplement to the culture medium gave optimal results. Higher concentrations had only a small additional effect on cell growth (Fig. 2). With NCS concentrations $<20\%$, the initial cell growth is slower. When the cell densities have reached 3×10^6 cells/g microcarrier beads, however, the doubling time of the cells is comparable in media supplemented with between 5% and 40% NCS (Fig. 2). This indicates that at this density, growth factors produced by the endothelial cells themselves become the major determinant for cell growth.

Direct Start of Microcarrier Culture

Instead of culturing endothelial cells first on dishes, they can be seeded on polystyrene microcarrier beads directly, and the use of trypsin can be completely avoided [6]. For this purpose endothelial cells are harvested from aortas as outlined above. The material isolated from ten aortas (approximately 1.5×10^7 cells) is suspended in 100 ml culture medium 2 and added together with 1 g microcarriers to a 500-ml spinner flask. The procedure is then started with step 4 of the procedure for microcarrier cultures. High concentrations of antibiotics as specified for culture medium 1 are used during the first 48 h. Typically, the cells reach confluency in 4 days, corresponding to 6×10^7 cells/g of microcarriers.

The doubling time of newly isolated cells grown directly on microcarriers is 30% slower than of cells that are first cultured on Petri dishes and subsequently trypsinized. The fact that cells trypsinized from dishes 2 and 7 days after isolation exhibit identical doubling times on microcarriers indicates that the cells grow better after a trypsinization step.

To start microcarrier cultures directly has disadvantages since the initial use of high concentrations of antibiotics and frequent media changes are more tedious and expensive than on culture dishes. Additionally, the number of viable cells in the material initially isolated is variable and difficult to determine. In contrast, the number of viable cells trypsinized from a primary dish culture can be easily and accurately determined.

Carriers Other than Polystyrene Beads

Apart from the polystyrene beads, whose properties have been described so far, a number of other microcarrier materials are available. They all have a specific gravity slightly higher than 1 g/ml so that they float easily in aqueous solutions. They differ, however, in their specific physical and chemical properties and are handled differently than polystyrene beads. Some of these materials, listed in Table 1, were tested in a comparative study.

Protocols: The same protocol as the one for polystyrene beads applies to glass microcarrier beads (Sigma). Other microcarriers, i.e., Ventregel II (Ventrex), Cytodex 3 (Pharmacia), CultiSpher-G (Percell Biolytica), must be hydrated with PBS before further use (≥ 3 h, 1 volume microcarrier plus 10 volumes PBS). They can either be autoclaved before use or incubated with high concentrations of antibiotics (e.g., as in culture medium 1) added to the PBS. After this step they are washed three times in culture medium 2. The protocol detailed for Biosilon microcarriers is then followed.

Glass Microcarriers (Sigma Type G 2767): These glass microcarrier beads from Sigma are solid beads with an average diameter slightly smaller than that of Biosilon beads. Per gram of beads the surface area is approximately 20% larger than of Biosilon beads. During exponential growth, porcine aortic endothelial cells grow on glass beads as rapidly as on Biosilon beads (Table 2).

Ventregel II (Ventrex, Type 79015): Ventregel II microcarriers consist of a solid polystyrene matrix coated with gelatin. Before use this gelatin coat must be hydrated. The physical dimensions of the hydrated beads are the same as those of Sigma glass beads type G 2767 (Table 1). The rate of growth of porcine aortic endothelial cells on these microcarriers is the same as on Biosilon and glass beads (Table 2). This agrees with the previous observation that aortic endothelial cells grow equally well on tissue grade polystyrene and gelatin [7].

Cytodex 3 (Pharmacia): Cytodex 3 microcarrier beads consist of a matrix of cross-linked dextran coated with denatured collagen I [5]. When hydrated the material swells extensively, resulting in spherical microcarriers with an average diameter 25% smaller than that of Biosilon beads. In contrast to plastic and glass beads, these microcarriers are porous.

The doubling time of porcine aortic endothelial cells on these microcarriers is the same as on Biosilon beads (Table 2). This doubling time (21 h) is nearly the same as that determined by another group [1]. Among the microcarrier types tested, Cytodex 3 is the only transparent material permitting phase-contrast microscopy of the cells growing on the microcarriers. The porosity of the bead material has been exploited in studies investigating transport through the monolayer of endothelial cells covering the bead (see, e.g., [4]). This internal fluid space is, however, poorly defined and therefore may represent a disadvantage in many applications.

CultiSpher-G (Percell Biolytica): These microcarriers are made from cross-linked gelatin. They swell extensively when hydrated. The surface of the swollen microcarriers is enlarged by large pores, into which the cells can extend. Microcarriers consisting wholly of gelatin have the useful property that they can be completely dissolved with trypsin or collagenase. Porcine endothelial cells fail, however, to grow on the commercial material CultiSpher-G even though they can grow on gelatin-coated dishes and microcarriers (Ventregel II, Table 2). Wissemann and Jacobson [9] described a procedure to prepare microcarriers of cross-linked gelatin that are suitable for endothelial microcarrier cultures.

Conclusions

A number of different microcarrier materials are suitable for establishing microcarrier cultures of porcine aortic endothelial cells. For many purposes solid polystyrene or glass beads are the materials of choice. These microcarrier materials are impermeable, neutral in most staining procedures, and easier to use than more complex materials (e.g., they do not need to be hydrated before use). For some purposes, the use of Cytodex 3 microcarriers might be advantageous since cells growing on them can be examined in phase-contrast microscopy.

When endothelial cell cultures are prepared from slaughterhouse material, frequent medium changes are required during the first days in order to remove both debris and microbial contaminants from the cultures. This is done most conveniently and with the minimal expense of media when the cells are grown on Petri dishes.

The main determinants for the growth of porcine endothelial cells on microcarriers are, first, medium composition, and, second, the initial ratio between cell number and number of microcarrier beads. Unless specified additions of growth factors are made to the culture media, porcine aortic endothelial cells are best grown with 20% NCS (FCS is not superior). When both the serum concentration and the cell-to-microcarrier ratio are low, the initial rate of cell growth is small. The results depicted in Fig. 1 indicate, however, that beyond a certain threshold density (3×10^6 cells/g Biosilon microcarriers) the velocity of cell proliferation becomes independent of the initial seeding density. With the exception of CultiSpher-G microcarriers, growth rates were nearly identical on all microcarrier materials tested.

References

1. Busch C, Cincilla PA, DeBault LE, Goldsmith JC, Owen WG (1982) Use of endothelium cultured on microcarriers as a model for the microcirculation. Lab Invest 47:498–504
2. Davies PF, Ganz P, Diehl PS (1985) Reversible microcarrier-mediated junctional communication between endothelial and smooth muscle cell monolayers: an in vitro model of vascular cell interactions. Lab Invest 85:710–718
3. Dickinson ES, Slakey LL (1982) Plasma-derived serum as a selective agent to obtain endothelial cultures from swine aorta. In Vitro 18:63–70
4. Killackey JFK, Johnston MG, Movat HZ (1986) Increased permeability of microcarrier-cultured endothelial monolayers in response to histamine and thrombin. A model for the in vitro study of increased permeability. Am J Pathol 122:50–61
5. Pharmacia Fine Chemicals (1981) Microcarrier cell culture. Principle and methods. Pharmacia Fine Chemicals, Uppsala
6. Ryan US, Mortara M, Whitaker C (1980) Methods for microcarrier culture of bovine pulmonary artery endothelial cells avoiding the use of enzymes. Tissue Cell 12:619–635
7. Schor AM, Schor SL, Allen TD (1983) Effects of culture conditions on the proliferation, morphology and migration of bovine aortic endothelial cells. J Cell Sci 62:267–285
8. Voyta JC, Via DP, Butterfield CE, Zetter BR (1984) Identification and isolation of endothelial cells based on their increased uptake of acetylated-low density lipoprotein. J Cell Biol 99:2034–2040
9. Wissemann KW, Jacobson BS (1985) Pure gelatin microcarriers: synthesis and use in cell attachment and growth of fibroblast and endothelial cells. In Vitro Cell Dev Biol 21:391–401

Growth Factors for Vascular Endothelial Cells

D. Gospodarowicz

Introduction

The search for the molecular mediators of growth control in vascular endothelial cells has led to the discovery of a family of endothelial cell growth factors (ECGFs) which differ widely in their ability to trigger endothelial cell proliferation and differentiation in an in vitro or in vivo context. Among the first vascular ECGFs to be described was the basic fibroblast growth factor (bFGF) isolated from the brain or pituitary [24, 33] and its closely related member, the acidic fibroblast factor (aFGF) [4, 58, 92]. At this date, at least nine ECGFs and angiogenic factors have been isolated and characterized (Table 1). Although all these factors have been shown to be angiogenic in vivo, they differ widely in their target cell specificity and by their ability to stimulate endothelial cell growth in vitro versus in vivo.

Growth Factors for Cultured Vascular Endothelial Cells

Basic and Acidic FGF

bFGF and aFGF are two closely related growth factors which differ in their isoelectric point, but have a high degree of structural homology [4, 13, 14, 43]. The primary translation product for both growth factors has a length of 155AA and an apparent MW of 18 kDa [1, 50]. It has been shown to be a growth factor for mesoderm- and neuroectoderm-derived cells [34, 46]. Related to bFGF, and to a lesser degree to aFGF, are a series of oncogene products: Int-2 [11], hst or k-FGF [9, 101], FGF-5 [102], and FGF-6 [8]. Of these, k-FGF and Int-2 have been shown to cross react with the FGF cell surface receptor and to trigger the proliferation of the same cell types as bFGF [10, 68].

bFGF is a potent mitogen for vascular endothelial cells derived from large vessels [33–35]. This effect was first described with vascular endothelial cells derived from bovine fetal aortic arch, or from fetal endocardium, bovine umbilical vein, and calf aortic arch [35]. In addition, the ability of FGF to act as a survival agent when cells were maintained at clonal density led to the establishment of the first clonal vascular endothelial cell strains which were strongly dependent on this mitogen [33] and exhibited increased lifespan when maintained in its presence [33–35]. bFGF is also mitogenic for human umbilical vascular endothelial cells [34]. These cells, in contrast to their bovine

Table 1. Angiogenic factors

Name	Target cells	Mode of action	In vitro effect	In vivo effect
Basic FGF	Many cell types	Direct	Strong	Strong
Acidic FGF	Many cell types	Direct	Weak	Weak
TNFα	Many cell types	Indirect (macrophages)	Cytotoxic	Strong
TGFβ	Many cell types	Indirect (macrophages)	Inhibitor	Strong
TGFα	Many cell types	Indirect-direct (macrophages)	Weak	Strong
EGF	Many cell types	Indirect-direct	Weak	Strong
PD-ECGF	Only vascular endothelial cells	Direct	Weak	Strong
FSdGF/ vasculotropin	Only vascular endothelial cells	Direct	Weak	Strong
Angiogenin		(?)	None	Strong

counterparts, are also sensitive to epidermal growth factor (EGF) [34, 60]. The proliferative effect caused by either EGF and FGF can also be potentiated by thrombin [34]. The response of capillary-derived endothelial cells to FGF is identical to that of endothelial cells derived from large vessels [44, 45]. This has led to the long-term establishment in culture of various vascular endothelial cell strains derived from the microvasculature of organs as different as the testis, fetal ovarian cortex, corpus luteum, adrenal cortex, brain, lung, thymus, and retina [44, 45].

In addition to being a mitogen for vascular endothelial cells, bFGF is also a morphogen for that cell type. This is a particularly interesting characteristic of FGF, since it has made possible the long-term culturing of cells, which otherwise would lose their normal phenotypes in culture when passaged repeatedly at low cell density [52, 97].

This biological effect of FGF has been best studied using vascular endothelial cell cultures cloned and maintained in the presence of FGF and then deprived of it for various periods [25, 26, 96, 97]. It has been related to the ability of FGF to control the synthesis of various basement lamina components [32, 36, 38, 39] which would in turn affect cell polarity and gene expression [29, 40, 41].

In the case of capillary-derived endothelial cells, bFGF has been reported to cause them to invade a three-dimensional collagen matrix and to promote the organization of the cells into tubules that resemble blood capillaries [64]. Concomitantly, bFGF stimulates vascular endothelial cells to produce a urokinase-type plasminogen activator (PA), a protease that has been implicated in the neovascular response [64, 70]. Thus, bFGF can stimulate processes that are characteristic of angiogenesis in vivo, including endothelial cell migration, invasion, and PA production [64, 70].

aFGF has much of the same effect as its basic counterpart. This reflects the interaction of both mitogens with the same cell-surface receptor [66]. Interestingly enough, they differ by their potency, which for aFGF, depending on the endothelial cell strains considered, can be 30- to 100-fold less than that of bFGF [26a, 43, 44].

Growth factors are known to require progression factors in order to support cell proliferation. In the case of vascular endothelial cells, the mitogenic effect of FGF was reported to be dependent on the serum or plasma concentration to which cells were exposed [34, 35]. This suggests that progression factors present in plasma or serum would complement the biological effects of FGF and that by limiting their concentration one would limit the effect of optimal FGF concentration. These progression factors are represented by transferrin and high density lipoproteins (HDLs) [89]. In their presence, vascular endothelial cells exposed to FGF can be serially passaged under serum-free conditions and exhibit all of the properties of counterpart cultures passaged in the presence of serum [90]. A requirement for transferrin is related to the fact that iron is an essential requirement of the enzyme ribonucleotide reductase, whose activity in turn is strongly related with the rate of RNA synthesis and is greatly increased during the S phase [91]. The requirement for HDL was shown in later studies to be related to its ability to stimulate the cellular enzyme HMG CoA reductase [41]. This results in increased mevalonate synthesis which is further used for the synthesis of isopentenyl adenosine, ubiquinone, dolichols, and as yet unidentified isoprenoid pyrophosphate compounds. Recent evidence suggests that cultured cells incorporate these unidentified isoprenoid groups into unique proteins that could play an important role in the maintenance of cell shape as well as in DNA synthesis [76, 83]. Thus, it is clear that changes in the synthesis of isoprenoid compounds could have an impact on cellular metabolism at many different levels. In addition to its ability to activate HMG CoA reductase, HDL may also support cell growth by providing choline and fatty acids [7].

Folliculostellate Derived Growth Factor (FSdGF)/Vasculotropin, and Platelet Derived Endothelial Cell Growth Factor (PD-ECGF)

FSdGF has been isolated from the condition medium of bovine pituitary derived folliculostellate cells [47], while vasculotropin has been isolated from the condition medium of mouse-derived pituitary cell line AtT-20 [69]. Both factors have been shown to be closely related and represent, respectively, the bovine and rodent form. This growth factor is a homodimer composed of 2 subunits with an apparent MW of 23 kDa [47, 69]. FSdGF is mitogenic in vitro for endothelial cells derived either from large vessels or from capillaries, and is angiogenic in vivo. In contrast to bFGF or aFGF, it does not stimulate the proliferation of cell types unrelated to vascular endothelial, not even that of vascular smooth muscle cells [47, 69]. As such, FSdGF seems to have a unique target cell specificity shared only by another vascular endothelial mitogen called PD-ECGF. This growth factor is an endothelial cell mitogen of relative molecular mass of 45 kDa purified to homogeneity from human platelets [49]. In contrast to other endothelial mitogens of the fibroblast growth (FGF) family [reviewed in 43, 57], PD-ECGF, which seems to be the only ECGF in human platelets, does not bind to heparin and does not stimulate the proliferation of fibroblasts [63]. In addition to being mitogenic for vascular endothelial cells, PD-ECGF is also chemotactic for vascular endothelial cells and is angiogenic in vivo [49]. The restricted target cell specificity of PD-ECGF and the fact that platelets are a main source of it, suggest that PD-ECGF has a role in maintaining the integrity of the blood vessels.

Epidermal Growth Factor (EGF)
and Transforming Growth Factor Alpha (TGFα)

Secreted TGFα exists as multiple species, ranging in apparent size from 5–20 kDa. The smallest form is 50 amino acids long and shares about 30% structural similarity with the 53 residue long EGF, including the conservation of all six cysteines [61]. This sequence relationship and the presumed formation of three similar disulfide bridges in both molecules provide a molecular explanation for the interaction of the two growth factors with the same cellular receptor [12, 62].

Both TGFα and EGF have been shown to stimulate the proliferation of vascular endothelial cells [77]. However, their effect differs, depending on the species from which vascular endothelial cells are derived, or by their territorial origin. For example, human umbilical endothelial cells respond to both TGFα and EGF [34, 35, 60, 77]. However, bovine vascular endothelial cells derived from either large vessels or capillaries do not respond to either EGF or TGFα [35, 43, 44], while mouse capillary endothelial cells seem to respond to both [72, 77]. These differences in cell responsiveness correlate with the presence of absence of functional EGF cell-surface receptors [34, 35].

Growth Potentiators and Inhibitors for Cultured Vascular Endothelial Cells

Growth Potentiators

Extracellular Matrix Components

The subendothelial extracellular matrix (ECM) is a determining factor in the modulation of the behavior of endothelial cells. It is, therefore, not surprising that the substrate upon which endothelial cells are maintained in vitro will modulate the response of vascular endothelial cells to serum growth factors or to their own autocrine growth factor [31, 80, 81]. The ECM produced by either vascular or corneal endothelial cells relieves FGF-responsive cells from their bFGF requirement [30]. This effect is now recognized to be due to the presence of bFGF-heparan sulfate-proteoglycan complexes which are an integral part of the ECM produced by cultured vascular or corneal endothelial cells [16, 98]. In addition, single isolated components of ECM have been shown to have a positive effect on cell proliferation. This includes gelatin (a denatured form of collagen), and fibronectin, which improved the growth rate of both human umbilical-derived endothelial cells as well as that of bovine capillary endothelial cells [42–44, 59]. In vitro fibronectin, alone or in combination with heparin, stimulates the migration of vascular endothelial cells [5, 95]. Similar observations were made by Nicosia and Madri [67] using plasma dots.

Heparin

Heparin and its closely related glycosaminoglycan, heparan sulfate, have both negative and positive effects on cell growth, depending on the cell type tested. Heparin has been shown to potentiate by 100-fold the mitogenic activity of aFGF when tested on HUE cells [23, 44], thus making it as potent as bFGF. The ability of heparin to potentiate the mitogenic effect of aFGF on HUE cells could reflect its ability to stabilize the tertiary structure of aFGF. It also reflects a lack of direct growth inhibitory effect of heparin on that cell type. However, when cells are growth inhibited by heparin, such as in the case of capillary endothelial cells, no such potentiation of the mitogenic effect of aFGF by heparin can be seen [44, 45]. Instead, strong inhibitory effects are observed, probably reflecting a direct effect of heparin itself on the cell metabolism.

In recent studies, heparin has been shown to protect both bFGF and aFGF from acid or heat inactivation [28]. It has been proposed that the high susceptibility of the FGFs to denaturing conditions could originate in the fact that none of the cysteine residues identified in their primary sequence [22] participate in disulfide bridging. Since bFGF and aFGF have extremely high affinities for heparin, one could speculate that at physiological salt concentration, one molecule of glycosaminoglycan could bind to the two heparin binding domains of FGF; this would result in cyclization of the FGF molecule, and would stabilize its tertiary structure, thus making it both heat and acid resistant [28]. The protective effect of heparin on the biological activity of FGF could have physiological significance. In previous studies, it was shown that even at 37°C, and at neutral pHs, the in vitro half-life of FGF is only 24 h [100]. It was surmised that the interaction of heparin or heparan sulfate with acidic or basic FGF should therefore result in greatly extending their biological half-life. It has [16].

When bFGF is bound to either heparin or heparan sulfate, it is also protected against degradation by trypsin or plasmin [74]. Heparin has also been shown to protect aFGF from plasmin degradation [86]. Heparin interaction may be important in stabilizing bFGF to proteases generated during wound healing and tumor growth: processes in which bFGF may play a role [74, 86].

Growth Inhibitors

Extracellular Matrix Components

Among the naturally produced ECMs which inhibit endothelial cell proliferation is the matrix produced by the teratocarcinoma cell line HR-9 and that isolated from EHS chondrosarcoma. In both cases, these matrices promote endothelial cell differentiation and reorganization in structures which look like capillary tubes (Martin, personal communication). Since those matrices have in common the presence of collagen type IV, laminin, and heparan sulfate PG, those components are likely to encourage cell differentiation at the expense of their proliferation.

Transforming Growth Factor Beta (TGFβ)

In the case of vascular endothelial cells, TGFβ is a potent inhibitor of both bFGF and aFGF [2, 19]. This effect is both dose dependent and characteristic of a noncompetitive interaction. TGFβ also inhibits bFGF induced motility of vascular endothelial cells in collagen matrices [65] and the FGF-dependent increase in plasminogen activator [73]. Subsequently, TGFβ blocks angiogenesis in vitro [65]. The above studies, which demonstrate that TGFβ and FGF can interact at the cellular level to modulate growth and differentiation, also suggest that many of the biological activities of FGF observed in vitro may be regulated by the presence of TGFβ and related proteins in the local cellular milieu.

Tumor Necrosis Factor Alpha (TNFα)

TNFα is known to induce hemorrhagic necrosis of solid tumors. When tested on capillary endothelial cells, TNF inhibited the proliferation of cells exposed to serum alone, but was not cytotoxic. In contrast, when added to cells stimulated by bFGF, considerable cytotoxic effect in addition to growth arrest was observed [20, 79]. TNF also exerts growth inhibitory and cytotoxic actions against vascular endothelial cells cultivated on the various basement membrane components, such as ECM secreted by bovine capillary endothelial cells, fibronectin, laminin, and type IV collagen [75].

Interferon (IFN)

IFN inhibits the proliferative response of vascular endothelial cells to aFGF [21]. This is in agreement with previous studies, which have shown that IFNγ is a potent antiproliferative agent for normal and transformed cells in vitro [94]. Interestingly, FGF has been reported to stimulate IFNγ production [51], suggesting that this factor could, in an autocrine fashion, autoregulate the proliferation of bFGF sensitive cells. It has been suggested that IFNγ could inhibit endothelial cell proliferation through a mechanism which involves growth factor receptor modulation [21].

Angiogenic Factors

The sprouting of capillaries from preexisting vessels, a phenomenon termed "angiogenesis" and distinct from vasculogenesis [71], involved a series of sequential steps. These are local degradation of basement membrane and outward migration of endothelial cells, followed by their proliferation and reassociation within a capillary structure [18]. All of the vascular ECGFs known to trigger endothelial cell proliferation and differentiation in vitro have also been shown to be angiogenic in vivo. This includes bFGF [37], aFGF [93], EGF and TGFα [37, 77], PD-ECGF [49], and FSdGF, also called vasculotropin [69]. Of these, only the FGFs are known to control, in an in vitro system, all sequential steps involved in angiogenesis [47]. Since the angiogenesis triggered by the factors cited above does not involve inflammation, it is likely to reflect their direct effect on endothelial cell proliferation. In addition to these factors, a number of other factors have been described which are angiogenic in vivo, although they have either no mitogenic effect in vitro on vascular endothelial cell proliferation, or are growth inhibitors or even cytotoxic [17].

Angiogenin, a polypeptide first isolated from the conditioned medium of a human adenocarcinoma cell line, is a potent stimulator of angiogenesis [15]. The polypeptide is a single chain with a molecular weight of 14.4 kDa and a pI of 9.5 [88]. Structural studies indicate that angiogenin has a 35% absolute sequence homology to a family of pancreatic ribonucleases. While angiogenin is inactive toward the more conventional substrates of ribonuclease such as wheat germ RNA, poly (C), poly (U), and RNA-DNA hybrids, it does cleave 28S and 18S ribosomal RNA to relatively large products of 100–500 nucleotides in length [82]. It is not known whether the ribonucleolytic activity of angiogenin is involved in the mechanism of angiogenesis.

TGFβ, an inhibitor of vascular endothelial cell growth in vitro, is also angiogenic in vivo [87]. This could, however, be an indirect effect of TGFβ, due to induced fibrosis, as well as to the chemotactic effect of TGFβ on human peripheral blood monocytes [99], which may subsequently release their own angiogenic factors, including bFGF [3]. Similarly, TNFγ, which in vitro, and depending on the status of vascular endothelial cell cultures, has either no mitogenic effect or is either a growth inhibitor or cytotoxic, is angiogenic in vivo [56, 65]. This angiogenic effect of TNF may be due to its ability to act as chemoattractant for leukocytes, as reflected by the inflammatory response that is seen in cornea after TNF implantation [56]. The apparently paradoxical effect of TGFβ and TNF in vitro versus in vivo can be resolved if there are multiple pathways for capillary formation. Neovascularization, for example, could be accomplished by a polypeptide such as FGF, which can "directly" stimulate capillary endothelial cells to undergo one or more of the key steps in capillary growth delineated above. Similarly, angiogenesis triggered by TGFβ

or TNF could be mediated through an intermediate cell, for which the macrophage is a good candidate [18].

Practical Aspects of the Use of Growth Factors to Culture Endothelial Cells

Growth Factor for Endothelial Cells

As seen in Table 1, growth factors for endothelial cells seem numerous. If one makes exception of those acting only in vivo, but not in vitro, they belong to three classes: the FGF, and EGF classes, and a new class mostly made up of PD-ECGF and vasculotropin (also called FSdGF).

Among these three classes, the FGF class has been the most used to grow vascular endothelial cells in vitro and to establish vascular endothelial cell strains. Although much confusion was originally created by the report that numerous ECGFs having the biological properties of FGF had in common the property of binding to heparin, it is now generally accepted that all of the heparin-binding endothelial cell mitogens described to date and other uncharacterized growth factors related to either bFGF or aFGF are the products of a single bFGF or aFGF gene [reviewed in 46].

Since both mitogens interact with the same cell-surface receptor, their biological properties are similar if not identical. Differences in specific activities have been observed, however, bFGF being 30- to 100-fold more potent that aFGF in triggering cell growth [26a]. The activity of aFGF can be potentiated by heparin. However, heparin itself is a highly heterogeneous compound, and there is considerable variability in the biological potency of different commercial preparations of heparin [6]. This, together with its intrinsic growth-inhibitory properties, make it difficult to use with specific vascular endothelial cell strains [44, 45]. Particular care should be taken in calibrating each particular lot of heparin. This has made the use of aFGF less popular than that of bFGF, the more so since recombinant aFGF has no activity unless heparin is present [56a]. Therefore, bFGF, which does not require the presence of heparin in order to be fully active (ED_{50}, 30–80 pg/ml), seems a better choice, if the ultimate goal is to grow vascular endothelial cells.

When cultured vascular endothelial cells, which themselves produce bFGF [80], are seeded at low or clonal density, they do not have time to condition their medium; it is then that the effect of added exogenous bFGF on their morphology, migration, and proliferation can best be seen.

The various FGF commercial preparations available at present range from crude to highly purified. The term "ECGF" covers a crude extract of brain which contains both bFGF and aFGF [55]. Of the two components, bFGF, because of its higher biological activity, represents the major component. It is used in the range of 50–100 µg/ml in contrast with purified brain-derived aFGF which is used at 30–100 ng/ml or pituitary derived bFGF which is used at 0.2–1 ng/ml. Although recombinant aFGF or bFGF are available in abundance, without the addition of heparin aFGF is totally inactive, while bFGF is 5- to 10-fold less potent than pituitary-derived bFGF. This likely reflects the difference in folding imposed on the molecule when synthesized by bacteria and yeast versus eukaryotic cells. This improper folding can be corrected by heparin to some extent. This is reflected by the increased potency of recombinant aFGF or bFGF when tested in the presence of heparin [56a]. It is likely that once they are expressed in bacteria or eukaryotic cells, the two FGF-related oncogenes, Int-2 and k-FGF, could also be used to grow endothelial cells.

Of the two other classes of endothelial cell growth factor, the EGF class has been the object of only cursory studies. Among the pitfalls in its use are (a) the small extent of the proliferative response as compared with FGF and (b) the variability in the response encountered between vascular territories and cell species. The last class of endothelial cell mitogens is composed of vasculotropin and PD-ECGF. These are the newest ECGFs as yet identified and they are still under evaluation by various groups. While less active than bFGF, their most attractive features are the uniqueness of their target cells, since only vascular endothelial cells seem to respond to them. Future research will tell whether this unique target cell specificity is reflected in some unique physiological properties or in tissue distribution and expression.

Substrate and Serum

The ability of bFGF to support cell proliferation is also a function of substrate and serum lot [31]. In some cases, batches of serum are strongly cytotoxic or do not support the mitogenic effect of FGF. This ability is also a function of substrate. For example, human umbilical endothelial cells respond better to FGF when maintained on gelatinized dishes, but not on plastic ones [42]; cell strains derived from bovine origin are fairly indifferent to this substrate, while rodent-derived endothelial cells require gelatinized or fibronectin-coated dishes in order to grow at optimal rates (unpublished observation).

Conclusions

Growth factors have been used to grow vascular endothelial cells in vitro for the past 14 years; the first report describing their use established a clonal vascular endothelial cell strain dating back to 1976 [33]. Although considerable concerns were raised about the possibility that by using growth factors one could select for a specific cell variant representing a subset of the original cell population [78], such concerns are no longer valid. Since vascular endothelial cells have the ability to produce bFGF [80], endothelial cells exposed to exogenous bFGF are, in fact, provided with their own autocrine growth factor. Such cells have, in their late passage, most of the phenotypic characteristics seen in primary culture. This is due in part to the fact that FGF is a morphogen as well as a mitogen for vascular endothelial cells, preventing them from entering a dedifferentiated fibroblastic status. These dual properties of FGF may reflect its biological functions, in early embryos. In blastulae, it has been shown to act as a morphogen, inducing early animal pole cells destined to form ectodermal structure to form ventromesodermal structures instead – among them, the vascular endothelium [48, 53, 54, 84, 85; reviewed in 27].

While FGF does represent the first identified ECGF, others may exist as well. These could control specific events taking place during vasculogenesis or angiogenesis. The rapid pace of research in the field of growth factors may enable us, quite soon, to reevaluate the role of specific ECGFs.

References

1. Abraham JA, Mergia Whang JL, Tumolo A, Friedman J, Hjerrild KA, Gospodarowicz D, Fiddes JC (1986) Nucleotide sequence of a bovine clone encoding the angiogenic protein, basic fibroblast growth factor. Science 233:545–548
2. Baird A, Durkin T (1986) Inhibition of endothelial cell proliferation by transforming growth factor: interactions with acidic and basic fibroblast growth factors. Biochem Biophys Res Commun 138:476–482
3. Baird A, Mormede P, Böhlen P (1985) Immunoreactive fibroblast growth factor in cells of peritoneal exudate suggest its identity with macrophage derived growth factor. Biochem Biophys Res Commun 126:358–364
4. Böhlen P, Esch F, Baird A, Gospodarowicz D (1985) Acidic fibroblast growth factor from bovine brain. Amino terminal sequence and comparison to basic fibroblast growth factor. EMBO J 4:1951–1956
5. Bowersox JC, Sorgente N (1982) Chemotaxis of aortic endothelial cells in response to fibronectin. Cancer Res 42:2547–2551
6. Castellot JJ, Choay J, Lormeau JC, Petitou M, Sache E, Karnovsky MJ (1984) Structural determinants of the capacity of heparin to inhibit the proliferation of vascular smooth muscle cells. J Cell Biol 102:1979–1984, 1986
7. Cohen DC, Gospodarowicz D (1985) Biotin and choline replace the growth requirement of Madin-Darby canine kidney cells for high density lipoproteins. J Cell Physiol 124:96–106
8. De Lapeyrière D, Marics I, Adelaïdem J, Raybaud F, Courlier F, Rosnet O, Benharrach D, Mattei MG, Birnbaum D (1989) Isolation and characterization of fibroblast growth factor genes. J Cell Biochem 13B:153
9. Delli-Bovi P, Curalota AM, Newman KM, Sato Y, Moscatelli D, Hewick RM, Rifkin D, Basilico C (1987) Processing, secretion, and biological properties of a novel growth factor of the fibroblast growth factor family with oncogene potential. Mol Cell Biol 8:2933–2941
10. Delli-Bovi P, Curalota AM, Kern FG, Greco A, Ittmann M, Basilico C (1988) An oncogene isolated by transfection of Karposi's sarcoma DNA encodes a growth factor that is a member of the FGF family. Cell 50:729–737
11. Dickson C, Peters G (1987) Potential oncogene product related to growth factors. Nature 326:833
12. Derynk R (1988) Transforming growth factor α. Cell 54:593–595
13. Esch F, Ueno N, Baird A, Hill F, Denoroy L, Gospodarowicz D, Guillemin R (1985) Primary structure of bovine brain acidic fibroblast growth factor (FGF). Biochem Biophys Res Commun 133:554–562
14. Esch F, Baird A, Ling N, Ueno N, Hill F, Denoroy L, Klepper R, Gospodarowicz D, Böhlen P, Guillemin R (1985) Primary structure of bovine pituitary basic fibroblast growth factor (FGF) and comparison with amino terminal sequence of bovine brain acidic FGF. Proc Natl Acad Sci USA 82:6507–6511
15. Fett JW, Strydom S, Lobb RR, Bethune JL, Alderman EM, Riordan JF, Vallee BL (1985) Isolation and characterization of angiogenin, an angiogenic protein from human carcinoma cells. Biochemistry 24:5480–5485

16. Flaumenhaft R, Moscatelli D, Saksela O, Rifkin DB (1989) Role of extracellular matrix in the action of bFGF: matrix as a source of growth factor for long term stimulation of plasminogen activator production and DNA synthesis. J Cell Physiol 140:75–81
17. Folkman J, Klagsbrun M (1987) Angiogenic factors. Science 235:442–447
18. Folkman J, Klagsbrun M (1987) A family of angiogenic peptides. Nature 329:671–672
19. Frater-Schröder M, Muller G, Burchmeier W, Böhlen P (1986) Transforming growth factor β inhibits endothelial cell proliferation. Biochem Biophys Res Commun 137:295–302
20. Frater-Schröder M, Risau W, Hallman R, Gautschi P, Böhlen P (1987) Tumor necrosis factor type a, a potent inhibitor of endothelial cell growth in vitro, is angiogenic in vivo. Proc Natl Acad Sci USA 84:5277–5281
21. Friesel R, Komoriya A, Maciag T (1987) Inhibition of endothelial cell proliferation by gamma interferon. J Cell Biol 104:689–696
22. Gautschi-Sova P, Muller T, Böhlen P (1986) Amino acid sequence of human acidic fibroblast growth factor. Biochem Biophys Res Commun 140:874–880
23. Gimenez-Gallego G, Conn G, Hatcher VB, Thomas KA (1986) Human brain derived acidic and basic fibroblast growth factors: amino terminal sequences and specific mitogenic activities. Biochem Biophys Res Commun 135:541–548
24. Gospodarowicz D (1974) Localization of a fibroblast growth factor and its effect alone and with hydrocortisone on 3T3 cell growth. Nature 240:123–127
25. Gospodarowicz D (1983) Control of endothelial cell proliferation and repair. In: Cryer A (ed) Biochemical interaction at the endothelium. Elsevier/North-Holland, New York, pp 366–396
26. Gospodarowicz D (1983) The control of mammalian cell proliferation by growth factors, basement lamina, and lipoproteins. J Invest Dermatol 81:405–505
26 a. Gospodarowicz D (1987) Purification of brain and pituitary FGF. Methods Enzymol 147 B:106–119
27. Gospodarowicz D (1990) Fibroblast growth factor and its involvement in developmental processes. Curr Top Dev Biol 24:57–93
28. Gospodarowicz D, Cheng J (1986) Heparin protects basic and acidic FGF from inactivation. J Cell Physiol 128:475–484
29. Gospodarowicz D, Greenburg G (1981) Growth control of mammalian cells, growth factors and extracellular matrix. In: Ritzen M, Aperia A, Hall K, Larsson A, Zetterberg A, Zetterström R (eds) The biology of normal human growth. Raven, New York, pp 1–21
30. Gospodarowicz D, Ill CR (1980) The extracellular matrix and the control of proliferation of vascular endothelial cells. J Clin Invest 65:1351–1364
31. Gospodarowicz D, Lui G-M (1981) Effect of substrata and fibroblast growth factor on the proliferation in vitro of bovine aortic endothelial cells. J Cell Physiol 109:69–81
32. Gospodarowicz D, Tauber J-P (1980) Growth factors and extracellular matrix. Endocr Rev 1:201–227
33. Gospodarowicz D, Moran JS, Braun DL, Birdwell C (1976) Clonal growth of bovine vascular endothelial cells: fibroblast growth factor as a survival agent. Proc Natl Acad Sci USA 73:4120–4124
34. Gospodarowicz D, Brown KS, Birdwell CR (1978) Control of proliferation of human vascular endothelial cells. I. Characterization of the response of human umbilical vein endothelial cells to fibroblast growth factor, and thrombin. J Cell Biol 77:774–788
35. Gospodarowicz D, Greenburg G, Bialecki H (1978) Factors involved in the modulation of cell proliferation in vivo and in vitro: the role of fibroblast and epidermal growth factors in the proliferative response of mammalian cells. In Vitro 14:85–118
36. Gospodarowicz D, Greenburg G, Birdwell CR (1978) Determination of cellular shape by the extracellular matrix and its correlation with the control of cellular growth. Cancer Res 38:4155–4171
37. Gospodarowicz D, Bialecki H, Thakral TK (1979) The angiogenic activity of the fibroblast and epidermal growth factor. Exp Eye Res 28:501–514

38. Gospodarowicz D, Vlodavsky I, Greenburg G, Alvarado J, Johnson LK, Moran J (1979) Cellular shape is determined by the extracellular matrix and is responsible for the control of cellular growth and function. Cold Spring Harbor Conf Cell Proliferation 9:561–592
39. Gospodarowicz D, Vlodavsky I, Greenburg G, Alvarado J, Johnson LK, Moran J (1979) Studies on atherogenesis and corneal transplantation using cultured vascular and corneal endothelia. Recent Prog Horm Res 35:375:448
40. Gospodarowicz D, Vlodavsky I, Savion N (1981) The role of fibroblast growth factor and the extracellular matrix in the control of proliferation and differentiation of corneal endothelial cells. Vision Res 21:87–103
41. Gospodarowicz D, Cohen DC, Fujii DK (1982) Regulation of cell growth by the basal lamina and plasma factors: relevance to embryonic control of cell proliferation and differentiation. Cold Spring Harbor Conf Cell Proliferation 9:95–124
42. Gospodarowicz D, Cheng J, Lirette M (1983) Bovine brain and pituitary fibroblast growth factors: comparison of their abilities to support the proliferation of human and bovine vascular endothelial cells. J Cell Biol 97:1677–1685
43. Gospodarowicz D, Neufeld G, Schweigerer L (1986) Fibroblast growth factor. Mol Cell Endocrinol 46:187–204
44. Gospodarowicz D, Massoglia S, Cheng J, Fujii DK (1986) Effect of fibroblast growth factor and lipoproteins on the proliferation of endothelial cells derived from bovine adrenal cortex, brain cortex, and corpus luteum capillaries. J Cell Physiol 127:121–136
45. Gospodarowicz D, Massoglia S, Cheng J, Fujii DK (1986) Effect of retina derived acidic and basic fibroblast growth factor and lipoproteins on the proliferation of bovine derived capillary endothelial cells. Exp Eye Res 63:459–476
46. Gospodarowicz D, Ferrara N, Schweigerer L, Neufeld G (1987) Structural characterization and biological functions of fibroblast growth factor. Endocr Rev 8:95–114
47. Gospodarowicz D, Abraham JA, Schilling J (1989) Isolation and characterization of a newly identified endothelial cell mitogen factor produced by pituitary derived folliculo stellate cells. Proc Natl Acad Sci USA 86:7311–7315
48. Grunz M, McKeehan WL, Knochel W, Born J, Tiedemann H, Tiedmann H (1988) Induction of mesodermal tissues by acidic and basic FGF. Cell Differ 22:183–190
49. Ishikawa F, Miyazone K, Hallman U, Drexler H, Wernstedt C, Hagiwara K, Usuki K, Takaku F, Risau W, Heldin C-H (1989) Identification of angiogenic activity and the cloning and expression of platelet derived endothelial cell growth factor. Nature 338:557–562
50. Jaye M, How KR, Burgess W, Ricca GA, Chiu IM, Ravera MW, O'Brien SJ, Modi WS, Maciag T, Drohan WN (1986) Human endothelial cell growth factors: cloning, nucleotide sequence, and chromosome localization. Science 233:541–545
51. Johnson HM, Torres A (1985) Peptide growth factors PDGF, EGF, and FGF regulate interferon production. J Immunol 134:2824–2826
52. Kato Y, Gospodarowicz D (1985) Sulfated proteoglycan synthesis by rabbit costal chondrocytes grown in the presence and absence of fibroblast growth factor. J Cell Biol 100:477–485
53. Kimmelman D, Kirschner M (1987) Synergistic induction of mesoderm by FGF and TGFβ and the identification of an mRNA coding for FGF in the early Xenopus embryo. Cell 51:869–877
54. Kimmelman D, Abraham JA, Haaparanta T, Palisi TM, Kirschner M (1988) The presence of fibroblast growth factor in the frog egg: its role as a natural mesoderm inducer. Science 242:1053–1056
55. Klagsbrun M, Shing Y (1985) Heparin affinity of anionic and cationic capillary endothelial cell growth factors: analysis of hypothalamus derived growth factors and fibroblast growth factors. Proc Natl Acad Sci USA 82:805–809
56. Leibovich SJ, Polverini PJ, Shepard HM, Wiseman DM, Shively V, Nuseir N (1987) Macrophage induced angiogenesis is mediated by tumor necrosis α. Nature 329:630–632

56a. Linemayer DL, Kelly LJ, Menke JG, Gimenez-Gallego G, DiSalvo J, Thomas KA (1987) Expression in *Escherichia coli* of a chemically synthesized gene for biologically active bovine fibroblast growth factor. Biotechnology 5:960–965

57. Lobb RR, Harper JW, Fett JW (1986) Purification of heparin binding growth factors. Anal Biochem 154:1–14

58. Maciag T (1984) Angiogenesis. Prog Hemost Thromb 7:167–182

59. Maciag T, Hoover GA, Stemerman MB, Weinstein R (1981) Serial propagation of human endothelial cells in vitro. J Cell Biol 91:420–540

60. Maciag T, Hoover GH, van der Speck J, Stemerman MB, Weinstein R (1982) Growth and differentiation of HUE cells in culture. Cold Spring Harbor Conf Cell Proliferation 9:525–540

61. Marquardt H, Huntkapiller MW, Hood LE, Todaro GJ (1984) Rat transforming growth factor type 1: structure and relation to epidermal growth factor. Science 223:1073–1082

62. Massagué J (1983) Epidermal growth factor-like transforming growth factor. II. Interaction with epidermal growth factor receptors in human placenta membranes and A431 cells. J Biol Chem 258:13614–13620

63. Miyazono K, Okabe T, Urabe A, Takaku F, Heldin C-H (1987) Purification and properties of an endothelial cell growth factor from human platelets. J Biol Chem 262:4098–4103

64. Montesano R, Vassali JD, Baird A, Guillemin R, Orci L (1986) Basic fibroblast growth factor induces angiogenesis in vitro. Proc Natl Acad Sci USA 83:7297–7301

65. Muller G, Behrens J, Nussbaumer U, Böhlen P, Birchmeier W (1987) Inhibitory action of transforming growth factor β on endothelial cells. Proc Natl Acad Sci USA 84: 5600–5604

66. Neufeld G, Gospodarowicz D (1986) Basic and acidic fibroblast growth factor interacts with the same cell surface receptor. J Biol Chem 261:5631–5637

67. Nicosia RF, Madri JA (1987) The microvascular extracellular matrix: developmental changes during angiogenesis in the aortic ring-plasma clot model. Am J Pathol 128:78–90

68. Paterno GD, Gillespie LL, Dixon MS, Slack JMW, Heath JK (1989) Mesoderm inducing properties of Int-2 and kFGF: 2 oncogene encoded growth factors related to FGF. Development 106:79–83

69. Plouët J, Schilling J, Gospodarowicz D (1989) Isolation and characterization of a newly identified endothelial cell mitogen produced by pituitary derived AtT-20 cells. EMBO 8:3801–3806

70. Presta M, Moscatelli D, Joseph-Silverstein J, Rifkin DB (1986) Purification from a human hepatoma cell line of a basic fibroblast growth factor-like molecule that stimulates capillary endothelial cell plasminogen activator production, DNA synthesis, and migration. Mol Cell Biol 6:4060–4066

71. Risau W, Lemmon V (1988) Changes in the extracellular matrix during embryonic vasculogenesis and angiogenesis. Dev Biol 125:441–450

72. Robinson RA, Teneyck CJ, Hart MN (1986) Growth control in cerebral microvessel derived endothelial cells. Brain Res 384:114–120

73. Saksela O, Moscatelli D, Sommer A, Rifkin DB (1987) The opposing effects of basic fibroblast growth factor and transforming growth factor beta on the regulation of plasminogen activator activity in capillary endothelial cells. J Cell Biol 105:957–963

74. Saksela O, Moscatelli D, Sommer A, Rifkin DB (1988) Endothelial cell derived heparan sulfate binds basic fibroblast growth factor and protects it from proteolytic degradation. J Cell Biol 107:743–751

75. Sato N, Fukuda K, Nariuchi H, Sagara N (1987) Tumor necrosis factor inhibiting angiogenesis in vitro. JNCI 79:1383–1392

76. Schnitzer-Polokoff, von Guten C, Logel J, Torget R, Sinensky M (1982) Isolation and characterization of a mammalian cell mutant defective in 3-hydroxyl-3-coenzyme A synthase. J Biol Chem 275:472–476

77. Schreiber AB, Winkler ME, Derynck R (1986) Transforming growth factor α: a more potent angiogenic mediator than epidermal growth factor. Science 232:1250–1253
78. Schwartz SM, Gajdusek CM, Selden SC III (1981) Vascular wall growth control: the role of the endothelium. Arteriosclerosis 1:107–126
79. Schweigerer L, Malerstein B, Gospodarowicz D (1987) Tumor necrosis factor inhibits the proliferation of cultured capillary endothelial cells. Biochem Biophys Res Commun 143:997–1004
80. Schweigerer L, Neufeld G, Friedman J, Abraham JA, Fiddes JC, Gospodarowicz D (1987) Basic fibroblast growth factor is expressed in capillary endothelial cells. Nature 325:257–259
81. Schweigerer L, Ferrara N, Neufeld G, Gospodarowicz D (1988) Basic fibroblast growth factor: expression in cultured cells derived from corneal endothelium and lens epithelium. Exp Eye Res 46:71–80
82. Shapiro R, Riordan J, Valee BL (1986) Characteristic ribonucleolytic activity of human angiogenin. Biochemistry 25:3527–3532
83. Sinensky M, Logel J (1985) Defective macromolecule biosynthesis and cell cycle progression in a mammalian cell starved for mevalonate. Proc Natl Acad Sci USA 82:3257–3261
84. Slack JMW, Isaacs HV (1989) Presence of basic fibroblast growth factor in the early Xenopus embryo. Development 105:147–153
85. Slack JMW, Darlington BG, Heath JK, Godsave SF (1987) Mesoderm induction in early Xenopus embryos by heparin binding growth factors. Nature 326:197–200
86. Sommer A, Rifkin DB (1989) Interaction of heparin with human basic fibroblast growth factor: protection of the angiogenic protein from proteolytic degradation by a glycosaminoglycan. J Cell Physiol 138:215–220
87. Sporn MB, Assoian DK, Smith JH, Roche NS, Wakefield LM, Heine UI, Liotta LA, Falanga V, Kehrl JH, Fauci AS (1986) TGFβ: rapid induction of fibrosis and angiogenesis in vivo and stimulation of collagen formation in vitro. Proc Natl Acad Sci USA 83:4167–4172
88. Strydom DJ, Fett JW, Lobb RR, Alderman EM, Bethune JL, Riodan JF, Vallee BL (1985) Amino acid sequence of human tumor derived angiogenin. Biochemistry 24:5486–5494
89. Tauber J-P, Cheng J, Gospodarowicz D (1980) The effect of high low density lipoproteins on the proliferation of vascular endothelial cells. J Clin Invest 66:696–708
90. Tauber J-P, Cheng J, Massoglia S, Gospodarowicz D (1981) High density lipoproteins and the growth of vascular endothelial cells in serum free medium. In Vitro 17:519–530
91. Thelander L, Reichard P (1979) Transferrin. Annu Rev Biochem 48:133–147
92. Thomas KA, Rios-Candelore M, Fitzpatrick S (1984) Purification and characterization of acidic fibroblast growth factor from bovine brain. Proc Natl Acad Sci USA 81:357–361
93. Thomas KA, Rios-Candelore M, Gimenez Gallego G, DiSalvo J, Bennett C, Rodkey J, Fitzpatrick S (1985) Pure brain derived acidic fibroblast growth factor is a potent angiogenic vascular endothelial cell mitogen with sequence homology to interleukin 1. Proc Natl Acad Sci USA 82:6409–6413
94. Ucer U, Bartsch H, Scheurich P, Pfizenmaier K (1985) Biological effects of gamma-IFN on human tumor cells: quantity and affinity of cell membrane receptors for gamma IFN in relation to cell growth inhibition and induction of HLA directed expression. Int J Cancer 36:103–110
95. Ungari S, Katari RS, Alessandri G, Gullino PM (1985) Cooperation between fibronectin and heparin in the mobilization of capillary endothelium. Invasion Metastasis 5:193–205
96. Vlodavsky I, Gospodarowicz D (1979) Structural and functional alterations in the surface of vascular endothelial cells associated with the formation of a confluent cell monolayer and with the withdrawal of fibroblast growth factor. J Supramol Struct 12:73–114

97. Vlodavsky K, Johnson LK, Greenburg G, Gospodarowicz D (1979) Vascular endothelial cells maintained in the absence of fibroblast growth factor undergo structural and functional alterations that are incompatible with their in vivo differentiated properties. J Cell Biol 893:468–486
98. Vlodavsky I, Folkman J, Sullivan R, Fridman R, Ishai-Michaeli R, Sasse J, Klagsbrun M (1987) Endothelial cell derived basic fibroblast growth factor: synthesis and deposition into subendothelial extracellular matrix. Proc Natl Acad Sci USA 84:2292–2296
99. Wahl SM, Hunt DA, Wakefield LM, McCartney FN, Wahl LM, Roberts AB (1987) Transforming growth factor β induced monocyte chemotaxis and growth factor production. Proc Natl Acad Sci USA 84:5788–5792
100. Westall FC, Rubin R, Gospodarowicz D (1983) Brain derived growth factor: a study of its inactivation. Life Sci 33:2425–2429
101. Yoshida TK, Miyagawa K, Odagiri H, Sakamoto H, Little PFG, Terada M, Sugimura T (1987) Genomic sequence of hst, a transforming gene encoding a protein homologous to fibroblast growth factors and the int-2 encoded protein. Proc Natl Acad Sci USA 84:7305–7309
102. Zhan X, Bates B, Xiaogao H, Goldfarb M (1988) The human FGF-5 oncogene encodes a novel protein related to FGFs. Mol Cell Biol 8:3487–3495

Note added in press: FSdGF/Vasculotropin has been shown to have the same NH_2 Terminal sequence as the Tumor vascular permeability factor (Connolly, DT, et al. J Clin Invest 84:1470–1478, 1989). Both factors have now been cloned (Keck et al. Science 246: 1309–1312 (1989), Leung et al. Science 246:1306–1309 (1989), Tascher et al. Biochem Biophys Res Commun 165:1198–1206 (1989))

Smooth Muscle Cells and Pericytes

Smooth Muscle Cells from Rabbit Aorta

P. Fallier-Becker, J. Rupp, J. Fingerle, and E. Betz

Introduction

The rabbit has been often used as an experimental animal in atherosclerosis research [51]. Early, in 1908, Ignatowski [34] studied the "influence of animal food on the organism of rabbits" and found an increase of lipids in artery walls when the rabbit was fed with a diet rich in fat and cholesterol. The production of atheromas by cholesterol rich diets became a standard technique and is still used when the effects of cholesterol on the arterial walls is studied. It is the basis of the lipid theory of atherogenesis [1]. A second theory for atherogenesis, the response-to-injury theory [44], is based upon experiments reported by Baumgartner [2]. He induced a thrombogenic reaction with subsequent fibromuscular intimal proliferates in the aortas of rabbits. Using a balloon catheter the endothelial lining of the aorta was removed and the vessel was distended by inflating the balloon. In 1966, Baumgartner and Studer [3] found a relationship between the distension of the aorta and a subsequent proliferation of cells in the intima. The mechanical stimulus caused smooth muscle cells (SMCs) of the media to migrate into the intima and to proliferate there to form a local intimal fibromuscular cushion. In 1979, another model to generate atherosclerotic proliferations was described by Betz and Schlote [7]. Again the rabbit was used as an experimental animal for local weak electrical stimulation of the carotid artery. Both models showed that in the early stages of atherogenesis, SMCs migrate from the media into the subendothelial space where they proliferate to form a focal thickening of the vessel wall termed a "fibromuscular plaque". In contrast, SMCs of the normal vessel wall reveal a low proliferation rate [18, 19]. In the normal vessel wall a balance exists between factors inhibiting and factors inducing proliferation of SMCs whereby the inhibitory factors dominate. These factors act in a complex manner in vivo and it is very difficult to study the mechanisms of inhibition or activation of SMC proliferation in detail. Since the rabbit was the first and most frequently used animal in experimental atherosclerosis research [51], in vitro systems consisting of cells from the rabbit may be helpful to analyze the interactions of factors affecting the proliferation of SMCs. This does not only save animal experiments [6] but also enables the study of therapeutic aspects which cannot be solved in in vivo experiments [5]. If, for example, a drug suppresses the development of a stenosing proliferate in an artery by inhibiting the migration or proliferation of SMCs in the intima, it cannot be concluded whether the drug causes this response because of its toxicity; it also cannot be concluded whether the drug inhibits the mitotic process directly or via an indirect mechanism, e.g., by inhibiting the permeability of the endothelium. Finally, the in vivo experiment does not allow the conclusion that the drug acts specifically on the SMCs in the artery walls. Therefore, cultures of various cell types are useful in order to study the action of antiatherogenic drugs. One of the most important cell type in this context is the arterial SMC. Cultures of SMCs enable migration and proliferation to be studied under controlled and standard conditions. In vitro it is even possible to observe either the migration or the proliferation as separate phenomena. It is useful to compare the responses of cultured SMCs of rabbits with the in vivo experiments. Thus, the effective doses of antiatherogenic drugs as well as their toxic concentration can be determined in vitro. Cultures of SMCs from the rabbit aorta can be easily established. The isolation of SMCs from the media can be achieved by enzymatic disaggregation [16] or using the primary explant technique [44]. In culture, SMCs reveal a characteristic behavior of proliferation, migration, differentiation, and growth pattern. Furthermore, more complex culture systems can be established which represent the in vivo situation better than mass cell cultures or clone cultures of SMCs.

The following is a description and a discussion of the materials and methods of cell culture techniques used in cardiovascular research.

Methods and Materials of Cell Culture Techniques in Cardiovascular Research

Isolation of Primary Cells [32]

The SMCs of the rabbit are mostly obtained from the aorta. The isolation of SMCs from this vessel will be described in detail.

Removal of the Aorta from the Rabbit

A male Chinchilla rabbit (special pathogen-free animals; weighing 2.5–3.0 kg; Thomae) is anesthetized with halothane. The slaughtered rabbit is allowed to bleed by cutting the carotid artery. The animal is hung upside down and skinned. The thorax is opened by cutting longitudinally along the median line with sterile scissors. After removal of the intestines the aorta is dissected out and transferred to a sterile bottle containing the transport medium (DMEM (Gibco) + 15 mM HEPES (Sigma) and 100 U/ml penicillin (Gibco); 0.1 mg/ml streptomycin (Gibco); pH 7.4).

Preparation of the Aorta

The aorta is transferred into a preparation dish (14 cm in diameter; Greiner) containing the preparation medium (transport medium: DMEM + 15 mM HEPES and penicillin/streptomycin; pH 7.4). The bottom of the dish is covered with silicone (Sylgard, Sasco, 1 cm thick). The vessel should always be covered with medium to avoid drying.

The aorta is cut in the middle into two pieces: the thoracic and the abdominal aorta. The vessel is carefully stretched and pinned onto the silicone-covered preparation dish, using insect needles. It is freed from fatty and loosely adhering connective tissue to avoid contamination with fibroblasts. The preparation medium is changed and the vessel is cut open longitudinally between the intercostal arteries (Fig. 1 a). Now the intimal layer consisting of endothelial cells and a few intimal smooth muscle cells normally present in the intima is gently removed with a scalpel. The intimal cells are removed with a pipette and the vessel surface is intensely rinsed with phosphate buffered saline (PBS; Oxoid). Then new preparation medium is poured onto the vessel.

Isolation of Media Pieces

The medial layer can be separated from the adventitial layer due to the differences in texture between the media and the adventitia. The SMCs are oriented helically around the longitudinal axis of the vessel [43]. Using two fine forceps, one having a curved tip, the medial layer is stripped off in small pieces (Fig. 1 b). The vessel is grasped at the lateral margin with the straight forceps and the medial layer is carefully detached from the adventitia with the curved forceps. The media pieces are placed in a culture dish containing a small volume of preparation medium.

Determination of the Wet Weight

A sterile, empty centrifuge tube (Greiner) is weighed. Then the media pieces are transferred to the upper margin of the tube, allowing the remaining medium to drain to the bottom of the tube (Fig. 1 c). The medium is sucked off. The tube is capped and the wet weight of the media pieces is determined. The total amount of media pieces of one thoracic aorta weigh about 200–300 mg, and of one abdominal aorta about 150–180 mg [31].

Enzymatic Disaggregation

After weighing the media pieces they are cut into smaller pieces (approx. 1 mm^2) using a pair of scissors (Fig. 1 d). Then the enzyme solution containing collagenase (Biochrom; Progen Biotechnik)/elastase (Boehringer Mannheim; see "Enzyme Concentration") in DMEM (pH 7.4) and soybean trypsin inhibitor (Serva) is poured into the tube rinsing the small media pieces to the bottom. The tube is closed with a cap and also with parafilm and is constantly shaken in a waterbath at 37 °C. Within 120 min disaggregation occurs. When the solution is turbid a microscopic control should be performed to decide whether the tissue is sufficiently digested. If single cells (and a few cell clusters) are visible, the digestion can be stopped and the suspension is gently pipetted

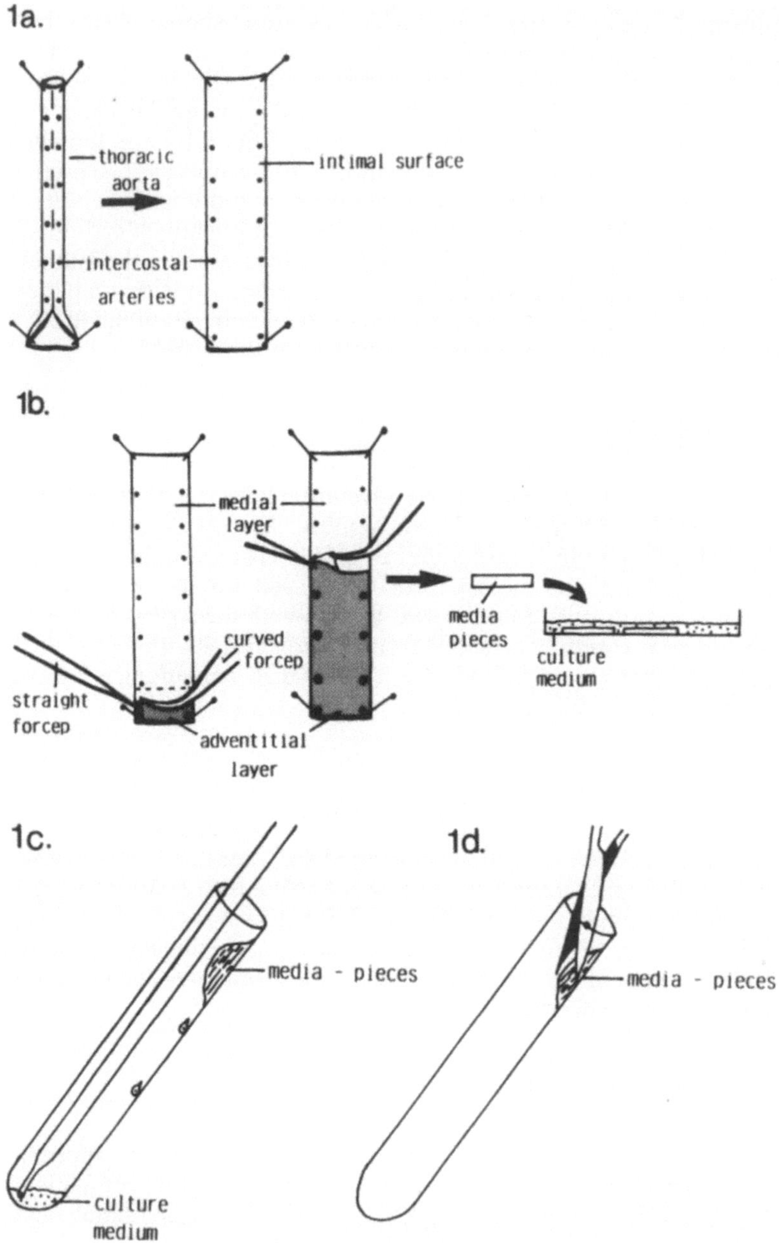

Fig. 1a–b. Preparation of the aorta. **a** The vessel is cut open longitudinally between the intercostal arteries. **b** After removing the intimal cells, the medial layer is stripped off the adventitial layer and the media pieces are transferred into a culture dish containing preparation medium. **c** Media pieces are transferred to the upper margin of a centrifuge tube in order to drain the pieces. **d** Media pieces are cut into smaller pieces using a pair of scissors

up and down until the tissue is completely dispersed. If some tissue is left undigested the cell suspension may be filtered through a sterile muslin or stainless steel mesh (100–200 μm), or larger pieces may simply be allowed to settle down. To achieve a good survival rate of SMCs a digestion time of 150 min should not be exceeded. An aliquot (1 ml) of the enzyme suspension is taken to determine the cell number ("Counting the Number of Cells"). Then the suspension is centrifuged to remove the enzyme. The pellet is resuspended in culture medium (20% DMEM/80% Ham's F12 (Gibco)+20% fetal calf serum (Sebak)+penicillin/streptomycin) and the cells are seeded in culture dishes at a seeding density of approximately 10000 cells/cm^2.

Enzyme Concentration

To decide whether the concentration of the enzyme solutions has to be increased, a microscopic control of the digestion should be performed after 90–120 min. Prolonged exposure to the enzyme should be avoided. Approximately 1 ml enzyme solution for every 100 mg of tissue is recommended.

Elastase

SMCs are tightly bound to their extracellular elastic material. The elastase enzyme has to break up these bindings to yield single cells. If in the microscopic control after 120 min elastin-SMC aggregates are still found the elastase concentration should be increased (0.5–1 mg/ml).

Collagenase

SMCs are also associated with collagenous fibers but not as tightly as to the elastic material. A collagenase enzyme should be used to digest the tissue without damaging the cells. A concentration of 1.0–1.8 mg collagenase/ml should be used.

Counting the Number of Cells

To determine the number of the isolated cells, an aliquot (1 ml) is taken from the enzyme solution. This suspension is centrifuged and the pellet is resuspended in 1 ml trypsin (Serva; 800 U/l; 0.02% in PBS without Ca^{2+} and Mg^{2+}, PBS$^-$) to obtain single cells.

Hemocytometer

After 5 min in trypsin the activity is stopped by adding soybean trypsin inhibitor or serum to the culture medium. The primary cells are then counted in the hemocytometer.

Electronic Cell Counter

After 5 min in trypsin the cells are suspended and diluted in a 0.9% NaCl solution and are counted immediately in an electronic cell counter (Schärfe).

Cell Yield

If a tissue is enzymatically digested to obtain single cells, only a small percentage of cells attach and can be therefore cultivated.

The cell yield is calculated by comparing the DNA content of freshly isolated cells with the DNA content of the undigested tissue. The wet weight of the media derived from a thoracic aorta is measured as previously described. An aliquot of 100 mg of the media tissue is used for the DNA measurement (see "Transfilter Cultures"). Another aliquot of 100 mg of the media tissue is enzymatically digested to obtain single cells. Afterwards, the DNA content of these freshly isolated SMCs is measured and compared with the DNA content of the starting material. The cell yield after enzymatic disaggregation amounts to approximately 10%–20%. For example, 100 mg of media tissue (wet weight) contains approximately 100 µg DNA; after enzymatic disaggregation, the DNA content of freshly isolated cells amounts to approximately 10 µg DNA.

Primary Explant Technique [30, 44]

To isolate primary cells from the medial layer of the rabbit aorta, the explant technique can be used instead of isolation by enzymatic disaggregation. As described in "Isolation of Media Pieces", the medial layer is separated from the adventitia using fine forceps. Media explants are obtained as shown in Fig. 2a. They are transferred to a culture dish (Falcon, Becton Dickinson) containing a small volume of preparation medium, then soaked with medium, and placed in a culture flask (Greiner). To prevent the explants being detached, the culture medium should not be poured immediately onto the explants. To make sure that the explants adhere to the bottom of the culture flask it is placed upright in an incubator. After 18–24 h the flask is placed horizontally and the explants are rinsed with medium (Fig. 2b). The medium is changed every 48 h.

Preferentially, the thoracic aorta is used to obtain media explants because it is easier to strip the media off the adventitia of this part of the aorta than of the abdominal aorta. An optimal size of media explants is approximately 0.25 cm^2. If this size is exceeded, detachment of explants occurs at a higher risk.

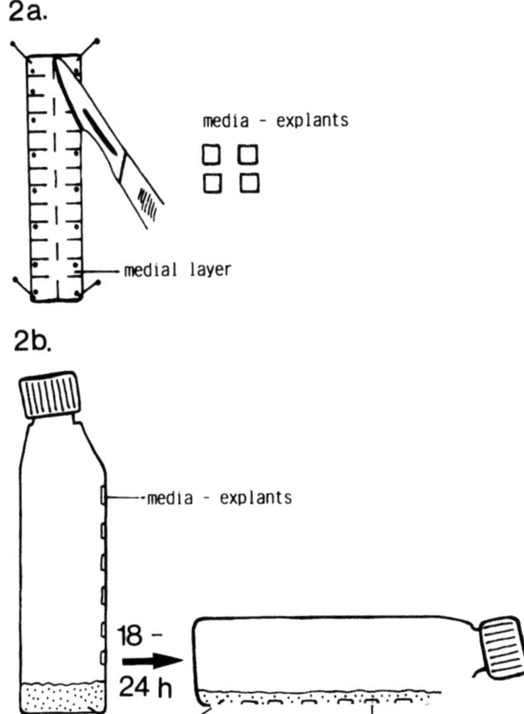

Fig. 2a, b. Primary explant technique. **a** Media explants are obtained detaching them from the adventitia from predetermined lines. **b** To minimize the risk of media explants being detached, the culture flask is placed upright in the incubator. After 18–24 h it is placed horizontally allowing the medium to cover the explants

Cultivation of Smooth Muscle Cells from the Rabbit Aorta

Mass Cell Cultures

Primary Cultures

Twenty-four hours after the isolation of SMCs by enzymatic disaggregation, the culture medium is changed. It should be microscopically controlled that the intact cells are attached to the bottom. The freshly isolated cells in culture are termed "primary cells" (P_0). The plating efficiency is normally 25%–75%. Dead cells and cell debris do not attach and have to be removed in order to avoid toxic effects on the intact cells. The cultures are kept in a 37 °C humidified incubator (Heraeus) with an atmosphere of 95% air and 5% CO_2.

The lag phase of a primary culture lasts approximately 3–7 days. Within this time the cells spread out but neither migrate nor proliferate. After this lag phase the SMC population starts to proliferate and the cell number increases logarithmically (log phase). When the SMC culture is confluent, the plateau phase is reached. The time until primary SMC cultures reach confluency depends on the seeding density and on the seeding efficiency. For example, if

SMCs are seeded at a density of 100 cells/cm^2 the culture reaches confluency after 13 days, and the final cell density then amounts to 40000 cells/cm^2; however, if 10000 SMC/cm^2 are seeded the culture is already confluent after 8 days at a final cell density of 140000 cells/cm^2.

Subcultures of Smooth Muscle Cells, Cell Lines

Seven to ten days after isolation, the primary cells are usually confluent and thus have to be transferred into new culture dishes. This transfer of cells into new culture dishes is termed "passaging" (P_0 to P_1, P_1 to P_2, etc.).

Procedure: The culture medium is removed and the attached cells are rinsed twice with PBS$^-$. Then 1 ml trypsin/55 cm^2 culture dish is added to the cells causing them to detach from the plastic bottom of the culture dish. After 1 min approximately 80% of the cells are detached (microscopically controlled). Then the trypsin is inactivated by adding culture medium containing 20% serum. The cells are suspended by repetitive pipetting. The cell number is determined as described above. To obtain a large number of cells the SMCs are seeded at a density of 10000 cells/cm^2. For example, a confluent SMC culture in a 55 cm^2 culture dish consists of approximately 4×10^6 cells; to cultivate a large number of SMCs the cells are seeded in eight 55-cm^2 dishes at a density of 10000 cells/cm^2 (ca. 5×10^5 cells/culture dish).

Cell line: After the first subculture – or passage – the primary SMC culture becomes a cell line and may be propagated and subcultured several times in an unaltered form for a limited number of cell generations beyond which they may either die out or give rise to a continuous cell line [23]. SMCs in the early subcultures show a slowly decreasing proliferation activity. At the same time the number of senescent and pleomorphous cells increases. This cell type does not proliferate further. Its fate is similar to that of the postmitotic fibroblast [4]. Up to the eighth cumulative population doubling, the SMC population reveals a large number of dividing cells (growth rate). From these cells a constant number is senescent and postmitotic. Parallel to an increase in in vitro age, the number of postmitotic cells increases and, as a consequence, the growth rate of the culture decreases. Therefore, after 8–12 population doublings the proliferation activity of the SMC culture is very low, but increases after spontaneous transformation. Then SMCs which did not undergo an aging process had survived. The speed of growth of these transformed cultures increases to 0.6–0.8 doublings/day which is higher than before [31]. To characterize a continuous cell line a chromosome analysis should be performed [53].

Cell Size Measurements

With increasing in vitro age the sizes of the single SMCs increase, however they vary considerably. If the volumes of primary SMCs are measured in a cell volume analyzer (Schärfe System) and compared with subcultured SMCs with a higher in vitro age, differences in the cell size distribution are observed. For these measurements SMC cultures are washed with PBS$^-$ and trypsinized at 37 °C for 10 min. The cells are resuspended in 0.9% NaCl before being measured.

Cell Freezing

For setting up a seed stock SMCs can be stored frozen in liquid nitrogen.

Procedure: SMC cultures are rinsed with PBS$^-$ and trypsinized (Section "Primary Cultures"). After detachment of the cells the trypsin activation is stopped by adding culture medium containing 20% serum. The SMCs are suspended and centrifuged to remove the culture medium. The cell pellet is resuspended in freezing medium (DMEM/20% fetal calf serum + 7.5% dimethylsulfoxide, DMSO; Serva). The freezing medium is stored at −20 °C in aliquots of 10 ml. It should be thawed only once. The temperature of the medium when it is used should be about 0°−1 °C. This can be achieved by thawing the freezing medium for a short time before use, whereby a small amount of ice remains in the medium. The following steps should be performed on ice: The cell pellet is suspended in the freezing medium and the suspension is filled in precooled 1.5 ml freezing ampules (glass or plastic, Lutz; Tecnomara). The ampules must be perfectly and quickly sealed. The SMCs should be frozen at a cell concentration of about $1-3 \times 10^6$ SMCs/1.5 ml freezing medium. The ampules are then transferred in a polystyrene foam box and kept in −80 °C for 24 h. For long-term storage the ampules are transferred in liquid nitrogen in a Dewar flask (Deutsche L'Air Liquide) at a temperature of −196 °C.

Thawing: To recover SMCs from the frozen storage, the ampules are retrieved and transferred to warm water. When the suspension starts to thaw, the ampule is opened and a small volume of warm medium is added. The cells are suspended in a pipette and transferred into a culture dish filled with warm medium. The dilution should be 1:10 or 1:20 (1 ml frozen cell suspension should be thawed in 10−20 ml medium).

Clone Cultures

If SMCs are seeded at a cell density of 10−20 cells/cm^2 the cultures are termed "clone cultures". This very low cell density enables single cells to attach to the plastic bottom with a relatively large distance between one another. Subsequently, the cells will grow as discrete colonies termed "clones" [23, 37].

Clone cultures are cultivated with a small amount of medium (for example, 5 ml medium/55-cm^2 culture dish) which requires changing only once a week. Individual clones may be separated using clone-rings. After 14 days in clone culture the cultures are fixed and stained (with, for example, Giemsa or Coomassie-Giemsa). The number of clones is evaluated using a colony counter

(Bender and Hobein). The results are expressed as the

Plating efficiency: (No. of colonies formed · 100)/No. of cells seeded

If it can be confirmed that each colony grew from a single cell, this term becomes the "cloning efficiency" [23].

Explant Cultures

Between day 5 and day 7 after the isolation of media explants, SMCs start to migrate out from the explants and the cell number increases by proliferation. After 14–21 days the plastic bottom is covered with SMCs. To passage the cells the media explants are removed using fine forceps. The primary SMCs are trypsinized ("Procedure" in "Subcultures of Smooth Muscle Cells, Cell Lines") and the cells are counted as in "Counting the Number of Cells". The SMCs are routinely seeded at a density of 10 000 cells/cm^2 and subcultured as described in "Mass Cell Cultures".

Cultures of Intimal Smooth Muscle Cells

Intimal SMCs are normally present in the rabbit aorta. They derive from the media and are located in small pockets of the internal elastic lamina beneath the endothelial lining. They can be activated to migrate and proliferate by mechanical stress in vivo [3] as well as in vitro [22]. Intimal SMCs are responsible for the formation of a neointima. Therefore, it may be interesting to isolate these intimal SMCs and cultivate them in order to compare their behavior in vitro with medial SMCs in culture. To obtain intimal SMCs from the rabbit aorta the enzymatic method is used.

Procedure: The aorta of a male Chinchilla rabbit is dissected and prepared as previously described. The vessel is cut open longitudinally and then transferred to a special holding device which covers the margins of the aorta to avoid contamination with fibroblasts. The uncovered area of the vessel extends to 6×52 mm and may be filled with the enzyme solution to isolate the intimal cells. The aorta is incubated with collagenase (1 mg/ml DMEM/HEPES) for 10–15 min at 37 °C. The intimal cells are then suspended by repetitive pipetting and centrifuged to remove the enzyme solution. The cell pellet is resuspended in culture medium and seeded in culture dishes. Endothelial cells that have also been isolated are overgrown by the intimal SMCs. These SMCs are able to be identified by staining the culture with an antibody against alpha smooth muscle (SM) actin ("Immunofluorescence"). After two to three passages all the cells are positive for alpha SM actin.

Identification of SMCs

After isolation and subculturing of SMCs it is necessary to determine that no other cell type (fibroblasts or endothelial cells) has contaminated the SMC cultures. With a light microscope SMCs can be identified morphologically and by their growth pattern. However, the most common and one of the best methods of characterization of SMCs currently available [42] is immunological staining of cytoskeleton proteins (SM myosin or alpha SM actin).

Morphology

Primary SMCs and subcultured SMCs with a low in vitro age ($<P_5$) have a spindle-shaped appearance. With a higher in vitro age ($>P_{10}$) SMCs spread and become larger with an increase in cell volume ("Cell Size Measurements"). Confluent SMC cultures show the characteristic "hill and valley" growth pattern (Fig. 3). They are able to form multilayers in culture, but there are also areas of monolayered SMCs [8]. With a higher in vitro age the monolayered areas become more dominant. If SMCs are cultivated for more than 2 weeks without passaging the cells, the formation of "nodules" may be observed (Fig. 3). These structures consist of cell debris in the center and intact SMCs on the surface. Intimal SMCs show a monolayered growth pattern devoid of the "hills and valleys" and are therefore different from medial SMCs.

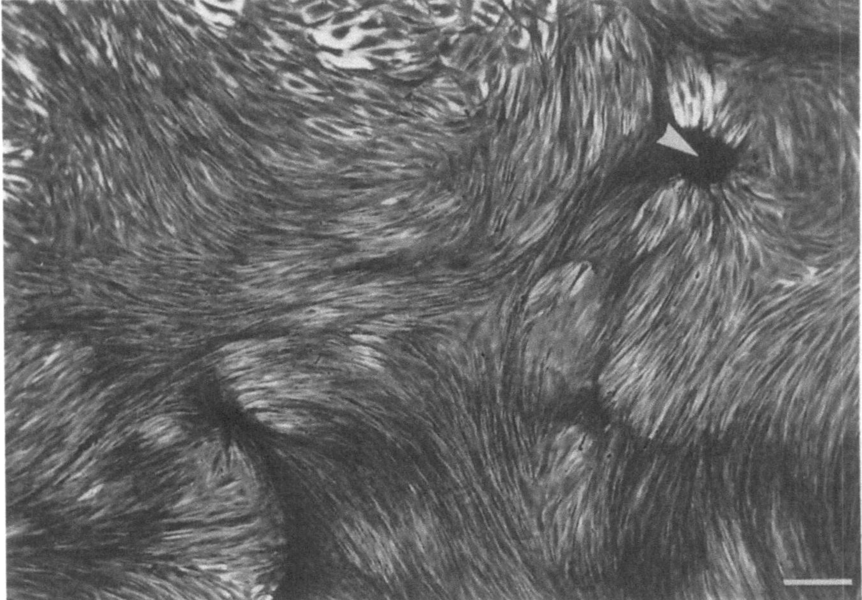

Fig. 3. Subcultures of SMC reveal the characteristic "hill and valley" growth pattern and "nodules" (▶). — 100 µm

Immunofluorescence

In a mixed culture of SMCs and endothelial cells (ECs) both cell types can be easily differentiated using light microscopy. ECs form clones of polygonal cells resembling a cobblestone pavement among the spindle-shaped SMC. However, it is nearly impossible to distinguish ECs from SMCs if they are cultured for a long time and have a high in vitro age. Identification of fibroblasts in a culture of SMCs is also very difficult under the light microscope. Therefore, immunological stainings are necessary to make sure that a culture of SMCs is not contaminated by another cell type. Differentiated SMCs present characteristic cytoskeletal proteins. SM myosin [28] and alpha SM actin, [49] for example, are marker proteins for differentiated SMCs and are able to be identified in individual cells by specific antibodies. These antibodies can be subsequently detected by second antibodies conjugated to fluorescein-iso-thio-cyanate or rhodamine. With a light microscope equipped for epifluoresence the stained cells can easily be identified.

For all immunological stainings the SMCs have to be cultured on coverslips. Before use the coverslips have to be treated as follows:

- Wash in hot 3% BM (Biomed Labordiagnostik; H_2SO_4 substitute) in water for 30 min
- Rinse in distilled water and sterilize at 150°C
- Place in culture dishes and incubate 30–60 min in sterile collagen in PBS^- (20 µl/ml)
- Remove PBS^- and seed with SMCs at a low density

SM Myosin

SM myosin [27] can immunologically be detected in SMCs using a polyclonal antibody. This antibody is cultivated in rabbits against purified chicken gizzard myosin. The production and specificity of the antiserum has been described by Gröschel-Stewart et al. [29]. The antiserum reacts specifically with SM myosin (heavy chain) but not with myosin isoforms of heart muscle cells, endothelial cells [16] and fibroblasts [28]. Western blot analysis showed a reaction of the antiserum with purified myosin derived from the media of swine carotid artery but not with myosin from skeletal muscle, nonmuscle tissue, heart muscle, and subcultured SMCs [46]. In cell cultures and histological sections this antiserum does not react with ECs or blood cells nor does it react with SMCs migrating out from media explants [32].

Procedure: The cells for immunological staining are grown on coverslips (Langenbrinck; "Immunofluorescence"). Coverslips are taken from the culture dishes and briefly rinsed with PBS with Ca^{2+} and Mg^{2+} (PBS^+), fixed with methanol (6 min, −20°C) and immediately used for immunological staining. Cells are rinsed with PBS^- and the antiserum against SM myosin is added subsequently at a concentration of 0.3 mg/ml PBS^- for 60 min at 37°C. After brief washings with PBS^- fluorescein- or rhodamine-conjugated goat-antirabbit IgG is added at a dilution of 1:80 for 60 min at 37°C. After several PBS^-

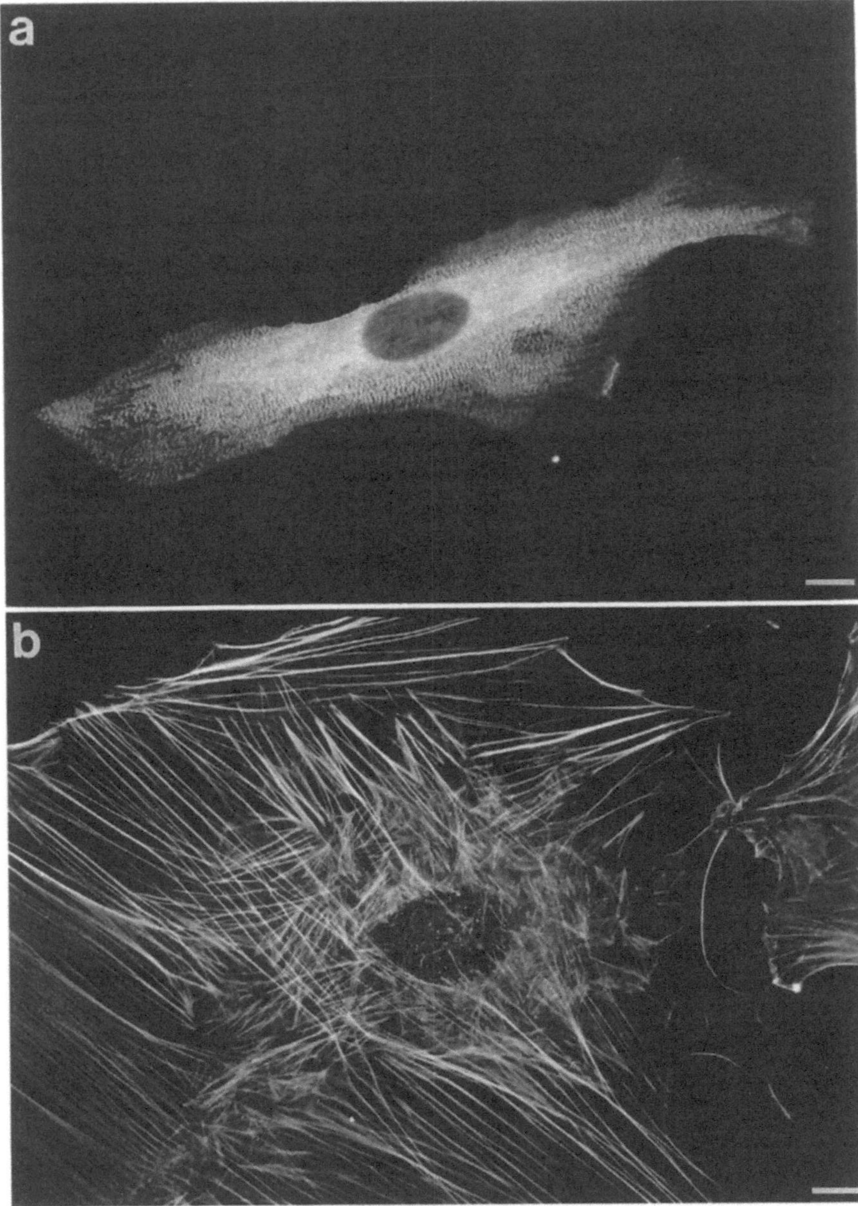

Fig. 4 a, b. Immunofluorescence of SM myosin (**a**) and alpha SM actin (**b**) of a SMC in primary culture. −1 µm

washings the coverslips are mounted in Mowiol (Hoechst) [40]. All procedures
are performed in a humid chamber. The coverslips are observed with a microscope equipped for epifluorescence (Fig. 4 a).

Alpha SM Actin

Alpha SM actin is expressed in differentiated SMCs and can be identified by
a monoclonal antibody from the mouse [48, 49]. The presence of alpha
SM actin is detected by indirect immunofluorescence with fluorescein- or
rhodamine-conjugated second antibodies.

Procedure: SMCs grown on coverslips ("Immunofluorescence") are rinsed
with PBS^+ and fixed with methanol (6 min, $-20\,°C$). The coverslips are washed
with PBS^- and incubated with the monoclonal antibody (anti alpha SM actin,
Renner) at a dilution of $1:300$ for 40 min at $37\,°C$. After washing with PBS^-
the second antibody goat-antimouse IgG (fluorescein- or rhodamine-conjugated) is added at a dilution of $1:25$ for 40 min at $37\,°C$. After several PBS^-
washings the coverslips are mounted in Mowiol [40]. All procedures are performed in a humid chamber. The coverslips are observed with a microscope
equipped for epifluorescence (Fig. 4 b).

Test Assays for Antiatherosclerotic Substances [6]

An important event which leads to the formation of a stenosing intimal proliferate during atherogenesis in vivo is the migration and proliferation of SMCs.
It is possible to observe these phenomena in several in vitro test systems
(assays). These assays are used to study the effects of antiatherogenic substances, e.g., heparinoids or various calcium antagonists. For this purpose
some test systems have been developed.

Proliferation Test Assay

Dose Response

Using this test assay, SMC growth is watched depending on different concentrations of antiatherogenic substances.

Procedure: SMCs are seeded at a density of 10000 cells/cm^2. The culture
medium is removed 24 h after seeding and the antiproliferative substances are
added to the cells in various concentrations. The medium containing the substance is changed once after 2 days. After 4 days incubation time, the SMCs
are trypsinized and counted in an electronic cell counter as already described.
If the substance tested is an efficient inhibitor of proliferation, the cell number
of the treated cultures should be lower than the untreated controls.

Viability Test: To make sure that the cells treated with a substance are not damaged at a certain concentration, a vital staining using fluorescein diacetate (FDA; Sigma) and/or ethidium bromide (EB; Serva) can be performed: The cultures are washed twice with PBS$^+$; then the cells are incubated for 2 min with PBS$^+$ containing 0.6% FDA stock solution (5 mg FDA/ml acetone) and 0.3% EB stock solution (1.25 mg/ml PBS$^+$). The staining solution is removed and PBS$^+$ is added. The cells are observed with a light microscope equipped for epifluorescence (with filters for blue/green and green/red). Living cells show a green fluorescence (FDA) and dead cells show a red fluorescence (EB).

Growth Speed

Furthermore, the growth speed of SMCs during 5–9 days is determined in a second test calculating the final cell number when the culture is confluent (plateau phase) in relation to the antiatherogenic substances. The experimental design is identical to that in "Proliferation Test Assay" except for the incubation time.

Migration Test Assay [11]

SMCs are cultured until they reach confluency. The antiatherosclerotic substance is then added, and 48 h later the confluent cell layer is injured by scraping away part of the cell sheet. The wound is made by pressing a razor blade down onto the bottom of a culture dish to cut the cells and to mark the dish. A sharply defined "wound" is produced and SMCs beyond this line are detached with a cotton wool bud and removed by soaking off the cells. The SMCs migrate across the wound edge into the wounded area devoid of cells and after 72 h the cells are fixed and stained. The migration distance and the number of cells that have migrated across the wounded edge are then determined.

Combined Migration-Proliferation Assay

This test assay has the potential of being able to study the combined effect of a substance on migration and proliferation of SMCs in the same experiment. The experimental design is identical to that in "Migration Test Assay". However, 18 h prior to fixation (methanol as described in "Procedure", "SM myosin" and "α-SM actin") the cultures are labeled with BrdU (30 μmol and 30 μmol dCyt). BrdU is incorporated into the DNA instead of thymidine in dividing cells and labels the S phase of cells in a similar way as ^{3}H-thymidine and can be detected using a monoclonal antibody against BrdU (1 : 50, 40 min, 37 °C; Bio Cell Consulting). With the Biotin-Avidin method the dividing SMCs can be identified (second antibody horse-antimouse biotinylated (Camon) 1 : 100, 10 min, 37 °C) and differentiated from the SMCs that have only migrated without proliferation. As test parameters, the migration distance and the growth fraction of the migrating cells are measured.

The described experiments can also be performed with explant cultures.

More Complex Culture Systems

Organ Cultures [21, 22, 41]

Whole vessel segments of the aorta from rabbits can be maintained in culture [9, 20, 24, 35, 38, 47] for more than 2 weeks without loss of endothelium and without loss of contractility. SMCs in organ culture can be stimulated to proliferate and to form a fibromuscular intimal proliferate in vitro. Cultures of whole segments of the rabbit aorta require a very careful preparation in order to maintain the endothelial lining without damaging the cells. The thoracic aorta of a rabbit is dissected off as described by Pederson and Bowyer [41]. To minimize trauma to the vessel, relaxation is achieved by injection of 30 ng papaverine/kg body weight i.v. and coagulation is prevented by 400 U heparin (Sigma)/kg body weight i.v. before excision of the vessel. Throughout the preparation of the vessel for organ culture it should be perfused via a catheter with culture medium (RPMI 1640, Gibco, supplemented with 15 mmol HEPES, 10% newborn calf serum and 20 µg/ml gentamicin). The thoracic aorta is then transferred to a preparation dish, opened longitudinally, and stretched out over six prepared silicone rubber pieces (1.5×1.2 cm) with minute pins. During this procedure the vessel is maintained at its original length and not allowed to shrink. The aorta is then divided into six pieces, each approximately 1.2 cm long and checked for endothelial injury with 0.01% trypan blue (Serva) in PBS$^-$. Each segment is then put into a 25 ml Pyrex glass bottle (Bender and Hobein) with 10 ml of culture medium (RPMI 1640, 30% newborn calf serum, 10 µg/ml gentamicin, 0.7 mmol glutamine, Gibco). The medium is changed every 48 h. To ensure a sufficient supply of medium, the cultures are mounted on a shaking holder inside the incubator.

Transfilter Cultures [26, 52]

In transfilter cultures, media explants from the rabbit aorta are cultivated on filters. In this system SMCs migrate out from the media explants and through the filter pores to the opposite side where they proliferate to form a SMC multilayer (Fig. 5). This process imitates the migration through the internal elastic lamina in experimental atherogenesis in vivo in which, during the initial phase, SMCs migrate from the medial layer into the subendothelial space passing the fenestrated internal elastic lamina and proliferating to form a thickening of the vessel wall. Thus, the transfilter culture serves as a model to study migration and proliferation of SMCs during atherogenesis.

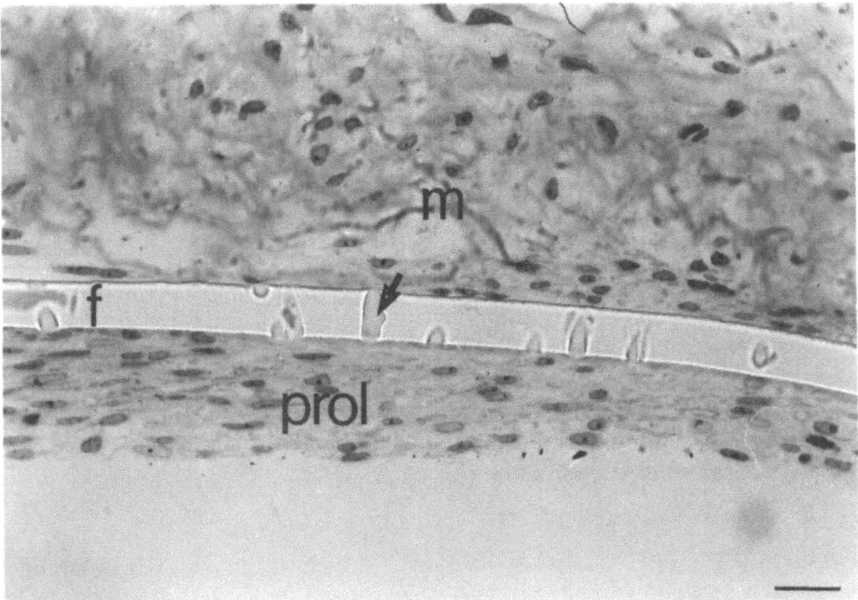

Fig. 5. Transfilter culture of a media explant (*m*) after 14 days incubation time. SMC have migrated from the explant and through the filter pores (→) to the other side of the filter (*f*) in order to form a proliferate (*prol*) of approximately 5–8 cell layers. — 10 µm

Filter and Pore Size

For establishing a transfilter culture it is necessary to choose the appropriate filter. To study migration of SMCs to the opposite side of the filter, a pore size not too large and not too narrow and a particular pore morphology is required. For this purpose polycarbonate filters (Nuclepore) or culture plate inserts for six-well plates (Nuclepore; Costar, Tecnomara) may be used. These filters are only 10 µm thick and have pores with a defined pore size. The choice of the pore size depends on the experimental design. If the study of SMC migration is desired, pore diameters of 3–5 µm are required. They allow the cells to pass through without difficulty. If, however, only substances secreted by cells or other chemically defined substances are required to be tested, then a smaller pore size should be used (0.2–0.45 µm) in order to prevent cell migration through the filter pores.

Coating the Filter with Collagen

The polycarbonate filters are coated with collagen following the method described by Cereijido et al. [13]. Coating facilitates the attachment of the media explants to the filter surface. Lathyrithic collagen type I from rat integumentum is dissolved in 0.1 *N* acetic acid at a concentration of 1 mg/ml. The filters are submerged in the collagen solution and dried in ammonia vapor for 1–3 min. Then the filters are transferred into 4% glutaraldehyde (Merck-Schuchardt) for 1–3 h and afterwards carefully washed with 0.9% NaCl several times and sterilized under UV in a solution of NaCl. The coated filters can be stored at 4°C for 4 weeks.

Media Explants

Aortic media explants are obtained as described in "Primary Explant Techniques". The media explants are placed on polycarbonate filters using a fine forceps. The internal elastic lamina is oriented towards the filter. Five media explants may be placed on one filter (culture plate insert). Only a small volume of culture medium is added to the explants to avoid detachment. After a few days the explants should be attached to the filter, culture medium may then be carefully poured onto the explants.

Incubation Time

Between day 5 and day 7, SMCs start migrating out from the media explants and through the filter pores. After 14 days, a cell multilayer of five to ten layers is formed and, after 28 days this multilayer resembles a fibromuscular plaque in vivo following balloon endothelial denudation.

Evaluation of Transfilter Cultures

Quantification of the Cell Multilayer

To quantify the SMC multilayer which has been formed within a certain incubation time, the DNA content of the multilayer formed on the opposite side of the filter is determined. For analysis, the method of Labarca and Paigen [36] is used.

Procedure: The transfilterd cultures are washed with sterile DNA buffer (2 mol NaCl, 50 mmol $NaH_2PO_4 \times H_2O$, 1 mmol EDTA, pH 7.4). Then the cell multilayer which has developed on the opposite side of the filter is detached by a rubber policeman. The cells are suspended in 2 ml DNA buffer and are sonicated for 1 min.

Sample: An aliquot of 100 µl from the cell suspension is taken and added to 100 µl Hoechst stain No. 33258 solution and to 1.8 ml DNA buffer. *Standard:* 100 µl Hoechst stain solution (Sigma) and 2 µl calf thymus DNA (or another DNA) are added to 1.898 ml DNA buffer.

The sample and standard solution are incubated for 1 h in darkness at room temperature. Then the DNA content of the sample is measured with a

DNA fluorometer (Atlanta). The DNA content of a cell multilayer formed within a certain incubation time on the other side of the filter can be used as an index for the SMC outgrowth. With this transfilter culture system, antiatherogenic substances can be tested. The DNA content of the cell multilayer of treated cultures may then be compared with untreated controls.

Immunohistological Stainings of Transfilter Cultures

For immunological studies, semithin sections of transfilter cultures are used. The transfilter cultures are fixed with 2% paraformaldehyde in PBS$^-$ (pH 7.4) for 3 h at room temperature, rinsed in PBS$^-$ several times, dehydrated stepwise in alcohol and alcohol-propylene oxide, embedded in Araldite (Serva) and polymerized at 60 °C for 3 days. The sections are mounted on poly-L-lysine (100 µg/ml; Serva) coated coverslips. After drying at 60 °C the Araldite is dissolved and removed from the sections using sodium methylate (Merck-Schuchardt) [38]. Sections are then immediately used for immunological staining. To demonstrate proliferation and the state of differentiation of SMCs, immunological staining with an anti-BrdU antibody and an anti-alpha SM actin antibody may be performed ("alpha SM Actin", "Combined Migration-Proliferation Assay"). SMCs that have migrated out from the media explants within 5–7 days in transfilter culture and that have passed through the filter pores, express no alpha SM actin. However, after 14 days incubation time SMCs of the multilayer reveal a weak fluorescence when they are stained with an antibody against alpha SM actin. After 28 days the fluorescence is intensive. This suggests that SMCs are able to re-express alpha SM actin in vitro as well as in vivo during formation of a fibromuscular plaque [25].

Discussion

SMCs of the rabbit aorta have been studied frequently and intensely to understand the events during pathogenesis of the vessel wall. In atherosclerosis research, several models to generate experimental lesions had been developed [3, 7]. However, models in vivo only allow studying the events occurring during migration, proliferation, and aging of SMCs but not the total differentiation pattern of SMCs. Therefore, SMC cultures have been established to enable observation of the behavior of SMCs in vitro continuously over a longer time, using, e.g., video techniques. These investigations revealed some problems concerning the validity of cultured cells as a model of physiological function in vivo [23]. A cell culture is obtained by the primary explant technique [43] or by enzymatic disaggregation [16] of the starting tissue. This isolation is a selective process, as it is not possible to take all the cells of the tissue in culture. During the subcultivation of the isolated cells dedifferentiation occurs and the cells which are able to proliferate and migrate are selected. Furthermore, in the later subcultures, a continuous cell line might be created by spontaneous transformation. These problems that generally arise if a cell culture is initiated, also apply to cultures of SMCs from rabbit aorta. In establishing a standardized SMC culture, the following difficulties have to be overcome:

1. SMC cultures should not be contaminated with fibroblasts from the adventitial layer; otherwise the SMCs are overgrown by fibroblasts.
2. Isolation and cultivation of SMCs are selective processes, since it is not possible to obtain a cell yield of 100%.
3. During cultivation, SMCs in culture are going to be adapted to the in vitro conditions. Dedifferentiation occurs and neither SM myosin nor alpha SM actin are found in the cells.

Contamination with Fibroblasts

To avoid contamination with fibroblasts the vessels should be carefully prepared in order to obtain media pieces that are free of adventitia. After the enzymatic disaggregation or in primary explant cultures, all the cells should be alpha SM actin positive to make sure that a pure SMC culture is obtained.

Selection

The problem of selecting a specific SMC type or subpopulation during the enzymatic isolation of cells can only be solved if the cell yield is 100%. As all the tested enzymes were not capable of achieving such a result, the selection problem can not be excluded. Increasing the cell yield depends on the convenient collagenase enzyme. The effective collagenase batches are crude preparations and beside the collagenase activity, they possess proteolytic and tryptic activities that are not specified. Therefore, no objective criteria exist for an optimal collagenase enzyme and this optimal enzyme has to be found by testing several batches from different suppliers. To minimize the problem of selection when cells are isolated from the tissue and taken in culture, a cell yield of approximately 10% should be achieved.

Concerning the primary explant technique, selection is also a problem. In this case, only cells which are capable of migrating and proliferating are taken in culture.

Dedifferentiation

When primary cultures are established, SMCs begin to proliferate after a lag phase of approximately 3 days. Within this time, the characteristic cytoskeleton proteins of differentiated SMCs, SM myosin and alpha SM actin, can be shown in the cells by immunological stainings. After 5–7 days incubation time, however, the number of SM myosin positive cells decreases. This suggests that SMCs in the log phase of primary cultures stop expressing SM myosin [15–17, 28]. When the cells reach confluency they re-express SM myosin [46]. In subculture, however, the number of SMCs which are SM myosin positive decreases and SMCs with a high in vitro age do not express SM myosin. In contrast, alpha SM actin is expressed in primary culture and also during the early subcultures but the number of alpha SM actin positive cells decreases. The expression of alpha SM actin in cultured SMCs is said to correlate with the growth phase [10]. With an increasing in vitro age, however, dedifferentiation of SMCs occurs. Electron micrographs of dedifferentiated SMCs in culture show an organelle-rich cytoplasm. This SMC phenotype differs from the differentiated SMCs in the primary culture and is termed "metabolic SMC phenotype". It derives from the differentiated contractile SMC phenotype by a modulation process [16]. The dedifferentiation process of SMCs in culture and the problem of identification of SMCs are correlated. Immunological stainings of SM myosin and alpha SM actin can be only performed on SMCs with a low in vitro age. Dedifferentiated SMCs can not be characterized and their presence is therefore difficult to prove.

Based upon the described problems of cultured SMCs, their validity in testing drugs is frequently under discussion. In atherosclerosis research, however, proliferation and migration of SMC are important events which lead to the formation of a fibromuscular intimal thickening. Both processes can be easily studied in vitro and, moreover, the effect of antiatherogenic substances on the migration and/or proliferation of SMCs can be investigated more exactly under controlled and standardized in vitro conditions [5].

To narrow the gap between in vivo experiments and single cell cultures, culture systems of a more complex nature have been established that imitate the in vivo situation. Organ cultures [21] for example, enable the maintenance of whole segments of the thoracic aorta in culture for more than 2 weeks with an intact endothelium and without any increase in the proliferation rate of the SMCs. If this system is disturbed by a mechanical stimulus comparable to an atherogenic stimulus in vivo, proliferation of SMCs and subsequent formation of an intimal thickening is induced [22].

Transfilter cultures [26, 52] of media explants enable the observation of the formation of a fibromuscular proliferate-like structure (Fallier-Becker et al., submitted). The SMCs of this proliferate are able to dedifferentiate during migration and proliferation and re-differentiate when the plaque formation is finished. Furthermore, the production of the extracellular matrix components, collagen and elastin, is increased. This is comparable to the in vivo situation in atherosclerotic vessels [33]. The transfilter culture system reveals many possibilities for modification. It is also used as a coculture system, when endothelial cells (ECs) are cocultivated with media explants on either side of the filter to study the interactions between the ECs and the SMCs of the media explants. If the ECs are confluent, they are able to prevent the formation of a SMC proliferate. Furthermore, macrophages and LDL, both playing a role in atherogenesis in vivo as atherogenic stimuli, can be added to generate an atheroma in vitro.

References

1. Assmann G (1982) Zur Ätiologie der Atherosklerose. Hämostaseologie 4:162–168
2. Baumgartner H-R (1963) Eine neue Methode zur Erzeugung von Thromben durch gezielte Überdehnung der Gefäßwand. Z Exp Med 137:227–247
3. Baumgartner H-R, Studer A (1966) Folgen des Gefäßkatheterismus am normo- und hypercholesterinaemischen Kaninchen. Pathol Microbiol 29:393–405
4. Bayreuther K, Rodemann HP, Hommel R, Dittmann K, Albiez M (1988) Human skin fibroblasts in vitro differentiate along a terminal cell lineage. Proc Natl Acad Sci USA 85:5112–5116
5. Betz E, Hämmerle H (1984) Arterienwandproliferate und Zellkulturen als Indikatoren für Hemmstoffe der Atherogenese. Funkt Biol Med 3:46–55
6. Betz E, Hämmerle H (1986) Die Wirkung von atherogenen Stoffen in Zellkulturen und Arterienwänden. In: Bundesministerium für Forschung und Technologie (ed) Ersatzmethoden zum Tierversuch. Bundesministerium für Forschung und Technologie, Bonn pp 253–266
7. Betz E, Schlote W (1979) Responses of the vessel walls to chronically applied electrical stimuli. Basic Res Cardiol 74:10–20
8. Björkerud S (1985) Cultivated human arterial smooth muscle displays heterogeneous pattern of growth and phenotypic variation. Lab Invest 53:303–310
9. Björnheden T, Bylock A, Hansson GK, Bondjers G (1983) A system for a long-term perfusion of rabbit aorta in vitro. Arteriosclerosis 3:366–382
10. Blank RS, Thompson MM, Owens GK (1988) Cell cycle versus density dependence of smooth muscle alpha actin expression in cultered rat aorta smooth muscle cells. J Cell Biol 107:299–306
11. Bürk RR (1973) A factor from a transformed cell line that affects cell migration. Proc Natl Acad Sci USA 70:369–372
12. Campbell GR, Chamley-Campbell JH, Burnstock G (1981) Differentiation and phenotypic modulation of arterial smooth muscle cells. In: Schwartz CJ, Werhessen NT, Wolf S (eds) Structure and function of the circulation, vol 3. Plenum, New York, pp 357–399
13. Cereijido M, Ehrenfeld J, Meza I, Martinez-Palomo A (1980) Structural and functional membrane polarity in cultured monolayers of MOCK cells. J Memb Biol 52:147–159
14. Chamley JH, Campbell GR (1974) Mitosis of contractile smooth muscle cells in tissue culture. Exp Cell Res 84:105–110
15. Chamley-Campbell JH, Campbell GR (1981) What controls smooth muscle phenotype? Atherosclerosis 40:347–357
16. Chamley-Campbell J, Campbell GR, Ross R (1979) The smooth muscle cell in culture. Physiol Rev 59:1–61
17. Chamley-Campbell JH, Campbell GR, Ross R (1981) Phenotype-dependent response of cultured aortic smooth muscle to serum mitogens. J Cell Biol 89:379–383
18. Clowes AW, Reidy MA, Clowes MM (1983) Kinetics of cellular proliferation after arterial injury. I. Smooth muscle growth in the absence of endothelium. Lab Invest 49:327–333
19. Clowes AW, Clowes MM, Reidy MA (1986) Kinetics of cellular proliferation after arterial injury. III. Endothelial and smooth muscle growth in chronically denuded vessels. Lab Invest 54:295–303
20. Ehrlich HP (1980) Culture of aorta. Methods Cell Biol 21A:117–134
21. Fingerle JK (1986) Organkultur der thorakalen Kaninchenaorta. Thesis, University of Tübingen
22. Fingerle J, Kraft T (1987) The induction of smooth muscle cell proliferation in vitro using an organ culture system. Int Angiol 6:65–72
23. Freshney RI (1983) Culture of animal cells. In: Freshney RI (ed) A manual of basic technique. Liss, New York
24. Fuchs JCA, Dixon SH Jr, Hagen P (1979) Lipid metabolism and accumulation in perfused rabbit aorta. J Surg Res 26:611–617

25. Gabbiani G (1987) The cytoskeleton of rat aortic smooth muscle cells: normal conditions, experimental intimal thickening and tissue culture. Acta histochem Suppl XXXIV:33–35
26. Grobstein C (1956) Trans-filter induction of tubules in mouse metanephrogenic mesenchym. Exp Cell Res 10:424–440
27. Gröschel-Stewart U, Drenkhahn D (1982) Muscular and cytoplasmic contractile proteins. Coll Relat Res 2:381–463
28. Gröschel-Stewart U, Chamley JH, McConnell JD, Burnstock G (1975) Comparison of the reaction of cultured smooth and cardiac muscle cells and fibroblasts to specific antibodies to myosin. Histochemistry 43:215–224
29. Gröschel-Stewart U, Schreiber J, Mahlmeister C, Weber K (1976) Production of specific antibodies to contractile proteins and their use in immunofluorescence microscopy. I. Antibodies to smooth and striated chicken muscle myosin. Histochemistry 46:229–236
30. Grünwald J, Haudenschild CC (1984) Intimal injury in vivo activates vascular smooth muscle cell migration and explant outgrowth in vitro. Arteriosclerosis 4:183–188
31. Hämmerle H (1987) Wachstum und Differenzierung von arteriellen glatten Muskelzellen bei der Atherombildung und in Zellkulturen. Thesis, University of Tübingen
32. Hämmerle H, Fingerle J, Rupp J, Grünwald J, Betz E, Haudenschild CC (1988) Expression of smooth muscle myosin in relation to growth kinetics of cultured aortic smooth muscle cells. Exp Cell Res 178:390–400
33. Hollander W, Colombo M, Faris B, Franzblau C, Schmid K, Wernli M, Bernasconi U (1984) Changes in the connective tissue proteins, glycosaminoglycans and calcium in the arteries of the cynomolgus monkey during atherosclerotic induction and regression. Atherosclerosis 51:89–108
34. Ignatowski AC (1908) Influence of animal food on the organism of rabbits. S Peterb Izviest Imp Voyenno Med Akad 16:154–173
35. Kagan HM, Milbury PE Jr, Kramsch DM (1979) A possible role for elastic ligands in the proteolytic degradation of arterial elastic lamellae in the rabbit. Circ Res 44:95–103
36. Labarca C, Paigen K (1980) A simple, rapid, and sensitive DNA assay procedure. Anal Biochem 102:344–352
37. Lindl T, Bauer J (1987) Zell und Gewebekultur. Einführung in die Grundlagen sowie ausgewählte Methoden und Anwendungen. Fischer, Stuttgart
38. Mauger JP, Worcel M, Tassin J, Courtois Y (1975) Contractility of smooth muscle cells of rabbit aorta in tissue culture. Nature 225:337–338
39. Mayer HD, Hampton JC, Rosario B (1961) A simple method for removing the resin from epoxy-embedding tissue. J Biophys Biochem Cytol 9:909–910
40. Osborne M, Weber K (1982) Immunofluorescence and immunocytochemical procedures with affinity purified antibodies: tubulin-containing structures. Methods Cell Biol 24:97–132
41. Pederson DC, Bowyer DE (1985) Endothelial injury and healing in vitro. Studies using an organ culture system. AJP 119:264–272
42. Raff MC, Fields KL, Hakomori SL, Minsky R, Pruss RM, Winter J (1979) Cell-type-specific markers for distinguishing and studying neurons and the major classes of glial cells in culture. Brain Res 174:283–309
43. Rhodin JAG (1980) Architecture of the vessel wall. In: Bohr DF, Somlyo AP, Sparks HV (eds) Handbook of Physiology, vol 2, sect 2. American Physiological Society, Bethesda, pp 1–31
44. Ross R (1971) The smooth muscle cell. II. Growth of smooth muscle in culture and formation of elastic fibers. J Cell Biol 50:172–186
45. Ross R, Glomset JA (1973) Atherosclerosis and the arterial smooth muscle cell. Proliferation of smooth muscle is a key event in the genesis of the lesions of atherosclerosis. Science 180:1332–1339
46. Rovner AS, Murphy RA, Owens GK (1986) Expression of smooth muscle and non-muscle myosin heavy chains in cultured vascular smooth muscle cells. J Biol Chem 261:14740–14745

47. Santillan GG, Schuh J, Chan SI, Bing RJ (1980) Binding and internalization of low density lipoproteins by perfused arteries. Biochem Biophys Res Commun 95:1410–1416
48. Spaet TH, Stemerman MB, Veith FJ, Lejnieks I (1975) Intimal injury and regrowth in the rabbit aorta. Medial smooth muscle cells as a source of neointima. Circ Res 36:58–70
49. Skalli O, Bloom WS Ropraz P, Azzarone B, Gabbiani G (1986) Cytoskeletal remodeling of rat aortic smooth muscle cells in vitro: relationships to culture conditions and analogies to in vivo situations. J Submicrosc Cytol 18:481–493
50. Skalli O, Ropraz P, Trzeciak A, Benzonana G, Gillessen D, Gabbiani G (1986) A monoclonal antibody against alpha-smooth muscle actin: a new probe for smooth muscle differentiation. J Cell Biol 103:2787–2796
51. Vesselinovitch D (1988) Animal models and the study of atherosclerosis. Arch Pathol Lab Med 112:1011–1016
52. Weber E, Hämmerle H, Vatti R, Berti G, Betz E (1986) Co-cultivation of endothelial and smooth muscle cells on opposite sides of a porous membrane. Appl Pathol 4:246–252
53. Worton RG, Duff C (1979) Karyotyping. Methods Enzymol 62:322–344

Smooth Muscle Cells from Adult Human Aorta

V. N. Smirnov and A. N. Orekhov

Introduction

Cultures of animal aortic cells have found wide application in the investigation of cellular aspects of atherogenesis. However, culture of human aortic cells is a more adequate model since atherosclerotic lesions occurring in human vessels differ from those induced in experimental animals.

This chapter deals primarily with the culture of human aortic smooth muscle cells (SMCs). The aorta was selected as a source of these cells, taking into account the high occurrence of atherosclerosis in this major vessel. In addition, the aorta is the best-studied human artery.

Special attention is paid to the primary culture of SMCs isolated by enzymatic dispersion. In primary culture, a differentiated state of the cells is retained for 5–7 days, during which the enzyme-isolated SMCs closely resemble their in vivo counterparts. However, the properties characteristic of differentiated SMCs are lost in long-term cultures [1]. Therefore, primary culture of aortic SMCs is proposed as an appropriate adequate model to investigate the cellular mechanisms of atherogenesis. To initiate cultures of differentiated aortic cells, the cell population must be obtained by enzymatic dispersion, but not by any other method (e.g., growing from explants) [1]. The elastase-collagenase mixture was employed to isolate SMCs from various blood vessels including adult human arteries [2–5].

Since initiation and development of atherosclerosis is confined primarily to the intima, a primary culture of subendothelial intimal SMCs is regarded as the most interesting in the investigation of the cellular mechanisms of atherogenesis. This chapter deals with human intimal cells, therefore, it seems reasonable to consider some of the properties of aortic intima.

Specific Features of Cellular Organization in Adult Human Aorta

The intima of adult human aorta is not homogeneous with the respect to its architecture. It is generally recognized that it is composed of two or, sometimes, three sublayers [6–9]. One may distinguish the musculoelastic layer adjacent to the media and the elastic-hyperplastic layer oriented towards the lumen. The elastic-hyperplastic layer is sometimes separated from the lumen by the innermost connective tissue sublayer. The precise borderline between the two innermost sublayers cannot be defined clearly. The morphological border between the intima and the media is represented by a well-marked

Fig. 1a, b. Human aortic sections stained for elastic fibres. **a** Internal elastic lamina (*upper*) and secondary elastic lamina (*lower*), indicated with *arrows,* × 200. **b** Changes in the sublayer thickness in atherosclerosis, × 250. hi, Hyperplastic intima; mi, muscular intima; m, media

internal elastic lamina (Fig. 1a). The second well-discernible elastic membrane is localized between the muscular and hyperplastic layers. This is the internal limited membrane or the secondary elastic lamina. Two elastic membranes, the internal elastic lamina and secondary elastic lamina, limit the muscular layer at both its sides. The thickness of the muscular layer changes insignificantly (Fig. 1b). By contrast, the thickness of the juxtaluminar hyperplastic layer varies considerably. At the site of an atherosclerotic lesion one may observe thickening of the hyperplastic but not of the muscular layer (Fig. 1b). Thus, intimal thickening at the site of an atherosclerotic lesion is due to thickening of the hyperplastic layer [6].

On a vertical section, most aortic cells are of similar elongated shape (Fig. 2a). However, on a horizontal section the cells exhibit really diverse shapes (Fig. 2b). In these preparations, as well as in three-dimensional pictures obtained by scanning electron microscopy, most cells in the muscular layer have an elongated bipolar shape characteristic of SMCs (Fig. 2c). These cells are densely packed and attach to each other by side surfaces. The cells of the muscular layer form dense rows of different orientations. Both in fatty streak and in atherosclerotic plaque, the structure of the muscular layer of human aortic intima is altered insignificantly. As in a grossly normal intima, the cells of the muscular layer of the plaque have an elongated bipolar shape resembling that of medial SMCs and they are very densely packed.

The hyperplastic intimal sublayer differs considerably from the muscular layer in respect to cellular organization. Most cells of the superficial hyperplastic layer are of stellate shape; they have multiple side processes (Fig. 2b, d). The cells of the hyperplastic layer are distributed rather sparsely but they contact each other with side processes. Long and thin processes of cells of the hyperplastic layer form a loose network. Cells are seen as the junctions of this network. In the hyperplastic layer there are more or less dense cell clusters (Fig. 2e). The cells are arranged in clusters with cell-free spaces between them.

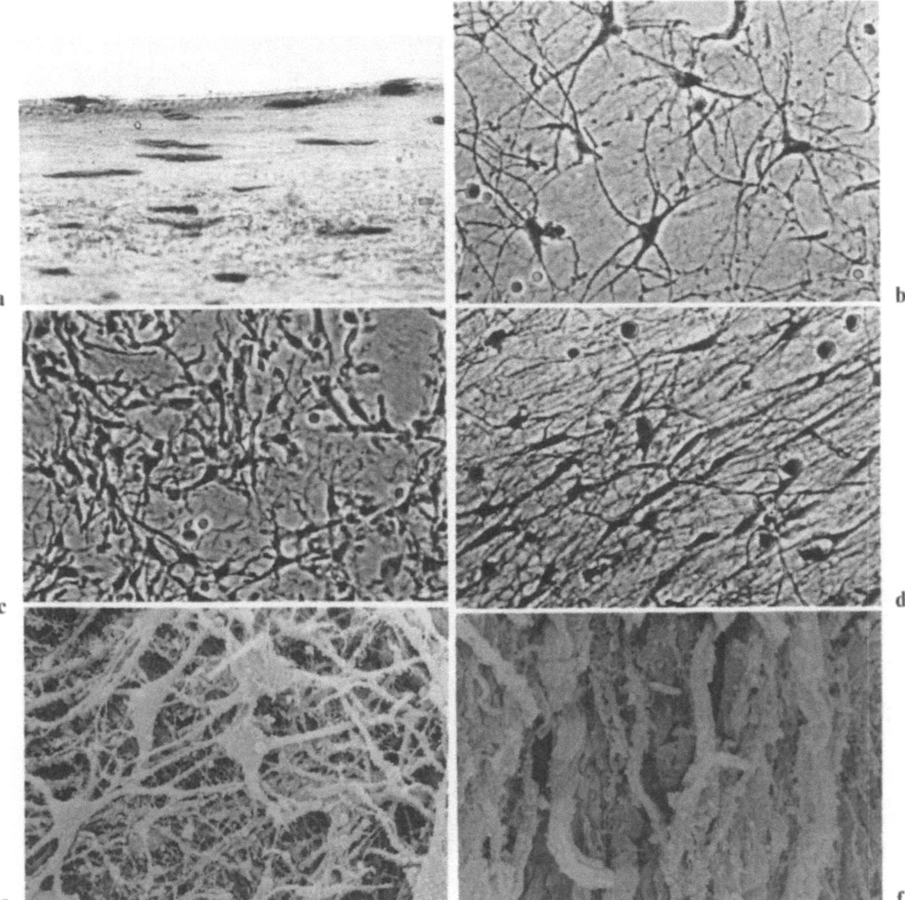

Fig. 2a–f. Subendothelial intimal cells. **a** Human aorta, vertical section (hematoxylin-eosin); elongated cells are clearly seen, × 400. **b** Hyperplastic intimal layer, horizontal section (phase contrast, hematoxylin), × 200. **c** Hyperplastic intimal layer, horizontal section (phase contrast, hematoxylin): cell clusters and areas of sparsely packed cells are seen, × 200. **d** Cellular network in the depth of the hyperplastic layer (phase contrast, hematoxylin): cell clusters and areas of sparsely packed cells are seen, × 200. **e** Hyperplastic layer (scanning electron microscopy), × 570. **f** Muscular layer (scanning electron microscopy), × 1050

Cells of the deeper parts of the hyperplastic layer localized at the muscular layer also form a loose network, however, these cells are more elongated (Fig. 2f).

Using scanning electron microscopy, it has been established that all cells of the hyperplastic layer contact each other by their processes (Fig. 2d). Cells have contacts not only in the horizontal but also in the vertical plane. Intercellular contacts were found between the cells localized at different levels of the hyperplastic layer. Thus, in the hyperplastic layer there is a three-dimensional structure made of horizontal and vertical networks.

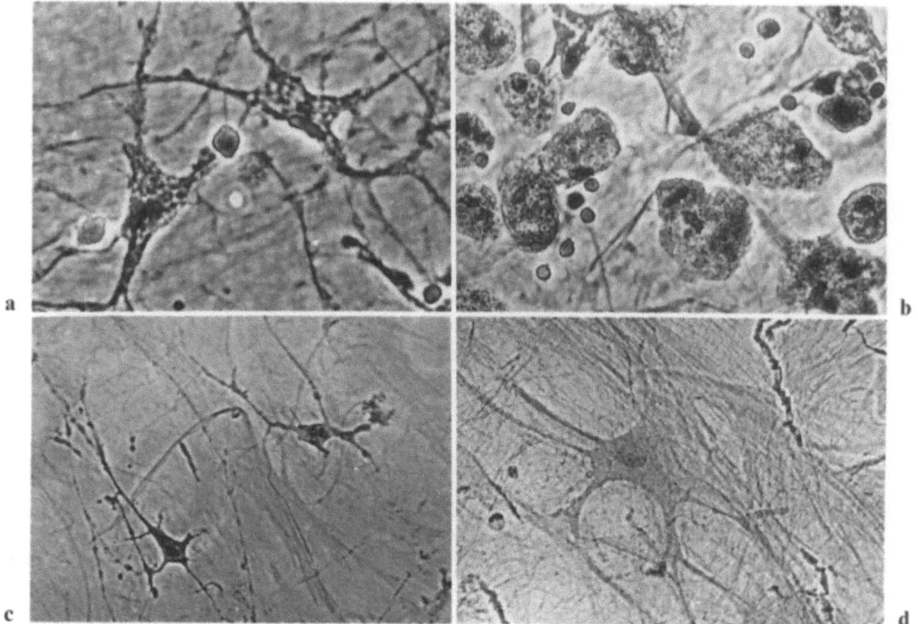

Fig. 3a–d. Atherosclerotic cells. **a** Stellate foam cell; **b** monocyte-like foam cells, (phase contrast, hematoxylin, × 480); **c** sparsely packed cells of fibrous plaque; **d** giant stellate cell; (phase contrast, hematoxylin, × 200)

In the superficial regions of the hyperplastic layer of fatty streak, the three-dimensional cellular network typical of unaffected intima is disintegrated. A great number of cells appear with the cytoplasm containing lipid inclusions (Fig. 3). These are so-called foam cells. One may distinguish foam cells of two major types: type 1, cells of stellate shape which form a three-dimensional network in the hyperplastic layer (Fig. 3a), and, type 2, round or ovoid cells resembling blood monocytes. Type 2 cells are few in normal vascular segments, while in the intima of a fatty streak their number increases (Fig. 3b). The greatest number of foam cells are found in the uppermost regions of the hyperplastic sublayer. At the deeper levels of this layer, the occurrence of foam cells gradually decreases, and finally at a certain depth one can find a cellular network made up of stellate cells which is characteristic of the hyperplastic layer of uninvolved intima.

In atherosclerotic plaque the changes in cellular organization of the hyperplastic layer are even more marked. However, in immediate proximity to the plaque, that is, in the marginal zone, subendothelial cells are essentially similar to those of unaffected areas. Subendothelial cells of the marginal zone, just as in unaffected areas, form a three-dimensional network. In the subendothelial zone of the central part of the plaque, the picture is dramatically different from the marginal zone and the normal intima. Directly beneath the endothelium of the plaque there is a thick fibrous layer called a fibrous cap. The cap consists of rough fibers of the connective tissue extracellular matrix; in the cap the cells are sparsely distributed, intercellular contacts are generally absent (Fig. 3c). In the deeper layers of the fibrous cap one can often see separate giant cells of stellate shape (Fig. 3d). The plaque core under the fibrous plaque cap is a result of the accumulation of fragments of disintegrated giant cells, blood cells, and extracellular matrix.

Thus, muscular and hyperplastic layers of the intima have considerable differences in respect to cellular organization. By shape and organization, the cells of the muscular layer resemble SMCs of the media. The cells of the hyperplastic layer have numerous processes and normally form a sparse three-dimensional network. In the atherosclerotic lesion characterized by lipid deposition, the cellular network of the hyperplastic layer is disintegrated.

To investigate the functions of cells populating the hyperplastic and muscular intimal layers, as well as the cells of the media, a method to obtain a primary cell culture from the given layer of human aorta has been developed. Isolation and culturing conditions for cells from different intimal layers are given in Table 1.

Table 1. Characteristics of cell isolation and cultivation

Tissue	Enzyme concentration (%)		Time of digestion (h)	Cell yield ($\times 10^{-6}$/g tissue)	Cell spreading (days)
	Collagenase	Elastase			
Hyperplastic intima	0.1	0	2–3	1–1.5	1–2
Muscular intima	0.15	0.01	3	1–3	2–3
Media	0.225	0.05	4	5–10	3–5

Culture of Smooth Muscle Cells

Primary Culture

Primary culture of aortic cells can be obtained by at least two methods: explant outgrowth and enzymatic dispersion. To isolate the cells of the hyperplastic layer 0.1% collagenase is used. Since the extracellular matrix of the muscular layer cannot be disintegrated only by collagenase, upon isolation from the muscular intima layer this enzyme is employed in a higher concentration and in combination with elastase. SMCs from the media are isolated using a mixture of collagenase and elastase taken at even higher concentrations. The ratio and concentrations of enzymes were specially selected for effective digestion of the matrix in each layer. Such a treatment enabled us to obtain the cells of the respective layer without significant contamination by the cells from other layers.

Enzyme Dispersion

Preparation of Enzymes for Tissue Digestion: Collagenase type II (Worthington), elastase type I and elastase type III (Sigma) were dissolved in Ca^{2+}- and Mg^{2+}-free Dulbecco's phosphate buffered saline (PBS; GIBCO Ltd.). All enzyme solutions contained 10% fetal calf serum, 0.1% glucose, 25 mM HEPES (pH 7.4), and 100 U/ml penicillin, 100 µg/ml streptomycin, and 2.5 µg/ml fungizone (all GIBCO). The enzymes can be dissolved in medium 199 (GIBCO) supplemented with serum and antibiotics. There was no difference in digestion efficiency, yield, and viability of cells isolated with enzymes dissolved either in medium 199 or PBS. Enzyme solutions were sterilized by filtration through a 0.45-µm filter, made into aliquots, frozen, and kept at $-20\,°C$ for up to 6 months.

To both collagenase and elastase solutions 10% fetal calf serum was added in order to inhibit nonspecific proteases present in the enzyme preparation [10]. This addition was found to have a positive effect on the ability of isolated cells to spread on a plastic substrate. The cells isolated in the presence of the serum attached and spread much faster than those isolated without serum. Sometimes the cells isolated in the absence of the serum did not spread at all or failed to attach to the substrate.

Preparation of Aortic Intima and Media for Enzymatic Digestion: Pieces of thoracic and abdominal aorta were taken aseptically from men aged 30–60 years within 1.5–3 h after sudden death. A piece of the vessel (3–5 cm) was ligated in situ at both ends, cut out, and placed in sterile Eagle's medium (150 ml) supplemented with kanamycin (100 µg/ml) and fungizone (2.5 µg/ml; GIBCO). Transportation of the material from a morgue to the laboratory usually lasted 45 min. The container was put into a laminar flow hood and all subsequent operations were carried out under sterile conditions. The vessel was cut longitudinally and the adventitia was separated from the adjacent media. The intima with internal media was placed into a 10×10 cm dish and washed with Eagle's medium containing antibiotics. The media was separated from the intima with microdissectional forceps. Separation along the internal elastic lamina was controlled under a light microscope. Separated media contained no fragments of intima or adventitia. However, during separation the intimal layer is stripped off together with small pieces of media. In the intima, the hyperplastic sublayer was mechanically separated from the muscular sublayer with forceps. Accuracy of separation can be assessed microscopically by the orientation of collagen and elastic fibers [11, 12].

Cells from normal and atherosclerotic aortic areas were isolated and cultured separately. Two types of atherosclerotic lesions were generally distinguished: an atherosclerotic plaque and a fatty streak. Plaques were cut out along their outline, the intima was separated from the media, and cells were isolated from each layer according to the procedure employed for uninvolved vascular segments. Intimal and medial cells from the areas with fatty streak congestions were also isolated and cultured separately.

Digestion of Media (Procedure I): The media was separated into fibers using microdissectional forceps. After dissection, the tissue was placed into a 50-ml tube with a lid and 0.2% solution of elastase type I was added in the ratio of 10 ml solution per 1 g of tissue. The tissue was incubated at 37 °C in a water-bath with agitation at 50 rpm. After a 2-h incubation, the elastase solution was discarded, the tissue was washed twice with PBS (pH 7.2), and 0.3% collagenase solution was added. The tissue was incubated with collagenase until practically complete dispersion, which usually required 3–5 h.

Digestion of Media (Procedure II): Dissected media was incubated with agitation at 37 °C in a mixture of 0.05% elastase type III and 0.225% collagenase. One gram of tissue was digested with 10 ml of enzyme solution. Incubation continued until complete dispersion of the tissue.

Cell yield, viability, and efficacy of tissue dispersion were independent of the procedure applied (i.e., procedure I or II).

Digestion of Hyperplastic Intima: Hyperplastic intima was placed in a 50-ml tube with screw cap and 0.1% collagenase was added (10 ml collagenase per 1 g tissue). The tissue was incubated at 37 °C with agitation until complete dispersion, which required 1–2 h.

Digestion of Muscular Intima: Muscular intima was separated into pieces with forceps, placed into a 50-ml cupped tube, and incubated with shaking at 37 °C with 0.15% collagenase and 0.01% elastase type III (10 ml of enzyme solution per 1 g of tissue) until practically complete dissolution (2–3 h).

In a special series of experiments, it was found that the mixture of enzymes used for the intimal cell isolation releases less than 0.5% of cells from the medial tissue during the time period required for complete digestion of the intima. Therefore, the conditions used for digestion of the muscular intima (0.15% collagenase and 0.01% elastase) essentially prevent contamination of the intimal cell suspension by medial cells.

Cell Culture: Using the technique for cell isolation described above, one can obtain $1–1.5 \times 10^6$ cells from 1 g of hyperplastic intima, $1–3 \times 10^6$ cells from 1 g of muscular intima, and $5–10 \times 10^6$ cells from 1 g of the media. Cell yield from the intima and media of normal vessel wall estimated on the basis of the DNA content was about 70%. Cell yield from the intimal layer of an atherosclerotic lesion was two- to three-fold lower, probably due to the large amount of extracellular matrix in atherosclerotic tissue. The viability of isolated cells determined by trypan blue staining was about 85%–90%.

Cell suspension obtained by enzymatic dispersion was filtered through gauze and centrifuged for 10 min at 200 g. Supernatant was discarded, and cells were resuspended in medium 199 containing 10% fetal calf serum, 2 mM glutamine, 100 U/ml penicillin, 100 µg/ml streptomycin, and 2.5 µg/ml fungizone (all GIBCO). Cells were seeded at a density of 10^4 cells per 1 cm^2 in tissue culture plastic, and cultured at 37 °C in a humidified atmosphere of 5% CO_2 and 95% of air.

Thirty to fifty percent of cells seeded into plastic flasks attach to the substrate and spread. Hyperplastic intimal cells attach and flatten after 1–2 days in culture and muscular intimal cells after 2–3 days, whereas medial cells require 3–5 days. The variable spreading rate cannot be explained by the differences in isolation techniques since intimal cells spread faster even when isolated under the conditions applied to medial cells.

Cellular Polymorphism

Light Microscopy of Culture: Cultured intimal and medial cells spread over the substrate and take on different shapes. Primary cultures of medial and intimal cells differ in their appearance. In cultures of media and muscular intima, one can see many elongated cells whereas most hyperplastic intima cells have several processes; these cultures also contain stellate cells absent from cell cultures initiated from the media (Fig. 4a, b).

The morphology of aortic primary cultures was analyzed after 7 days in culture since by this time most of the cells attached to the substrate had flattened. Based on morphological criteria, four major cellular types in cultures of cultured human aortic cells can be distinguished: (1) polygonal, (2) stellate, (3) asymmetric, and (4) elongated. It should be pointed out that this classification is rather tentative, and is employed to describe the polymorphism of cultured aortic cells.

Elongated cells (Fig. 4c) have an elongated endoplasm which occupies a larger part of the cell. Lamellar ectoplasm is usually smaller in area and is localized to one of the cell ends. Side lamellar protuberances are usually one-sided.

Asymmetric cells (Fig. 4d) have an irregularly shaped endoplasm with one or several thread-like processes. Cells with processes always have lamellar protuberances on the side opposite to the processes. In all cells, these processes lack any regular form their edges being very thin and indistinctly shaped.

Stellate cells (Fig. 4e) differ from the other cell types in the endoplasm divided into long radial processes. The nucleus is localized in the center of the cell. The ectoplasm is not clearly defined. Lamellae can be found at the end of the processes and in the form of a web at their basis.

In *polygonal cells* (Fig. 4f), the endoplasm is localized in the central part of the cell and is evenly surrounded by lamellar ectoplasm. In most cases, lamellar edges are definitely shaped into a characteristic polygonal figure. Between the vertices of the polygon the lamellar edges are smoothly concave. The number of vertices in polygonal cells usually varies from three to five.

After spreading, stellate cells take on the appearance of polygonal cells. The ectoplasm of stellate cells is expanded between the processes, thus forming a characteristic lamella. Therefore, stellate and polygonal cells may be regarded as a single cell type varying in their degree of spreading.

In addition to varying shape, the cell types described have other morphological differences which can be revealed only by a detailed analysis [13, 14]. Thus, three to five nucleoli of irregular shape are characteristic of asymmetric cells. Other cells usually have one to two large round nucleoli. The ectoplasm of elongated and polygonal cells generally exhibits more or less regular striation. In polygonal cells it is present in the form of intersecting or adjacent files of parallel fibrils. In elongated cells the fibrils are longitudinally oriented. The

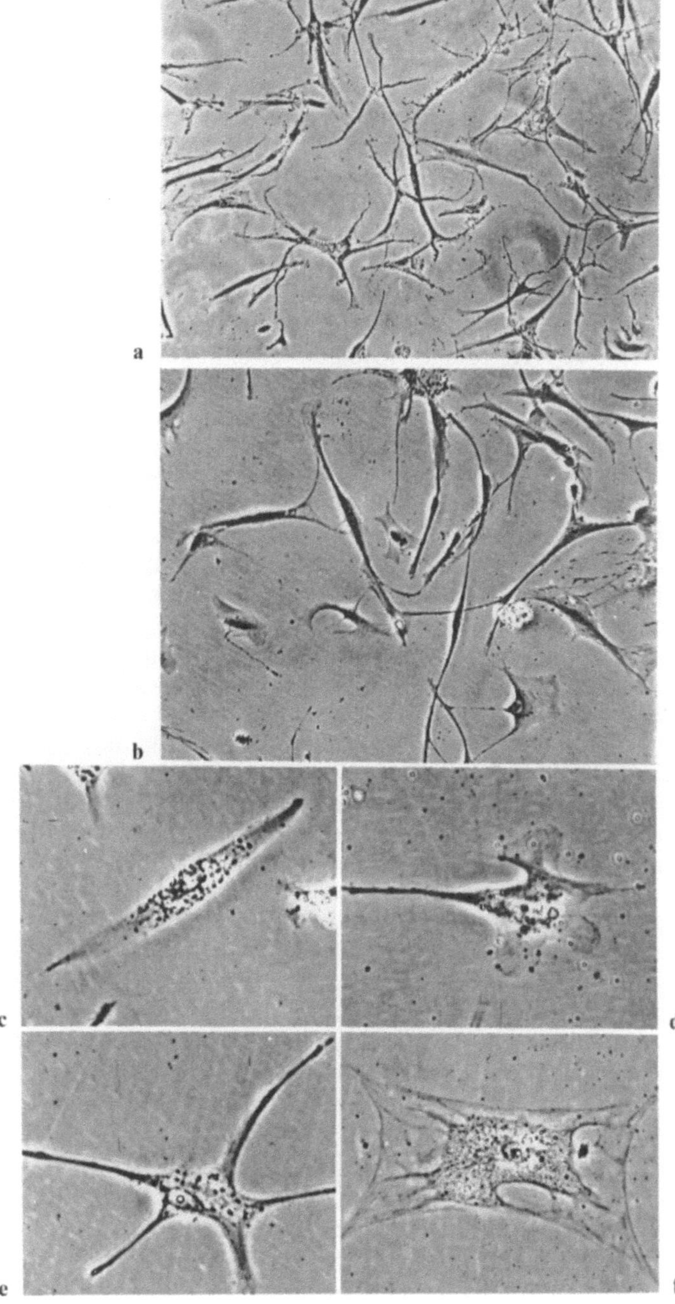

Fig. 4a–f. Aortic cells in primary culture. **a** Culture of hyperplastic intimal cells; **b** culture of medial cells; **c** elongated cell; **d** asymmetric cell; **e** stellate cell; **f** polygonal cell. (Phase contrast, ×600)

cytoskeleton of asymmetric cells is constituted by short and thin fibrils, while in asymmetric cells they are localized without a visible regularity.

Indentification of Cultured Cells: Immunofluorescence was used to identify the principal cell types present in primary cultures of human aorta. As known, SMCs in primary culture may be distinguished from fibroblasts and endothelial cells using antibodies to smooth muscle myosin [1, 15]. Four morphological cell types, namely, stellate, polygonal, asymmetric, and elongated, are brightly stained with antimyosin antibodies [13, 14]. From these observations it may be concluded that the cells which make up the population of aortic culture are primarily of the smooth muscle origin.

Ultrastructural analysis of isolated cells in suspension showed that almost all cells derived from the medial layer of human aorta have a similar structure [14, 16]. In suspension, all medial cells have a rounded shape and heavily deformed nucleus. All subcellular organelles are in the central part of the cell, while a broad marginal zone is filled with densely packed microfilaments. The presence of multiple filaments suggests that these cells are SMCs.

Most of the cells isolated from muscular intima have a structure similar to that of medial cells. Suspensions of hyperplastic intimal cells occasionally contain some lymphocyte-like cells and a small number of nondifferentiated cells. Cells that exhibit certain ultrastructural characteristics of SMCs form the major part of the cell population isolated from hyperplastic intima.

Investigation of the ultrastructure in the cells cultured from the media and muscular intima showed that most of the cultured cells have all the necessary characteristics to permit them to be regarded as SMCs [2, 7, 17]. Specifically, they have the basal lamina, microfilament bundles with dense bodies, and characteristic micropinocytotic vesicles [14, 16]. However, many cultured cells from the hyperplastic intima and some medial cells failed to exhibit the whole set of smooth muscle characteristics and some even lack them completely. This fact may reflect the process of dedifferentiation or modification of SMCs in culture. Furthermore, the possibility that the cells underwent modification while still in the vessel wall cannot be excluded.

In cultures derived from atherosclerotic lesions, one can often find cells with vesicle-like inclusions which in some cases fill the whole cytoplasm. As determined by oil red 0 staining, these inclusions are of lipid nature. The cells with lipid inclusions were found among all four morphological types. Using an immunofluorescent technique, it was demonstrated that lipid-overloaded cells cultured from atherosclerotic aorta have a smooth muscle nature [14].

Explant Method and Long-Term Culture

For isolation of human aortic SMCs with the explant method and for long-term culturing. Dulbecco's modified Eagle's medium supplemented with 10% human serum, 10 mM HEPES, 100 µg/ml sodium pyruvate, 50 µg/ml ascorbic acid, 2 mM L-glutamine, 100 µg/ml streptomycin, and 2.5 µg/ml fungizone was routinely used. All components except the human serum were obtained either from GIBCO or Flow Laboratories (U.K.). Human serum was prepared from pooled donor blood and heat-inactivated at 56 °C for 20 min.

To isolate the cells from human aortic intima and media, pieces of the tissue were finely minced with scissors and scalpel blades and transferred to the tissue culture flask in the minimal amount of complete medium, just enough to cover the pieces. The initial density of the explants was approximately 1 piece/cm^2. The flasks were incubated at 37 °C in an atmosphere of 5% CO_2:95% air overnight, then fresh medium was added to the necessary volume, and detached pieces of the tissue were removed and discarded. The medium was changed twice a week, and the cells usually began to migrate from explants between days 7 and 10 of culturing. After 21–25 days in culture, the islets of outgrowing cells were trypsinized with 0.25% trypsin (GIBCO), and seeded into 35-mm tissue culture dishes. After confluency, cells were trypsinized, resuspended in complete medium, and split 1:3. The cells could be cultured up to the 18th–20th passage when the signs of quiescence were obvious.

It should be mentioned that the cells growing out from explant lack certain features of differentiated (or "contractile") SMCs, and express the "synthetic" phenotype [1]. Thus, the cells prepared using the explant method even in primary culture are quite similar to subcultured SMCs. When a large amount of subcultured cells is required (immunization or drug screening) the explant method seems to be preferable since it is less expensive.

Atherosclerotic Parameters of Cultured Cells

Since subendothelial intimal cells appear to be the principal cells involved in atherogenesis, their culture seems to be the most appropriate model to study cellular aspects of atherosclerosis. Primary culture of enzyme-isolated intimal cells which retain their main in vivo characteristics is of paramount interest.

Increased proliferation and intracellular lipid accumulation are the earliest signs of atherosclerosis [18, 19], therefore we will present these parameters for primary cultures of intimal cells.

Proliferative Activity

Cells isolated from intimal sublayers preincubated with [^{3}H]thymidine did not differ in the DNA synthesis. However, in the intimal cells cultured from fatty streak [^{3}H]thymidine incorporation was considerably higher compared with that in cells from noncompromised intima (Table 2).

Table 2. [^{3}H]Thymidine incorporation in cells isolated from human aorta

Tissue	[^{3}H]Thymidine incorporation (dpm/10^3 cells)		
	Normal	Fatty streak	Plaque
Hyperplastic intima	6.1 ± 0.3	14.2 ± 0.3*	5.2 ± 0.4
Muscular intima	8.4 ± 0.3	15.4 ± 0.2*	7.2 ± 0.5

* Significant difference from normal ($P < 0.05$)

Among intimal cell cultures, the highest value of thymidine index (the relative number of cells incorporating thymidine into the synthesized DNA) was in those derived from fatty lesions [20, 21]. Their mean thymidine index value in fatty streak cultures was 2.5-fold higher than that in the cultures established from unaffected areas [20]. On the other hand, the mean thymidine index value for the plaque cells was slightly lower compared with the cells of unaffected intima [20]. Similar results were obtained with radiometric determination of thymidine incorporation [21].

Thus, the cells derived from fatty streak exhibited the highest proliferative activity, while that of the cells cultured from atherosclerotic plaque was essentially the same as in the cells from uninvolved intima.

Lipid Content

In intimal cells isolated from fatty streak and atherosclerotic plaque intracellular lipid contents were considerably higher than in normal intimal cells [22]. Lipid content in the cells from the muscular sublayer was higher than in cells from the hyperplastic sublayer (Table 3). Generally, intracellular lipid content of both sublayers was 1.5- to 2-fold higher than that in grossly normal intima. In fatty streak, cells from both the hyperplastic and muscular layers contain more lipids than those in the correspondent layers of fibrous plaque (Table 3).

More than 99% of lipids in human aortic cells are phospholipids, triglycerides, and free and esterified cholesterol. Intracellular lipid accumulation occurring in atherosclerosis is associated with an increase in both esterified or free cholesterol and, to a lesser degree, trigylceride contents [22]. In cells cultured from unaffected intima and fatty streak, the lipid content did not change significantly within 2 weeks of culturing and was similar to the lipid content in suspension of freshly isolated cells [21, 22]. The lipid content in cells cultured from plaque at the beginning of culturing was slightly lower than in the suspension of freshly isolated cells. This level was retained until day 10–12, and by the 14th day the lipid content decreased by 20% [22]. On day 3 in culture, the ratio of phospholipids, triglycerides, cholesterol, and cholesteryl esters in cultured cells was virtually the same. The proportion of triglycerides and free cholesterol in cells did not change significantly from day 3 to 14 in culture. The proportion of phospholipids and cholesterol esters was constant until day 12, and by the 14th day the proportion of the former slightly increased, while the share of the latter dropped [21, 22].

Table 3. Lipid content in cells isolated from human aorta

Tissue	Lipid content (μg/10^5 cells)		
	Normal	Fatty streak	Plaque
Hyperplastic intima	45 ± 9	77 ± 14	70 ± 9
Muscular intima	69 ± 14	135 ± 15*	122 ± 22*

* Significant difference from normal ($P<0.05$)

Table 4. Drugs tested on a cellular model

Drug	References
Antiatherosclerotic	
Cyclic AMP elevators	Orekhov et al. [21], Tertov et al. [23, 27]
Prostacyclin	Orekhov et al. [21, 24, 28], Kudryashov et al. [25], Akopov et al. [30]
Prostaglandin E_2	Orekhov et al. [21], Kudryashov et al. [25]
Artificial HDL	Orekhov et al. [26]
Antioxidants	Orekhov et al. [21]
Calcium antagonists	Akopov et al. [30], Baldenkov et al. [32], Orekhov et al. [33–35]
Trapidil and trapidil derivatives	Giessler et al. [29]
Lipoxygenase inhibitors	Tertov et al. [31]
Atherogenic	
Beta-blockers	Tertov et al. [27], Orekhov et al. [35]
Thromboxane A_2	Akopov et al. [30], Baldenkov et al. [32]
Indifferent	
Nitrates	Orekhov et al. [35]

Use of Cell Cultures for the Study of Atherosclerosis

This section will consider the use of primary culture of human aortic intimal cells in atherosclerosis research. The investigations of antiatherosclerotic effects of drugs in cell culture will be given as one of the illustrations. Findings of the study on the mechanisms of cellular lipid accumulation, which is the main cellular manifestation of atherosclerosis, will serve as another example.

Atherosclerosis-Related Effects of Drugs

Cells cultured from atherosclerotic lesions of human aorta retain their main properties characteristic of atherosclerotic cells. This allows one to regard a culture of atherosclerotic cells as a convenient model to assess antiatherosclerotic activity of various agents [21]. The effects of different drugs and chemicals were examined using this model [23–36]. The drugs tested are summarized in Table 4. Some of them produced antiatherosclerotic effects on cell culture, others appeared to be ineffective or even capable of stimulating the development of atherogenic processes.

Among the number of tested drugs three classes of cardiovascular drugs (calcium antagonists, beta-blockers, and nitrates) have been tested. All calcium antagonists lowered both the lipid level and [^{3}H]thymidine incorporation in cultured atherosclerotic cells [33–35]. Verapamil and nifedipine proved to be the most effective. Thus, calcium antagonists manifested direct antiatherosclerotic action at the cellular level. Nitrates had no effect on intracellular cholesterol level and practically did not affect the proliferative activity of atherosclerotic cells [35]. On the other hand, all beta-blockers examined increased atherosclerotic manifestations, i.e., all of these drugs exhibited atherogenic activity in culture [35, 36]. To confirm these observations in vivo, the blood plasma of patients who received calcium antagonists and beta-blockers was analysed. After oral administration of a calcium antagonist, the patients' plasma acquired antiatherosclerotic properties, i.e., the intracellular cholesterol level was lowered and the proliferation of atherosclerotic cells inhibited [35]. In contrast, the plasma of patients who received propranolol was atherogenic. Its atherogenic properties manifested themselves at the arterial cell level in the rise of intracellular cholesterol and stimulation of cell proliferation [35, 36].

The model can be used not only to test drugs but foodstuffs as well. Specifically, the patients were given canned meat of a mollusc belonging to genus *Buccinum*. After this single dietary load, the patients' blood plasma exhibited marked antiatherosclerotic properties: the addition of this plasma to cultured cells caused a decrease in intracellular cholesterol level (Table 5). Another patient had initially atherogenic plasma (see next section) causing a more than two-fold increase in the cholesterol level of cells derived from normal intima. This patient received a single dietary load of Antarctic krill meat. Two hours later the atherogenicity of his plasma had decreased and 4 h later it was practically absent (Table 5). This, krill meat possesses a preventive antiatherogenic action on arterial cells.

General Mechanisms or Intracellular Lipid Accumulation

Lipid accumulation in subendothelial intimal cells is the earliest noticeable sign of atherosclerosis. The increase in cholesterol content of intimal aortic cells stimulates their proliferation and synthesis of the extracellular matrix which are the main manifestation of atherosclerosis at the cellular level [37, 38]. Therefore, it was suggested that intracellular lipid accumulation is an initiating factor of atherosclerosis at the level of vascular cells. In this section two mechanisms of intracellular lipid accumulation are considered.

Atherogenic Properties of Blood Plasma

It has been found that incubation of intimal cells cultured from unaffected human aorta with blood serum or plasma of patients with angiographically assessed coronary atherosclerosis led to accumulation of intracellular cholesterol [39–42]. It has been demonstrated that the deposition of intracellular cholesterol is accompanied by stimulation of cell proliferation, and enhanced synthesis of total protein, collagen, sulfated glycosaminoglycans, and hyaluronic acid [37, 38]. Low density lipoprotein (LDL), derived from atherogenic plasma, causes atherogenic manifestations in cultured cells [38, 40–42]. The sialic acid content in atherogenic LDL was threefold lower than that in nonatherogenic LDL [43]. On the other hand, the addition to cultured cells of initially nonatherogenic native LDL desialylated with neuraminidase caused a

Table 5. Effects of sea products on atherosclerosis-related properties of patients' plasma

Product	Intracellular cholesterol content (µg/mg cell protein)			
	Control	0 h	2 h	4 h
Buccinum	196 ± 17	204 ± 16	181 ± 19	150 ± 16*
Krill	37 ± 5	86 ± 4	71 ± 7	49 ± 4*

* Significant difference from 0 h ($P < 0.05$)

Table 6. Affinity constants of lipoprotein-anti-LDL interaction

Lipoprotein	Affinity constant $(\times 10^{-7}, M^{-1})$
Healthy donors' LDL	2.4
Cu^{2+}-oxidized LDL	2.5
Glycosylated LDL	2.6
Acetylated LDL	2.8
Lp(a)	3.6
Malondialdehyde-treated LDL	10.9
Patients' LDL	11.3
Desialylated LDL	89.4

substantial increase in intracellular cholesterol [43]. The atherogenic plasma of patients contains a nonlipid factor, immunoglobulin G (IgG), interacting with LDL and capable of increasing the atherogenicity of LDL obtained from this plasma [40]. The formation of an LDL complex with these IgG is accompanied by the binding of the C1q component of complement. F(ab')$_2$ fragments of these IgG interact with LDL. Table 6 shows the affinity constants of anti-LDL IgG to native and modified lipoproteins. The antibodies showed high affinity for LDL isolated from the atherogenic blood plasma of patients. Desialylated LDL (initially native LDL treated with neuraminidase) had the highest affinity among modified lipoproteins. One may assume that autoantibodies in the blood are produced exactly against modified (desialylated) LDL.

Thus, it has been established that the plasma of atherosclerotic patients contains two atherogenic factors. There are modified (desialylated) LDL, which is itself atherogenic, and antibodies against modified LDL, which are capable of increasing the atherogenicity of modified LDL and converting initially nonatherogenic lipoproteins into atherogenic lipoproteins.

General Mechanisms of LDL-Induced Lipid Accumulation

Unlike native LDL, modified LDL induces intracellular lipid accumulation. LDL can be modified either chemically [44–50] or by insolubilization (formation of an insoluble complex with any agent) [51–56]. The general mechanisms leading from various lipid modifications to cellular lipidosis are unknown.

Table 7. Microaggregation of native and modified LDL and its effect on cholesteryl ester content in cultured aortic cells

Lipoprotein	Cholesteryl ester content (% above control)	Average size of particle (relative units)
Healthy donors' LDL	17 ± 6	6.8
LDL + lipopolysaccharide	108 ± 15	32.5
Glycosylated LDL	117 ± 11	30.0
Patients' LDL	122 ± 14	22.5
LDL + smoke[a]	135 ± 11	29.3
Lp(a)	154 ± 10	39.0
LDL + anti-LDL IgG	169 ± 21	48.4
Desialylated LDL	184 ± 14	39.5
Cu^{2+}-oxidized LDL	186 ± 17	45.2
Lp(a) + LDL	202 ± 21	56.4
Malondialdehyde-treated LDL	217 ± 28	52.3

[a] Cigarette smoke extract was prepared according to Yokode et al. [57]

Using primary culture of human aortic intimal cells, it was demonstrated that the formation of lipoprotein aggregates is a general moment in all cases of chemical modification. It was found that LDL modified in vitro, but not native LDL, form aggregates consisting of three to five or even more particles (Table 7). At the same time, there is a direct correlation between the degree of aggregation and intracellular lipid content ($r = 0.88$, $P < 0.01$). Removal of LDL aggregates from the culture medium prevented the increase in intracellular lipid content. These results indicate that LDL aggregation plays a main, if not the primary, role in intracellular lipid accumulation.

Cultured intimal cells were also used to elucidate the nature of LDL atherogenicity resulting from insolubilization. It was found that formation of an insoluble LDL-containing complex is the principal step of this modification [56]. It is noteworthy that the degree of lipid accumulation is independent of the nature of the insolubilizing agent. Presumably, both after aggregation and insolubilization lipoprotein is internalized not via the specific LDL receptor associated with regulation of intracellular lipid metabolism, but rather via nonspecific phagocytosis. LDL internalized in this nonspecific manner probably cannot trigger the cellular mechanisms for its utilization, which results in LDL accumulation in the cell.

References

1. Chamley-Campbell J, Campbell GR, Ross R (1979) The smooth muscle cell in culture. Physiol Rev 59:1–61
2. Gimbrone MA, Cotran RS (1975) Human vascular smooth muscle in culture. Growth and ultrastructure. Lab Invest 33:16–27
3. May JF, Paule WJ, Rounds DE, Blankenborn DH, Zemplenyi T (1975) The induction of atherosclerotic plaque-like mounds in cultures of aortic smooth muscle cells. Virchows Arch[B] 18:205–211

4. Ronnemaa T, Jarvelainen H, Lehtonen A, Tammi M, Larjava H, Saarni H, Vihersaari T, Viikari J (1980) Serum lipoprotein composition, hormones, and the synthesis of glycosaminoglycans by human aortic smooth muscle cells. Artery 8:323–328
5. Thyberg J, Nilsson J, Palmberg L, Sjolund M (1985) Adult human arterial smooth muscle cells in primary culture. Modulation from contractile to synthetic phenotype. Cell Tissue Res 239:69–74
6. Orekhov AN, Andreeva ER, Tertov VV (1987) The distribution of cells and chemical components in the intima of human aorta. In: Chazov EI, Smirnov VN (eds) Human atherosclerosis. Harwood, Chur, pp 75–100 (Soviet medical reviews, sect A: Cardiology reviews, vol 1)
7. Geer JC, Haust MD (1972) Smooth muscle cells in atherosclerosis. Karger, Basel
8. Gross L, Epstein EZ, Kugel MA (1980) Histology of the coronary arteries and their branches in the human heart. Am J Pathol 10:253–274
9. Movat HZ, More RH, Haust MD (1958) The diffuse intimal thickening of the human aorta with aging. Am J Pathol 24:1023–1031
10. Brattain MG (1979) Tissue disaggregation. In: Melamed MR, Mullaney PF, Mendelson MI (eds) Flow cytometry and sorting. Wiley, New York, pp 193–205
11. Orekhov AN, Karpova II, Tertov VV, Rudchenko SA, Andreeva ER, Krushinsky AV, Smirnov VN (1984) Cellular composition of atherosclerotic and uninvolved human aortic subendothelial intima. Light-microscopic study of dissociated aortic cells. Am J Pathol 115:17–24
12. Orekhov AN, Andreeva ER, Tertov VV, Krushinsky AV (1984) Dissociated cells from different layers of adult human aortic wall. Acta Anat (Basel) 119:99–105
13. Orekhov AN, Andreeva ER, Krushinsky AV, Smirnov VN (1984) Primary cultures of enzyme-isolated cells from normal and atherosclerotic human aorta. Med Biol 62:255–259
14. Orekhov AN, Krushinsky AV, Andreeva ER, Repin VS, Smirnov VN (1986) Adult human aortic cells in primary culture: heterogeneity in shape. Heart Vessels 2(4):193–201
15. Hofman W, Goeger D (1977) Immunofluorescence in the identification of differentiating arterial smooth muscle cells in culture. Prog Biochem Pharmacol 13:52–54
16. Krushinsky AV, Orekhov AN (1982) Morphological analysis of cells isolated from the intima and media of human aorta. In: Chazov EI, Smirnov VN (eds) Vessel wall in athero- and thrombogenesis: studies in the USSR. Springer, Berlin Heidelberg New York, pp 41–51
17. Ross R (1971) The smooth muscle cell. II. Growth of smooth muscle in culture and formation of elastic fibers. J Cell Biol 50:172–186
18. Ross R, Glomset J (1973) Atherosclerosis and arterial smooth muscle cell. Science 180:1332–1339
19. Smith EB (1974) The relationship between plasma and tissue lipids in human atherosclerosis. Adv Lipid Res 12:1–49
20. Orekhov AN, Kosykh VA, Repin VS, Smirnov VN (1983) Cell proliferation in normal and atherosclerotic human aorta. II. Autoradiographic observation on deoxyribonucleic acid synthesis in primary cell culture. Lab Invest 48(6):749–754
21. Orekhov AN, Tertov VV, Kudryashov SA, Khashimov KA, Smirnov VN (1986) Primary culture of human aortic intima cells as a model for testing antiatherosclerotic drugs. Effects of cyclic AMP, prostaglandins, calcium antagonists, antioxidants, and lipid-lowering agents. Atherosclerosis 60(2):101–110
22. Orekhov AN, Tertov VV, Novikov ID, Krushinsky AV, Andreeva ER, Lankin VZ, Smirnov VN (1985) Lipids in cells of atherosclerotic and uninvolved human aorta. I. Lipid composition of aortic tissue and enzyme isolated and cultured cells. Exp Mol Pathol 42(1):117–137
23. Tertov VV, Orekhov AN, Repin VS, Smirnov VN (1982) Dibutyryl cyclic AMP decrease proliferative activity and the cholesteryl ester content in cultured cells of atherosclerotic human aorta. Biochem Biophys Res Commun 109(4):1228–1233

24. Orekhov AN, Tertov VV, Smirnov VN (1983) Prostacyclin analogues as antiatherosclerotic drugs. Lancet 2(8348):521
25. Kudryashov SA, Tertov VV, Orekhov AN, Geling NG, Smirnov VN (1984) Regression of atherosclerotic manifestations in primary culture of human aortic cells: effects of prostaglandins. Biomed Biochim Acta 43(8/9):S284–S286
26. Orekhov AN, Misherin AY, Tertov VV, Khashimov KA, Pokrovsky SN, Repin VS; Smirnov VN (1984) Artificial HDL as an antiatherosclerotic drug. Lancet 2(8412): 1149–1150
27. Tertov VV, Orekhov AN, Smirnov VN (1986) Agents that increase cellular cyclic AMP inhibit proliferative activity and decrease lipid content in cells cultured from atherosclerotic human aorta. Artery 13(6):365–372
28. Orekhov AN, Tertov VV, Mazurov AV, Andreeva ER, Repin VS; Smirnov VN (1986) "Regression" of atherosclerosis in cell culture: effects of stable prostacyclin analogues. Drug Dev Res 9(3):189–201
29. Giessler C, Fahr A, Tertov VV, Kudryashov SA; Orekhov AN, Smirnov VN, Mest H-J (1987) Trapidil derivatives as potential antiatherosclerotic drugs. Arzneimittelforschung 37-I(5):538–541
30. Akopov SE, Orekhov AN, Tertov VV, Khashimov KA, Gabrielyan ES, Smirnov VN (1988) Stable analogues of prostacyclin and thromboxane A_2 display contradictory influences on atherosclerotic properties of cells cultured from human aorta. The effect of calcium antagonists. Atherosclerosis 72(2/3):245–248
31. Tertov VV, Panosyan AG, Akopov SE, Orekhov AN (1988) The effects of eicozanoids and lipoxygenase inhibitors on the lipid metabolism of aortic cells. Biomed Biochim Acta 47(10/11):S286–S288
32. Baldenkov GN, Akopov SE, Li HR, Orekhov AN (1988) Prostacyclin, thromboxane A_2 and calcium antagonists: effects on atherosclerotic characteristics of vascular cells. Biomed Biochim Acta 47(10/11):S324–S327
33. Orekhov AN, Tertov VV, Khashimov KA, Kudryashov SA, Smirnov VN (1986) Antiatherosclerotic effects of verapamil in primary culture of human aortic intimal cells. J Hypertens [Suppl 6] 4:S153–S155
34. Orekhov AN, Tertov VV, Khashimov KA, Kudryashov SA, Smirnov VN (1987) Evidence of antiatherosclerotic action of verapamil from direct effects on arterial cells. Am J Cardiol 59(5):495–496
35. Orekhov AN, Baldenkov GN, Tertov VV, Li Hwa Ryong, Kozlov SG, Lyakishev AA, Tkachuk VA, Ruda MY, Smirnov VN (1988) Cardiovascular drugs and atherosclerosis: effects of calcium antagonists, beta-blockers, and nitrates on atherosclerotic characteristics of human aortic cells. J Cardiovas Pharmacol [Suppl 6] 12:S66–S68
36. Orekhov AN, Ruda MY, Baldenkov GN, Tertov VV, Khashimov KA, Li Hwa Ryong, Lyakishev A, Kozlov SG, Tkachuk VA, Smirnov VN (1988) Atherogenic effects of beta blockers on cells cultured from normal and atherosclerotic aorta. Am J Cardiol 61(13):1116–1117
37. Tertov VV, Orekhov AN, Li Hwa Ryong, Smirnov VN (1988) Intracellular cholesterol accumulation is accompanied by enhanced proliferative activity of human aortic intimal cells. Tissue Cell 20(6):849–854
38. Orekhov AN, Tertov VV (1988) Atherogenic low density lipoprotein isolated from the blood of patients with coronary atherosclerosis. In: 8th International symposium on atherosclerosis. Satellite meeting: Modified lipoproteins. Oct 7–8, Venice, pp 145–149
39. Chazov EI, Tertov VV, Orekhov AN, Lyakishev AA, Perova NV, Kurdanov KA, Khashimov KA, Novikov ID, Smirnov VN (1986) Atherogenicity of blood serum from patients with coronary heart disease. Lancet 2(8507):595–598
40. Orekhov AN, Tertov VV, Pokrovsky SN, Adamova IY, Martsenyuk ON, Lyakishev AA, Smirnov VN (1988) Blood serum atherogenicity associated with coronary atherosclerosis. Evidence for nonlipid factor providing atherogenicity of low-density lipoproteins and an approach to its elimination. Circ Res 62(3):421–429
41. Chazov EI, Orekhov AN, Tertov VV, Pokrovsky SN, Adamova IY, Lyakishev AA, Gratsianski NA, Nechaev AS, Perova NV, Khashimov KA, Kurdanov KA,

Kukharchuk VV, Smirnov VN (1988) Atherogenicity of blood plasma from patients with coronary atherosclerosis and its correction. Atherosclerosis Rev 17:9–20

42. Tertov VV, Orekhov AN, Martsenyuk ON, Perova NV, Smirnov VN (1989) Low density lipoproteins isolated from the blood of patients with coronary heart disease induce the accumulation of lipids in human aortic cells. Exp Mol Pathol 50 (in press)

43. Orekhov AN, Tertov VV, Mukhin DN, Mikhailenko IA (1989) Modification of low density lipoprotein by desialylation causes lipid accumulation in cultured cells. Discovery of desyalylated lipoprotein with altered cellular metabolism in the blood of atherosclerosic patients. Biochem Biophys Res Commun 162:206–211

44. Goldstein JL, Anderson RGW, Buja LM, Basu SK, Brown MS (1977) Overloading human aortic smooth muscle cells with low density lipoprotein-cholesterol esters reproduced features of atherosclerosis in vitro. J Clin Invest 59:1196–1202

45. Goldstein JL, Ho YK, Basu SK, Brown MS (1979) A binding site on macrophages that mediates the uptake and degradation of acetylated low density lipoprotein, producing massive cholesterol deposition. Proc Natl Acad Sci USA 76:333–337

46. Shechter I, Fogelman AM, Haberland ME, Seager J, Hokom M, Edwards PA (1981) The metabolism of native and malondialdehyde-altered low density lipoproteins by human monocyte-macrophages. J Lipid Res 22:63–71

47. Weisgraber KH, Innerarity TL, Mahley RW (1978) Role of the lysine residues of plasma lipoproteins in high affinity binding to cell surface receptors on human fibroblasts. J Biol Chem 253:9053–9062

48. Traber MG, Defendi V, Kayden KJ (1981) Receptor activities for low-density lipoprotein and acetylated low-density lipoprotein in a mouse macrophage cell line (IC21) and in human monocyte-derived macrophages. J Exp Med 154:1852–1867

49. Fogelman AM, Haberland ME, Seager J, Hokom M, Edwards PA (1981) Factors regulating the activities of the low density lipoprotein receptor and the scavenger receptor on human monocyte-macrophages. J Lipid Res 22:1131–1141

50. Brown MS, Basu SK, Falck JR, Ho YK, Goldstein JL (1980) The scavenger cell pathway for lipoprotein degradation: specificity of the binding site that mediates the uptake of negatively-charged LDL by macrophages. J Supramol Struct 13:67–81

51. Basu SK, Brown MS, Ho YK, Goldstein JL (1979) Degradation of low density lipoprotein-dextran sulfate complexes associated with deposition of cholesteryl esters in mouse macrophages. J Biol Chem 254:7141–7146

52. Vijayagopal P, Srinivasan SR, Jones KM, Radhakrishnamurthy B, Berenson GS (1985) Complexes of low-density lipoproteins and arterial proteoglycan aggregates promote cholesteryl ester accumulation in mouse macrophages. Biochim Biophys Acta 837:251–261

53. Salisbury BGJ, Falcone DJ, Minick CR (1985) Insoluble low-density lipoprotein-proteoglycan complexes enhance cholesteryl ester accumulation in macrophages. Am J Pathol 120:6–11

54. Falcone DJ, Mateo N, Shio H, Minick CR, Fowler SD (1984) Lipoprotein-heparin-fibronectin-denatured collagen complexes enhance cholesteryl ester accumulation in macrophages. J Cell Biol 99:1266–1274

55. Orekhov AN, Tertov VV, Mukhin DN, Koteliansky VE, Glukhova MA, Khashimov KA, Smirnov VN (1987) Association of low-density lipoprotein with particulate connective tissue matrix components enhances cholesterol accumulation in cultured subendothelial cells of human aorta. Biochim Biophys Acta 928(3):251–258

56. Orekhov AN, Tertov VV, Mukhin DN, Koteliansky VE, Glukhova MA, Frid MG, Sukhova GK, Khashimov KA, Smirnov VN (1989) Insolubilization of low density lipoprotein induces cholesterol accumulation in cultured subendothelial cells of human aorta. Atherosclerosis (in press)

57. Yokode M, Kita T, Arai H, Kawai C, Narumiya S, Fujiwara M (1988) Cholesteryl ester accumulation in macrophages incubated with low density lipoprotein pretreated with cigarette smoke extract. Proc Natl Acad Sci USA 85:2344–2348

Coculture of Endothelial and Smooth Muscle Cells*

P. F. Davies

Introduction

Coculture of endothelial cells (ECs) and smooth muscle cells (SMCs) is an experimental approach appropriate to investigations of vascular cell biology and pathophysiology. It is an attempt to recapitulate the relationships that exist in all blood vessels except capillaries (although it should be noted that pericytes, with characteristics of rudimentary SMCs, may interact with the capillary endothelium). A number of important mechanisms of endothelial-smooth muscle interactions have recently been elucidated, including the regulation of vessel tone by endothelial-derived relaxing and contracting factors [1–3], endothelial-derived mitogens in the regulation os smooth muscle proliferation [4, 5], and cholesterol homeostasis [5, 6]. In contrast to humoral interactions, direct contact between endothelium and SMCs which occurs in arteries [7, 8] and the microcirculation [9] probably involves gap junctional channels between the cells (as demonstrated in culture) [10, 11]. The presence of these structures suggests that efficient communication via electrical coupling and the intercytoplasmic diffusion of second messengers may occur between the cells in intact blood vessels. This chapter summarizes the practical alternatives currently available for establishing either humoral or contact-mediated communication between endothelium and SMCs in the culture dish. Variations of the basic procedures are limited only by the ingenuity of the investigator.

Outline of Cellular Interactions

As shown in Fig. 1, there are two general mechanisms of interactions between cell types; (a) humoral and (b) direct contact. Three general pathways of humoral communications are shown for endothelial products acting on smooth muscle. Activities in the opposite direction (smooth muscle to endothelium) are equally valid in theory, although less likely in situ because diffusion against a net convective flow of plasma filtrate will inhibit the rapid transfer of smooth muscle products to the endothelium. The most relevant form of contact-mediated communication, the formation of gap junctional

* This study was supported by grants HL 36049 and HL 36028 from the National Heart, Lung and Blood Institute of the NIH

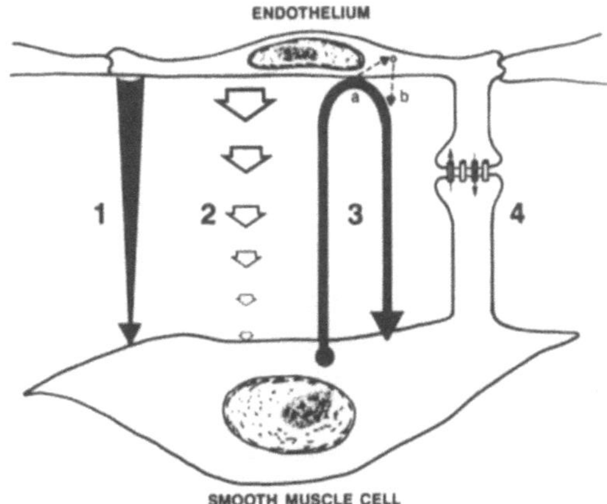

Fig. 1. Humoral and contact-mediated general interactions between endothelial cells (ECs) and smooth muscle cells (SMCs). Constitutive or regulated production of stable (*1*) or short-lived (*2*) factors acting on receptors of the target cell, e.g. mitogens, endothelin, prostaglandins, relaxing factors. The possibility of bidirectional humoral communication is also shown (*3*). A nonhumoral mechanism is the formation of gap junctions (*4*) providing direct electrical and metabolite coupling between the cells

channels (myoendothelial junctions), is shown. These are sites of coupling where intramembrane protein channels composed of specific connexin proteins [12, 13] allow electrical and chemical communication [12, 14].

In addition to the presence of both endothelium and adjacent smooth muscle, there is a complex extracellular matrix in vivo whose role in regulating cell differentiation, accessibility of interstitial solutes (e.g., lipoproteins) to the SMCs, and regulation of growth of arterial cells is poorly understood. Few attempts at coculture to date have considered the importance of the role of extracellular matrix other than that synthesized and secreted by the cocultured cells during the course of the experiment. This has occurred perhaps because investigations involving the use of biochemical tracers are complicated by the presence of large amounts of extracellular material which can bind radiolabelled metabolites by unpredictable and poorly understood mechanisms. Certain experiments, however, (e.g., morphology, cell migration), lend themselves to more complex arrangements of multiple cell types and can include extracellular matrices as part of the coculture design [15].

Experimental Approaches to Coculture

Coplating

This is the simplest approach to the culture of endothelial cells with smooth muscle. Following trypsinization of the two cell populations, they are mixed in the desired ratio and plated. Hemocytometry, Coulter counting, or determination of cell protein or DNA in the homogeneous cultures before mixing can be used to estimate the final ratio of the mixed cell population. Variations in plating efficiency and speed of cell attachment should be considered if larger numbers of cells are to be plated to establish a confluent coculture; there is the possibility that one of the cell types will preferentially attach and spread over all of the available dish surface to the exclusion of the second cell type. Consequently, the actual ratio of cells in the coculture will not accurately reflect the ratio at time of plating. When this method is used, therefore, it is a good idea to check the final disposition of the coculture by identifying the different cells in a representative dish using specific antibodies, fluorescent di-I-Ac low-density lipoprotein (LDL) uptake (for ECs), or by prelabelling one of the cell types with fluorescent latex beads (e.g., Covaspheres, Duke Scientific).

In coplating experiments, both contact and humoral interactions can occur, and, indeed, the presence of both mechanisms may modify cell function compared with a humoral only or contact only configuration. Large differences in cyclic nucleotide levels in SMCs have been reported using this kind of coculture compared with monotypic culture and the response appears to be related to the length of time during which the cells interact with each other [16]. The intimacy of the association between the cell types may be important for the magnitude of the response.

An extension of this simple approach is to add the ECs after establishment of the smooth muscle culture. ECs, however, tend not to form a confluent monolayer over the smooth muscle monolayer; instead they form islands of growth-inhibited cells. To obtain an endothelial monolayer, SMCs can be grown in a collagen gel matrix and the ECs subsequently plated onto the gel surface where they quickly grow to confluence [17]. The gel provides a 3-dimensional environment for the SMCs which more closely approximates vascular tissue.

Use of Microcarriers in Coculture

As outlined elsewhere in this book, SMCs and ECs like other anchorage-dependent cells, are readily grown on microcarriers. Each microcarrier bead becomes a microculture of up to approximately 200 cells which can be transferred from place to place quickly and easily, a major advantage for setting up a coculture [5, 18]. We have used this technique extensively to study endothelial-derived mitogens, smooth muscle lipoprotein metabolism, cyclic nucleotide levels, and gap junction communication [5, 8, 16, 18]. In most instances, it is convenient to grow ECs (rather than SMCs) on the microcarriers and combine them with conventional dish culture of SMCs. A routine procedure for microcarrier culture of arterial and venous ECs is first presented.

Microcarrier Culture of Endothelium

The microcarriers used were Biosilon solid polystyrene beads (Nunc; distributed by Vangard, Neptune, NJ). The stirrer system was a Techne MCS-104 Unit (Techne); the flasks were Techne 1-l stirrer flasks; the volume of culture medium, 500 ml; and the gas space, 500 ml. A maximum of four flasks per stirrer unit will produce $\geq 10^9$ cells at confluence.

1. The stirrer unit is kept in a standard incubator and its exposed surfaces occasionally washed down with disinfectant.

2. Prior to use, the inside of each flask is coated with a 2% solution of silicone in distilled water (Prosil-28, SCM), drained and allowed to dry. The flask and stirring rod is loosely assembled and autoclaved. All cell and flask handling procedures are conducted in a laminar flow hood.
3. 250 ml of complete culture medium is added to each flask 6-g microcarriers (keep sterile) are weighed out and added to each flask. Smaller amounts may be used proportionally. The solution is rocked gently to sink the beads, and the flasks placed in the incubator at 37 °C for 30 min.
4. ECs in suspension are added to the flask. A seeding density of at least 5 cells/microcarrier bead is recommended. There are approximately 2×10^6 microcarriers (6 g) in each flask. Therefore, 10^7 cells are added per flask. NB: if only a smaller number of cells are available for seeding, use 1 g microcarriers and seed 1.5×10^6 cells.
5. The cell/bead mixture is agitated by turning on the stirring mechanism for 2 min. The beads and cells are allowed to settle undisturbed for several hours. The agitation is repeated for 2 min, then again left undisturbed for several hours. Finally, 250-ml additional medium is carefully added to each flask and the stirrer unit is switched on to run continuously. A typical routine would be to seed the cells in the morning, agitate around noon, and leave undisturbed until late afternoon, at which time, continuous stirring is activated.
6. The MCS-104 unit has a slow start-up mechanism which minimizes shearing forces at the cell/microcarrier interface. A stirring speed of 40–50 rpm works well to maintain the microcarriers in suspension.
7. A weekly change of medium is recommended for 6-g quantities of microcarriers; less frequent changes are permissible for smaller quantities of cells.
8. A small aliquot of microcarriers can be removed periodically (use wide-bore pipettes to avoid cell damage) for observation using phase-contrast microscopy and for cell staining. (See Chap. 22 for details of the hematoxylin staining procedure.)

Humoral Communication

In this technique, the two cell populations are kept separated yet allowed to share the same medium. An advantage of microcarriers for this approach is the relatively large numbers of ECs that can be introduced into coculture, a property of the large surface area/volume ratio of the beads. A simple apparatus (Fig. 2) uses a plastic cylinder attached to the underside of the culture dish lid (by sterile silicon grease) and bounded on the lower end by a nylon net (spectramesh, Cole Parmer), pore size 1 µm. It is relatively easy to construct a large number of coculture adapters of this kind in the laboratory. (Details of their construction can be found in [18].) Recently, the Transwell system (Costar) has become commercially available and works well for microcarrier coculture.

An important factor in the design of humoral coculture geometry is the distance between the two cell populations. Unstable, short-lived cell products are unlikely to act effectively on the other cell population if they break down faster than the time required to diffuse to the target cell. Thus, the biological

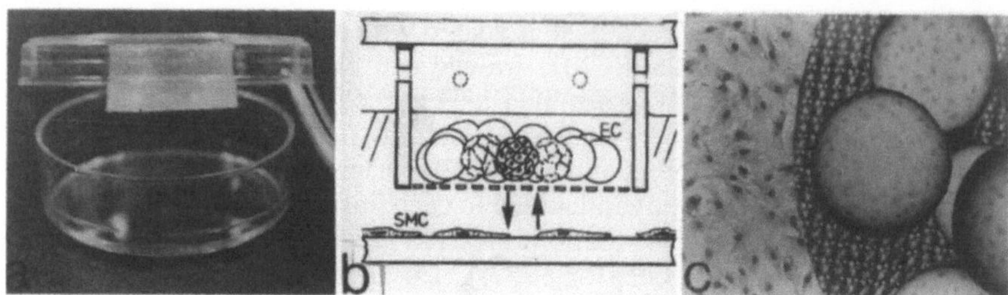

Fig. 2a–c. A microcarrier coculture system for studies of humoral communication between vascular cells. **a** A shallow plastic cylinder, the lower end closed by a porous nylon net (1- to 10-µm pores), is reversibly attached to the underside of a culture dish lid. As shown in **b** ECs on microcarriers are loaded into the cylinder which is partially submerged in culture medium bathing SMCs on the dish surface. Humoral exchange can occur between the cell populations without contact. **c** A photomontage reconstructed from three separate photographs of the smooth muscle monolayer, the nylon net, and the microcarrier-bound ECs stained with hematoxylin for visualization. [From 18]

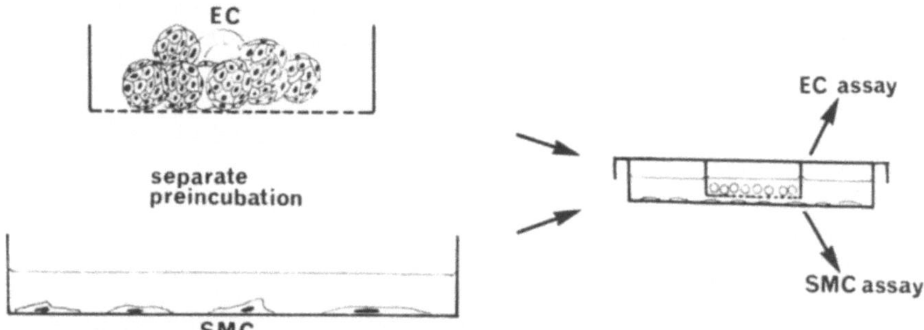

Fig. 3. Example of the use of coculture for EC and SMC studies. After growing the two cell populations separately, they can be separately preincubated if desired (e.g., for labelling, or induction of receptors). The cell populations are then brought together in coculture for the desired length of time, separated, and assayed individually

effects measured in these systems are more likely to be the result of stable cellular products, e.g., endothelium-derived mitogens and endothelin, than molecules with short half-lives, e.g., relaxing factors, prostaglandins.

The advantages of this approach include the ease and speed in which cells can be brought together and separated for subsequent analysis (Fig. 3) and the relatively large numbers of microcarriers in the system which allows concentration of cellular products.

Nonmicrocarrier Alternatives

Humoral communication can be investigated without using microcarriers. Alternative systems include:

1. ECs are grown on one surface of a Rose chamber; SMCs are cultured on the opposite surface [19].
2. Mixed slide cultures of each cell population can be constructed [20].
3. Physical division of the available culture dish surface by a strip of material to which the cells will not adhere, e.g., application of a narrow width of Hydron NCC (Interferon Sciences) as a 1.2% solution in 95% ethanol. After the Hydron has dried, it will inhibit cell attachment [21]. ECs can be plated on one side of the strip, SMCs on the other.

As with the microcarrier techniques, some thought should be given to the spatial relationships of the cell populations in analyzing coculture-dependent biological effects in these systems.

Contact-Mediated Interactions in Coculture

In addition to simple coplating of mixed cell populations, contact-mediated interactions between ECs and SMCs can be investigated using microcarriers [8]. A major advantage is the separation of the two cell populations after short periods of coculture without the use of proteolytic enzymes. During this procedure, ECs and SMCs rapidly establish bidirectional gap junctional communication [8]. A simple method is as follows:

1. Establish a monolayer culture of SMCs in dishes or culture flasks.
2. Establish a confluent microcarrier culture of ECs.
3. Transfer the microcarriers to the smooth muscle monolayer. The number of microcarriers can be adjusted to provide a single "layer" of microcarriers as observed under the microscope. If left undisturbed, the microcarriers attach within 1 h when EC-SMC contact is established on the underside of each bead (Fig. 4). We estimate that 4–12 cell pairs are involved in attachment after 3 h of coculture [8]. To reverse the attachment, the culture vessel is sharply tapped with a finger; the microcarriers detach and can be harvested by pipette. The re-isolated cultures are greater than 95% pure after 3 h of coculture. However, some ECs will migrate off the microcarriers over a period of 24 h or longer, causing an increase in contamination of the smooth muscle monolayer.

It is useful in this system to compare populations of cells in contact with each other to cells in a similar geometry, but without cell contact, i.e., humoral. This can be set up by gentle agitation of the culture vessel after addition of the microcarriers. An orbital shaker in the incubator is set to oscillate at a speed just fast enough to maintain movement of the microcarriers. Thus, a direct comparison between humoral and contact-mediated effects can be made at equal spatial distances between microcarriers and SMCs. There still remain some difficulties of interpretation, however, the most obvious of which is that the agitated system is stirred.

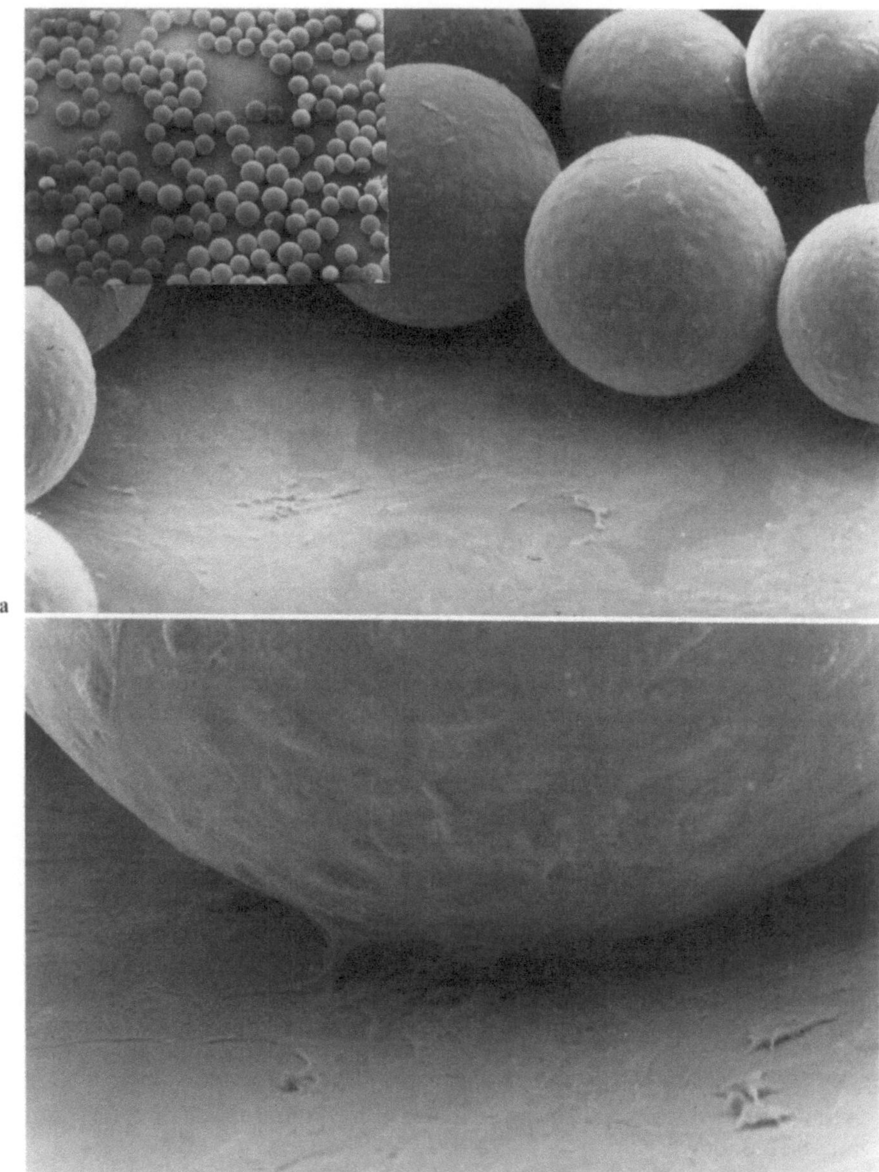

Fig. 4a, b. Microcarriers used in a model of myoendothelial gap junction formation. **a** Contact coculture of ECs on microcarriers laid over a monolayer of SMCs. Approximately 150–200 ECs are on each microcarrier (inset: low power). **b** Heterocellular contact at the attachment site of the microcarrier. Gap junctions from between the two cell types at these locations. Scanning electron microscopy. [From 11]

Isolated Pairs of Cells

The sophistication of modern cell biological and physiological techniques enables single cell analyses of biological effects. For example, using fluorescent dyes, not only can single cell intracellular free calcium levels be determined, but calcium distribution throughout the cell can be quantitated by image analysis. Similarly, the refinement of single cell recording techniques in electrophysiology has also narrowed the acquisition of biological information to the single cell or membrane patch. Techniques such as these are likely to assume increasing use in studies of cell interactions.

As an extension of recent electrophysiological studies [22, 23], we are experimenting with techniques to isolate heterotypic pairs of vascular cells. Essentially, the procedures involve repetitive dilution and plating of an EC SMC mixed suspension until a stage is reached when the probability of finding mixed pairs of cells is achieved. A 1:1 mixed suspension of ECs and SMCs is plated into multiwells starting at a density of approximately 5×10^4 cells/cm^2 surface area (approximately half confluence) and serially diluted by a factor of 2 into adjacent wells. The aim is to find pairs of cells in contact with each other, but separated from other pairs, groups, and single cells. To identify the cell types, one of the cell populations is prelabelled (before it is trypsinized) with fluorescent Covaspheres. One or two of the latex beads are taken up per cell where they remain sequestered in the perinuclear region in secondary lysosomes. After washing the monolayer, the cells are trypsinized and mixed with the second trypsinized cell population. The mixed culture is plated and pairs of cells are examined in the fluorescence microscope where Covasphere fluorescence is used as an indication of cell identity.

Conclusion

The above techniques are relatively straightforward in execution; interpretation of results from the various systems, however, is often quite difficult, especially when precise mechanisms of interactions are sought. Careful thought should be given to the appropriate controls, and often it is not possible to conduct unambiguous control experiments, especially for the separation of contact-mediated versus humoral-mediated mechanisms. However, the use of coculture reestablishes in a nominal fashion some of the interactions which must exist in vascular tissue in vivo, and provides important clues about vascular communication relevant to normal physiology and the pathogenesis of vascular diseases.

References

1. Furchgott RF, Zawadzki JV (1980) The obligatory role of endothelial cells in the relaxation of arterial smooth muscle by acetylcholine. Nature 288:373–376
2. Peach MJ, Loeb AL, Singer HA, Saye JA (1984) Endothelium-derived vascular relaxing factor. Hypertension [Suppl 1] 7:I94–I100

 3. Yanagisawa M, Kurihara H, Kimura S, Tomobe Y, Kobayashi M, Yazaki Y, Goto K, Masaki T (1988) A novel potent vasoconstrictor peptide produced by vascular endothelial cells. Nature 332:411−415
 4. DiCorleto PE, Bowen-Pope DF (1983) Cultured endothelial cells produce a platelet-derived growth factor-like protein. Proc Natl Acad Sci USA 80:1919−1923
 5. Davies PF, Truskey GA, Warren HB, O'Connor SE, Eisenhaure BH (1985) Metabolic cooperation between vascular endothelial cells and smooth muscle cells in co-culture: changes in low density lipoprotein metabolism. J Cell Biol 101:871−879
 6. Hajjar DP, Marcus AJ, Hajjar KA (1987) Interactions of arterial cells: studies on the mechanisms of endothelial cell modulation of cholesterol metabolism in cocultured smooth muscle cells. J Biol Chem 262:6976−6981
 7. Spagnoli LG, Villaschi S, Neri L, Palmieri G (1982) Gap junctions in myo-endothelial bridges of rabbit carotid arteries. Experientia 38:124−125
 8. Davies PF (1986) Biology of disease: vascular cell interactions with special reference to the pathogenesis of atherosclerosis. Lab Invest 55:5−24
 9. Rhodin JA, (1967) The ultrastructure of mammalian arterioles and precapillary sphincters. J Ultrastruct Res 18:181−223
10. Larson DM, Sheridan JD (1985) Junctional transfer in cultured vascular endothelium. I. Dye and nucleotide transfer. J Membr Biol 83:157−168
11. Davies PF, Ganz P, Diehl PS (1985) Methods in laboratory investigation. Reversible microcarrier-mediated junctional communication between endothelial and smooth muscle cell monolayers: an in vitro model of vascular cell interactions. Lab Invest 53:710−718
12. Loewenstein WR (1981) Junctional intercellular communication: the cell-to-cell membrane channel. Physiol Rev 61:829−913
13. Goodenough DA (1976) In vitro formation of gap junction vesicles. J Cell Biol 68:220−231
14. Flag-Newton J, Simpson I, Loewenstein WR (1979) Permeability of the cell-to-cell membrane channels in mammalian cell junction. Science 205:404−407
15. Navab M, Hough GP, Stevenson LW, Drinkwater DC, Laks H, Fogelman AM (1988) Monocyte migration into the subendothelial space of a coculture of adult human aortic endothelial and smooth muscle cells. J Clin Invest 82:1853−1863
16. Ganz P, Davies PF, Leopold JA, Gimbrone MA, Alexander RW (1986) Short and long-term interactions of endothelium and vascular smooth muscle in co-culture: effects on cyclic GMP production. Proc Natl Acad Sci USA 83:3552−3556
17. Schor AM, Schor SL (1986) The isolation and culture of endothelial cells and pericytes from the bovine retinal microvasculature: a comparative study with large vessel vascular cells. Microvasc Res 32:21−38
18. Davies PF, Kerr C (1982) Cocultivation of vascular endothelium and smooth muscle cells using microcarrier techniques. Exp Cell Res 141:455−457
19. Campbell JH, Campbell GR (1981) What controls smooth muscle phenotype? Atherosclerosis 40:347−354
20. Harel L, Jullien M, DeMonti M (1978) Diffusible factors controlling density inhibition of 3T3 cell growth: a new approach. J Cell Physiol 96:327−333
21. Folkman J, Moscona A (1978) Role of cell shape in growth control. Nature 273:345−349
22. Olesen SP, Clapham DE, Davies PF (1988) Haemodynamic shear stress activates a K+ current in vascular endothelial cells. Nature 331:168−170
23. Olesen SP, Davies PF, Clapham DE (1988) Muscarinic activated K+ current in bovine aortic endothelial cells. Circ Res 62:1059−1064

Culture and Study of Pericytes

P. A. D'Amore

Introduction

The microvasculature consists of two cell types, the capillary endothelial cell and the pericyte (also called the Rouget cell or mural cell). The earliest description of the pericyte is credited to Rouget [47] and an extensive structural study was presented by Zimmerman [66] in which he described the existence of these cells in nearly all vertebrates. Though early studies suggested that the pulmonary circulation was devoid of pericytes [36, 66], Weibel [65] in a very thorough study reported the existence of pericytes on pulmonary alveolar capillaries. In this report he commented that pericytes were less frequent in the pulmonary circulation than in the systemic circulation and had rare and small processes. Further, he pointed out that there appear to be species differences; pericytes are frequently seen in human, dog, and guinea pig lungs, are rarer in rat lungs, and have not been found in shrew lungs: in fact, that "the smaller the animal, the rarer the pericytes".

The origin of pericytes has not been definitively determined. Clark and Clark [11], in their investigations of vessel formation in tadpoles, indicated that the pericytes arose from "wandering mesenchymal" cells. From their study on developing retinal vessels in kittens, Shakib and de Olivera [51] concluded that vessels form as solid cords of cells immediately in front of vascular beds and that it is from these cords that the endothelial cells and pericytes derive. As the basement membrane appears in the maturing capillaries, the outer cells differentiate into pericytes while the inner cells become endothelial cells. Using the quail-chick nuclear marker in a study of bone differentiation, Jotereau and LeDourain [27] found that in chick limb buds grafted onto quail embryos the vascular endothelium in the graft was of host origin (quail) whereas the pericytes were of donor origin. These observations indicate distinct lineages for the endothelial cell and pericyte and make it unlikely that the endothelial cell could serve as a precursor for the pericyte and vice versa.

To date pericytes have been defined primarily by their periendothelial location in the microvasculature. However, similarity in ultrastructure and location have led to their comparison to other cells. For instance, the perivascular location and the presence of both vimentin and desmin in the Ito cells of the hepatic sinus and the reticular cells of the splenic sinusoid make these cells analogous to the pericyte [20]. The most common comparison that is made is between pericytes and kidney mesangial cells. Like pericytes, mesangial cells have been reported to contract in response to vasoactive agents [50], shown to phagocytize particulate matter [42], and likened to smooth muscle cells [3].

Until the advent of its tissue culture, little was known about the function of the pericyte. In spite of this inability to assign a definite function(s) to the pericyte, the literature is filled with speculations [for review see 53]. A number of early investigators made observations which suggested that pericytes were capable of independent contraction [for review see 10]. A majority of these reports were based on alterations of capillary morphology in which shortening and thickening of the endothelial cells and/or pericytes was noted. Clark and Clark [10] studied the relation of the pericyte to capillary contractility using the transparent tails of tadpoles. They reported that the chloretone used to anesthetize the larvae had the effect of narrowing the capillaries and that removal of the drug led to "expansion" of the capillaries. Contraction of pericapillary cells in response to electrical stimulation was observed in capillaries of frog tongue [4]. Structural studies have also been conducted at the electron-microscopic level. Tilton et al. [59] used planimetry to quantify alterations in endothelial morphology (buckling), a reflection of pericyte contractility. Perfusion of rat cardiac and skeletal muscle with various vasoactive agents led to increased buckling of the endothelial plasma membrane that was apposed to the pericyte.

Some have suggested a phagocytic role for the pericyte. The observations of phagokinetic activity by skeletal muscle pericytes [40] and by dermal pericytes [13] seem to provide evidence for this role. Yet, the finding of few lysosomal elements [65] does not support this concept.

The fact that "pericytes are carried everywhere by growing capillaries in the embryo, in wound healing, and during bone formation" [46] has led many to suggest that the pericytes serve as precursors to other cells. Ultrastructural study of pericytes in injured rat brain led Boya [5] to speculate that pericytes are capable of transforming into microglial cells. A number of investigators have drawn analogies between pericytes and vascular smooth muscle cells [46, 65] and some have suggested that the pericyte may differentiate into smooth muscle cells [11]. Rhodin [46] observed in his ultrastructural studies that pericytes were similar to the mesenchymal cells of the embryo, leading him to conclude that "one of the most important functions of the pericyte is to serve as a precursor to smooth muscle cells when there is a functional demand for this differentiation".

Materials

The most widely used tissues for the culture of microvascular cells include brain [15, 22], skin [52], retina [6, 21], adrenal cortex [18], and adipose tissue [63]. A major consideration with regard to the selection of tissues for the culture of pericytes is the ratio of endothelial cells to pericytes. Although not systematically studied, this ratio is known to vary from one microvascular bed to another. Since the retina has the highest ratio of pericytes to endothelial cells [37], it has been our tissue of choice.

The following protocol for the isolation of pericytes from bovine retina may be applied to the isolation of pericytes from the microvessels of any tissue. Modifications may be necessary in: (a) the concentration of collagenase used,

(b) in incubation time in the collagenase, and/or (c) the addition of other enzymes (e.g., elastase, dispase, or trypsin).

Required materials include:
(a) Betadine-phosphate buffered saline (PBS); (20% v/v)
(b) Sterile PBS
(c) 0.2% collagenase stock (type II, Worthington) with 0.2% bovine serum albumin in calcium-magnesium free-PBS (CMF-PBS)
(d) Dulbecco's modified Eagle's medium (1000 mg/L glucose)
(e) Fetal bovine serum or calf serum
(f) Autoclavable funnels
(g) 110-µm nylon mesh (Nitex)
(h) 100-mm disposable Petri dishes
(i) 25-cm tissue culture flasks
(j) Conical centrifuge tubes (15 ml and 50 ml)
(k) 25-ml Erlenmeyer flasks (sterile) with stirring bars
(l) Amber corks for flasks
(m)Disposable 5- and 10-ml pipettes
(n) Sterile dissecting equipment: straight-toothed large forceps, curved medium scissors, blunt probe, small straight scissors, and scalpel blade handles with blades (no. 21)
(o) Rotary shaker

Methods

Isolation of Tissue: General Description

The tissue should be sterilely removed from the donor. When this is not possible, as is the case when obtaining tissues from a slaughterhouse, it is advisable to incubate the tissue in a 20% betadine-PBS solution for 15 min prior to dissection. Follow the betadine soak with two thorough rinses in saline. If the tissue of interest is protected from the betadine, such as the retina in the eye, or the cortex of the adrenal gland, these tissues may be dissected out. If, on the other hand, cardiac tissue, mesentery, or brain is the tissue source then one should dissect out a portion of the tissue that has not been exposed to the betadine. On the basis of our experience with retinal tissue, we utilize approximately 0.8–1.0 g of tissue (wet weight; two to three retinas) for every 5 ml of collagenase. If the tissue is not clean, dissect away unwanted material. For example, in dissecting adipose tissue from dermal tissue be careful not to remove any of the dermal cells and in the case of the retina, adherent pigmented epithelial cells can become a contamination problem if not removed before collagenase treatment.

Dissection of Retina from the Eye

Obtain an intact eye and soak in 20% betadine for 10 min and follow with two 10-min PBS washes. Clear away excess connective tissue and wash again thoroughly in sterile PBS.

Using a sharp pair of scissors or a scalpel blade, puncture the eye about 5 mm posterior to the limbus (the line of contact between the cornea and the sclera of the eye) and cut around the globe at this level. Remove the vitreous body using a toothed forceps.

With a blunt probe, gently separate the retina from the eye cup, adding a small amount of PBS if the tissue is dry and adherent. When the entire retina is free, cut it at its connection to the optic nerve and transfer it to a 100-mm Petri dish that contains 10 ml of PBS.

Using a small pair of scissors, cut away any major areas that appear black (adherent pigmented epithelium) and wash the retina thoroughly by adding and carefully aspirating 10-ml aliquots of PBS. Remove any additional retinal pigmented epithelial cells (black areas) by teasing them away from the retinal tissue with two sterile Pasteur pipettes. Aspirate these along with the PBS washes.

Capillary Isolation and Pericyte Culture

Mince the tissue thoroughly using two scalpel blades. The final tissue pieces should be small enough to fit into the bore of a 5-ml pipette. Wash the minced tissue well with PBS by suspension and gentle centrifugation ($800 \times g$). Suspend the minced tissue in the 0.2% collagenase-BSA solution, transfer to an Erlenmeyer flask and cap with a sterile amber cork. Shake at 40 rpm on a rotary shaker at room temperature for 1 h. Pipette the solution against the side of the flask three times to free the capillary fragments which have been loosened by the enzyme treatment. Pipette the solution onto Nitex mesh, which is loosely stretched over the top of a funnel and secured to the sides with autoclave tape. Wash the tissue through the mesh with 20 ml of DMEM with 5% calf serum. The smaller vessel fragments (two to six cells) will pass through the mesh along with any single cells and the mesh will retain larger vessel fragments and pieces of parenchyma as well as connective tissue. The supernatant (which has passed through the mesh) is pelleted by gentle centrifugation ($800\ g$), then resuspended in DMEM with 5% calf serum and washed twice with the same medium.

Plate the washed pellet (from two to three retinas) into one T-25 flask with DMEM plus 10% calf serum. The following day wash the cells gently, refeed with the same medium and then refeed every 3 days thereafter. The pericytes do not achieve confluence, but will grow to a density that represents about 80% coverage of the flask within 3–4 weeks.

Subculture

In our studies we have been very conservative with respect to the subculture of pericytes; we generally use first passage cells, that is, cells that have been exposed to trypsin only once. Our reasons for this are twofold. The first is practical; the plating efficiency of trypsinized pericytes is poor, averaging 50%. Pericytes require an extended trypsinization time to be detached from the flask (5–10 min as apposed to 1–2 min for endothelial cells). We do not know the reason for this relative trypsin resistance; it may be due to a unique composition of the extracellular matrix or it may simply be because the cells have synthesized an extensive amount of basement membrane in the 3–4 weeks required for them to grow to a usable density. The second, and more important reason for using very early passage cells, is the concern that the pericytes, like smooth muscle cells, may "dedifferentiate" with passage. The phenotype of smooth muscle cells has been shown to change with increasing passage and this alteration has been interpreted as a loss of their differentiated phenotype [9, 44] (Antonelli-Orlidge and D'Amore, manuscript submitted). Since it seems clear that the pericyte is "related" to the smooth muscle cell, we speculate that the same dedifferentiation process might apply.

Characterization

Morphology and Growth Rate

On the first day following the establishment of the primary pericytes cultures very few cells will be spread onto the culture dish. Clusters of refractile cells, the attached capillary fragments, can be seen. In the following week irregularly shaped cells will be observed migrating from the capillary fragments and some single cells will become evident as they spread onto the dish. Occasional colonies of endothelial cells will be noted during the first week of the primary culture. These colonies are characterized by the classic endothelial morphology of polygonal, contacted cells which in the case of capillaries tend to grow as colonies (versus single cells). These endothelial colonies will not grow beyond about 100 cells and will within these first 2 weeks actually lift. We speculate that the lack of growth is due to the presence of pericytes which we know inhibit endothelial proliferation when grown in contact (see Discussion for the details). Pericytes in culture are large irregularly shaped cells, which unlike endothelial cells never form a confluent monolayer (Fig. 1). This growth pattern is similar to that seen *in vivo* where "the cell processes of one pericyte rarely contact those of a neighboring pericyte" [46]. Pericytes cultured under the conditions described above have a very slow doubling time. Once the pericytes have passed a lag phase, which is most likely due to the low cell density of the cells (Fig. 2) they enter their log phase of growth and thereafter display a doubling time of 3–4 days. Both fibroblast growth factors (both acidic and basic) and platelet-derived growth factors are potent mitogens for the pericytes (D'Amore et al., manuscript submitted).

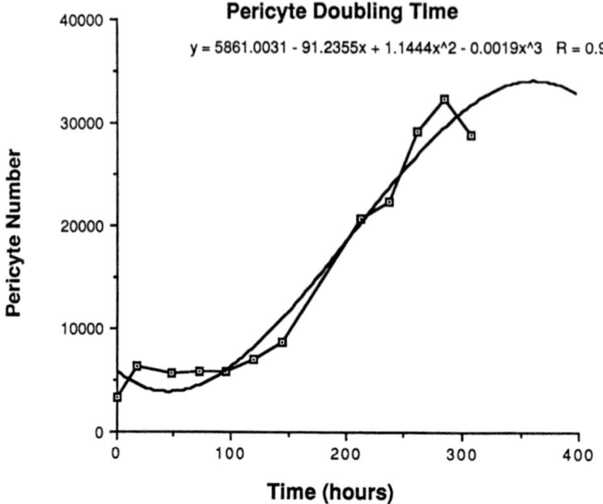

Fig. 1. Growth rate of pericytes in culture. Pericytes were isolated from bovine retinas as described in Methods. Cells were plated at 5000 cells/2.1 cm^2 well. Cell numbers were determined electronically at the indicated times. Pericytes grew slowly until they reached a density of approximately 10,000 cells/well at which time they went into a more rapid growth phase. (We speculate that this is due to the cells' ability to condition their media). Calculations during this phase indicate that the doubling time of the pericytes is about 72 h

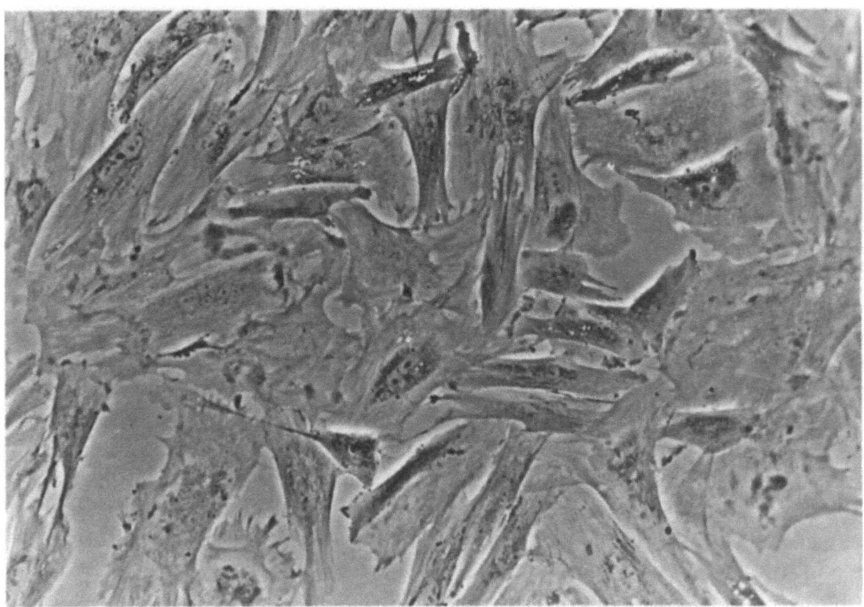

Fig. 2. Bovine retinal microvessel pericytes. Pericytes isolated as described in Methods and grown for approximately 3 weeks. Note the irregular shape and the presence of phase-dense fibers. The pericytes will never form a confluent monolayer. Rather, after reaching this density the cells will begin to form nodules, which represent multiple layers of the cells

Cell-Specific Markers

No one specific marker has been identified that characterizes pericytes alone (e.g., such as von Willebrands factor (VWF) is for endothelial cells). As a result, the identification of a cell population as pericytes depends on demonstrating the presence of a variety of characteristics along with the absence of the most likely contaminant cells. In a comparison of the actin isoforms in endothelial cells, pericytes, and smooth muscle cells we have found that, whereas endothelial cells contain nonmuscle actin and smooth muscle cells contain predominantly muscle actin, the pericytes contain both actin isotypes [23]. Thus, the presence of muscle actin has come to be a useful (if not entirely specific) marker for pericytes, particularly where likely contaminating cells are not of muscle lineage. For example, pericytes cultured from retina are more likely to be contaminated by endothelial cells and/or retinal pigmented epithelial. The identity of the endothelial cells can be assessed by assaying for the presence of VWF or for the ability to take up acetylated low density lipoprotein (acyl-LDL) [62]. The epithelial cell contamination is more problematic. Early contaminants can be identified by the presence of black lipofuscin granules that characterize the pigmented epithelial. However, epithelial cells without pigment may be present and even those with pigment lose the granules with cell passage.

Krause et al. [35] have reported the characterization of a monoclonal antibody that recognizes a 140 kDa protein that is specific for cerebral pericytes of the rat. They have demonstrated that the antigen is found only in areas of the brain that possess a blood-brain barrier. It is also present on the apical surfaces of transporting epithelial and as such may be involved in a transport function of the blood brain barrier. This antigen has not yet been demonstrated on pericytes cultured from these areas of the brain, so its usefulness as a marker for cultured cells is uncertain. Another monoclonal antibody (3G5) directed against a surface glycolipid has been shown to be a useful marker for pericytes [43]. Using immunofluorescent microscopy it has been shown that these antibodies do not react with cultured smooth muscle cells, endothelial cells, or retinal pigmented epithelial cells. Further, using flow cytometry this antibody was used to purify pericytes from trypsinized mixed retinal cells. In cultures where the pericytes represented 8% of the original cell mix, the pericytes were enriched to greater than 70% of the cells.

Discussion

Several investigators have reported the culture of microvascular pericytes [6, 7, 21, 49]. The retina has been the most popular tissue source and it is very likely that the high density of pericytes on the retinal capillaries accounts for this choice. For the most part, these procedures employ a collagenase digestion step [49], although the concentration of the enzyme and the length of the incubation varies. Some permutations have been introduced in an effort to increase efficiency of the process. Schor and Schor [49] report that low oxygen tension increases pericyte plating efficiency.

Why one is able to obtain virtually pure cultures of pericytes from obviously heterogeneous mixed primary cultures (containing both endothelial cells and pericytes) is not totally clear, but is most likely due to the inhibitory influence that the pericytes have on endothelial cell growth (see below). Others have reported this phenomenon and in this regard Schor and Schor [49] noted: "In primary mixed cultures, pericytes grew vigorously, eventually becoming the only cell type apparent in confluent cultures and in subsequent passages." This success in the culture of pericytes has allowed for the characterization and investigations into some of the functions of pericytes.

Metabolism

As was the case for vascular endothelium, the ability to culture pericytes has led to insight into the pericytes' metabolic activities. The pericytes have been shown to contribute to the basement membrane which envelopes them *in situ* and have been shown to synthesize collagen [12] and glycosaminoglycans (predominantly heparan sulfate) [57].

Using both isolated retinal vessels and cultured cells, pericytes have been shown to contain aldose reductase, an enzyme which reduces hexose sugars to their corresponding sugar alcohols [1, 33]. The intracellular accumulation of these alcohols has been suggested to contribute to the selective loss of pericytes that is known to occur in association with diabetic retinopathy. Cultured pericytes have also been shown to transport ascorbate by facilitated diffusion, a function which the authors speculate may be vulnerable to impairment by hyperglycemia [34]. In addition, prostaglandin synthesis has been demonstrated by pericytes *in vitro* and it has been suggested that pericytes may influence thrombi formation via this product [26].

Contractility

The evidence for a contractile role for the pericyte has come in a number of forms. The presence of a variety of contractile proteins led a number of investigators to postulate contractile functions for the pericyte. Early ultrastructural studies reported the presence of actin- and myosin-like filaments in rat brain pericytes [39, 64]. Using immunohistochemical methods we have demonstrated that pericytes *in vivo* and *in vitro* contain both muscle and nonmuscle actin isotypes, whereas endothelial cells possess only nonmuscle, and smooth muscle cells express predominantly muscle actin [23], supporting the hypothesis that pericytes are the capillary and venular correlate of smooth muscle cells. Other contractile proteins have been similarly localized in the pericyte. Immunoperoxidase localization of tropomyosin [29] and myosin [30] in the pericyte along with the presence of cyclic GMP-dependent protein kinase (not found in endothelial cells) [28] led to the conclusion that the pericytes are "contractile elements related to smooth muscle cells" and involved in "the regulation of blood flow through the microvasculature." Quantitatively the level of tropomyosin in the pericytes was less than that in smooth muscle cells but higher than that in the endothelial cells, and unlike the endothelium, the pericytes contained the muscle isoform [29]. The amounts of myosin were also found to vary with microvascular bed and within segments of the same bed [30]. In pericytes of smaller capillaries the nonmuscle isomyosin predominated, whereas in pericytes associated with larger capillaries and postcapillary venules the muscle isoform was the primary myosin found. Immunocytochemical studies revealed the presence of both vimentin and desmin in pericytes of various chicken microvascular beds as well as in the cells apposed to endothelial cells in beds with no pericytes (e.g., the Ito cell of the hepatic sinus and the reticular cell of the splenic sinusoid) [20]. There is an intriguing consistency among all of these results with the pericyte expressing a profile of contractile proteins that is intermediate between that of the smooth muscle cells and endothelial cells. This set of observations lends support to Rhodin's early characterization of the pericyte as an "undifferentiated" cell (see Introduction).

More recently DeNofrio et al. [16] have shown that actin isoforms are functionally sorted in the pericyte. Nonmuscle actin was localized in the membrane ruffles, pseudopods, and on the stress fibers, whereas the muscle actin was only seen in the stress fibers and not in the motile areas of cytoplasm. Biochemical analysis of the proportions of the various actin isoforms in the pericyte revealed a ratio of 1 alpha-muscle actin to 2.75 beta-nonmuscle actin to 3 gamma-nonmuscle actin. Similarly, Skalli et al. [54] have reported finding alpha actin in both pericyte and smooth muscle cell cytoplasm where it was localized in microfilament bundles.

Another level of significance for the contractile potential of the pericytes comes from a comparison of pericyte distribution in normotensive and hypertensive rat brain by Herman et al. [24, 25]. Using short-term cultures of cerebral capillaries they observed that pericytes remained tightly associated with the endothelial cells of capillaries from hypertensive-prone and hypertensive animals but not from normotensive animals [25]. Further, the number of pericytes was increased two to five times in the capillaries from the hypertensive brains. Results obtained when pericytes were localized *in situ* using immunohistochemical methods corroborated these *in vitro* findings; four times as many "pericyte-rich" capillaries were found in the motor cortex of hypertensive animals as compared with their normotensive controls [24]. The authors postulate a role for the increased density of pericytes in the capillaries of hypertensive animals in regulating blood flow and vascular repair following injury *in vivo*.

Functional evidence for the contractile ability of the pericytes was demonstrated in two separate culture systems. Using a collagen lattice and a silicone rubber sheet, Kelley et al. [31] compared the contractility of vascular cells and found that all of the vascular cells could contract the substrata in the following order of efficiency: smooth muscle cells > pericytes > aortic endothelial cells. Schor and Schor [49] observed that smooth muscle cells and pericytes plated onto a collagen gel contracted the gel whereas aortic and retinal capillary endothelial cells did not. Using the silicone rubber model, Kelley et al. demonstrated that the pericyte contractile response was altered by certain vasoactive agents and cAMP agonists [32]. Specifically, histamine or serotonin contracted the pericytes whereas isoproterenol relaxed the cells. In investigating the mechanism of the contraction, dibutyryl cAMP and forskolin were both found to induce pericyte relaxation and to elevate cAMP levels. Further, staining of the cells with rhodamine-phalloidin to visualize actin-containing filaments revealed that relaxation of the pericytes was associated with the disassembly of stress fibers.

Pericyte-Endothelial Cell Interactions

Ultrastructural analyses have revealed a high degree of physical interaction between the endothelial cells and pericytes *in vivo*. A number of transmission electron microscopic studies reveal the close association between the cells (Fig. 3) and allow the visualization of sites of contact between the two cell types. The pericyte arrangement along skeletal muscle capillaries was studied in capillary preparations in which basement membrane components were enzymatically removed [41]. Observations in these studies revealed the pericytes aligned along the long axis of the capillaries, covering an average 82% of the circumference. A comparison of pericyte "coverage" of the endothelial cell surface in the retina and brain revealed a ratio of 0.41 (pericyte plasma membrane in contact with outer vascular circumference/outer endothelial circumference) in the retinal microvessels and 0.22–0.30 for five regions of the cerebral cortex [19]. These authors suggest that this degree of association between the endothelium and pericytes and the significant differences from one microvascular bed to the next has important implications for the control of microcirculatory function.

Tilton et al. [60] have conducted a number of careful morphometric studies that shed some light on the nature of these interactions. In one study they examined the differences in the structure and distribution of pericytes in various vascular beds of the eye. They found, for instance, that the percentage of capillary circumference covered by pericytes (46%–58%) and the percentage of capillary sections with pericyte nuclei (12%–16%) were similar in retina, iris, and cililary processes, whereas it was 50% of those values in the choriocapillaris. The number of pericyte processes per capillary was also measured to vary among the different vascular beds with 3–4 for the capillaries of the retina and choroid versus 9–11 for the capillaries of the iris and cililar process. Tilton et al. [60] have categorized the endothelial cell-pericyte interactions into three groups: (a) pericyte processes in close apposition to the endothelium, (b) pericyte processes that protude into the endothelium, and (c) endothelial processes that protrude into pericytes. The significance of the differences in the shape, frequency, and distribution of pericytes in various microvascular beds is not known but is likely to be related to differences in capillary functions and hemodynamics.

Using freeze-fracture to examine the junctions of choroidal vessels, Spitznas and Reale [56] demonstrated the presence of gap junctions between the endothelial cells and pericytes. With the techniques of dye and radiolabeled uridine transfer, Larson et al. [38] examined co-cultures of bovine brain microvascular endothelial cells and pericytes for evidence of junctions. While dye

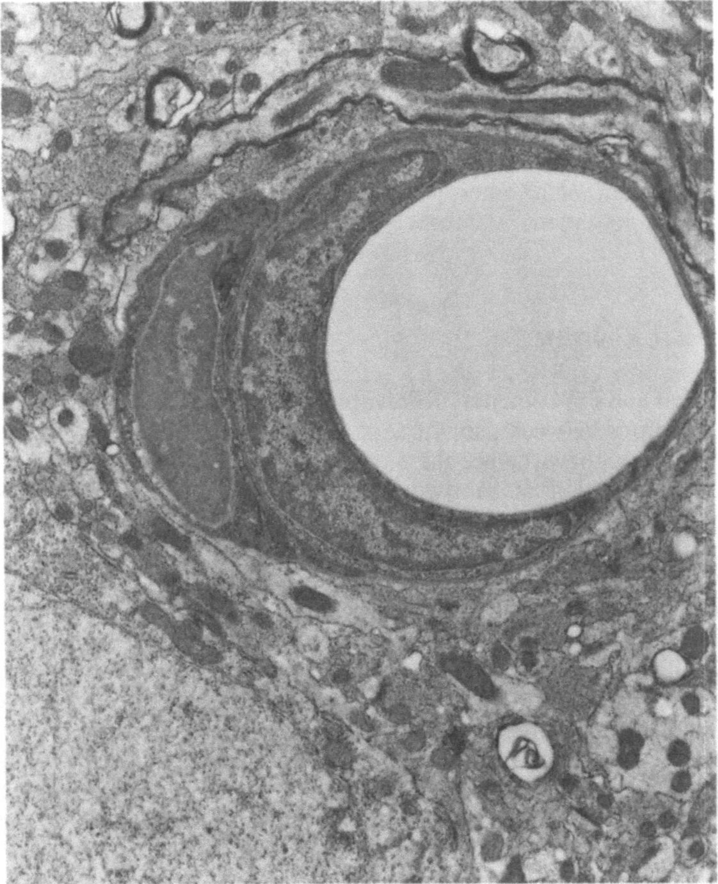

Fig. 3. Cross-sectional view of a renal capillary. This capillary consists of a single endothelial cell (EC) with its associated abluminal pericyte. This fortuitous section contains both endothelial and pericyte nuclei. (Micrograph courtesy of Dr. Ira M. Herman, Tuft's Medical School, Boston, MA)

transfer between cells in homogeneous cultures of endothelial cells and pericytes was high, it was low between cells in cocultures of the two cell types. On the other hand, labeled nucleotide transfer was extensive both in the homogeneous cultures as well as in the cocultures of endothelial cells and pericytes, demonstrating the presence of functional junctions. Using electron probe microanalysis we have similarly demonstrated ionic coupling between endothelial cells and pericytes in cocultures [58]. Endothelial cells in pure cultures had a sodium/potassium ratio that was approximately eight-fold higher than that of pericytes, whereas cocultured endothelial cells and pericytes had ion ratios that were intermediate to the cells cultured alone, indicating the existence of functional gap junctions.

In an ingenious series of studies Carlson [8] has used scanning (SEM) and transmission electron microscopy (TEM) to ultrastructurally examine retinal capillaries and their basement membranes. TEM revealed sites of cell contact where membranes of pericytes and endothelial cells appeared to fuse. Detergent solubilization of isolated capillaries allowed examination of acellular basement membranes and revealed the presence of numerous fenestrations in the basement membranes, which he comments "correlate well with the subendothelial basement membrane discontinuities occupied by periendothelial junctions."

Growth Control

Clinical and experimental observations regarding the frequency and nature of associations between pericytes and endothelial cells led to the suggestion that pericytes might influence the growth of the endothelium. One of the primary observations that has motivated this concept is the fact that there is a selective loss of pericytes from the retinal microvasculature of diabetics prior to the onset of neovascularization (diabetic retinopathy) [37, 55], i.e., the loss of the pericytes appears to be permissive for the subsequent vasoproliferation. Similarly, hemangioendotheliomas, vascular tumors, were noted to have a paucity of pericytes [17]. Additional evidence that the presence of the pericyte might "stabilize" the capillary comes from the ultrastructural observations of Crocker et al. [14] who noted that the arrival of the pericyte at the newly forming capillary marked the cessation of growth. They hypothesized that "it is the incorporation of intersititial cells in the basement membrane of newly formed capillaries with the subsequent close association of areas of their cytoplasmic membranes that is responsible for inhibiting endothelial proliferation."

Motivated by these findings, we established coculture systems that would permit us to investigate the possibility of pericyte-mediated endothelial cell growth control. Using this model we found that when endothelial cells were cocultured in the presence of pericytes the endothelial growth was totally inhibited [45]. The growth inhibition was reversible and dependent on contact between the two cell types, was observed at endothelial cell to pericyte ratios of up to and including 10:1 and was cell-specific. Subsequent studies have revealed that the pericyte-mediated growth inhibition in the cocultures is mediated by transforming growth factor type beta (TGF-β) [2]. The latter finding is significant for at least two reasons. First, although a number of cultured cells produce TGF-β, the factor is generally secreted in a biologically latent form. This has led to the speculation that a major site of regulation with respect to TGF-β focuses on the manner and site of its activation. Sato and Rifkin [48] have shown that inhibitors of plasmin can block TGF-β activation in the cocultures. Second, there is ample reason to believe that the interactions that occur in the coculture system reflect those that take place *in vivo*. As was discussed above, pericytes and endothelial cells appear to make frequent contact through discontinuities in the basement membrane [8]. In the coculture system we have observed endothelial-pericyte contacts (Fig. 4) and we and others [38] have demonstrated that the cells are coupled by gap junctions [58].

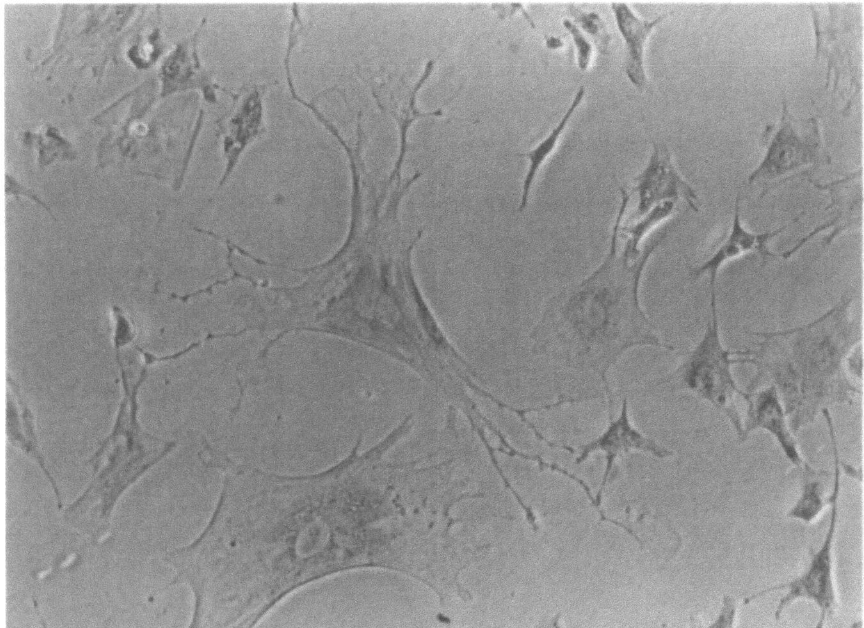

Fig. 4. Coculture of ECs and pericytes. Cocultures of ECs and growth-arrested pericytes were established as described in Orlidge and D'Amore [44] at a ratio of about one pericyte to five ECs. In this field the central pericyte is seen extending neurite-like processes to contact a number of surrounding ECs. (Micrograph courtesy of Alicia Antonelli, Children's Hospital, Boston, MA)

Taken together, these observations strongly indicate that the coculture system that we have developed accurately reflects the interactions between these cells *in situ*.

Clinical observations and our data strongly suggest a role for the pericyte in the control of vascular endothelial cell growth. This suggestion, however, must be taken in context of the fact that there are likely to be many levels of growth control in a tissue as important as the vasculature. More specifically, we speculate that it is not loss of the pericytes in the retinal microcirculation alone that leads to neovascularization. This speculation is supported by the finding of pericyte degeneration and increased pericyte turnover in other tissues (primarily skeletal muscle) without associated angiogenesis [61]. It seems more likely that growth control is the result of a balance between inhibitory and stimulatory forces. Thus, loss of the pericytes might instead create a "permissive" environment for the stimulation of new vessel growth that would leave the retinal circulation more vulnerable to angiogenic stimuli. The development of areas of ischemic retina that are known to be associated with diabetic retinopathy might lead to the release of angiogenic factors from the injured tissue. The loss of the pericytes would result in a decrease in local TFG-β concentrations. This coupled with increases in local concentrations of stimulatory factors might then lead to new vessel growth.

References

1. Akagi Y, Kador PF, Kuwabara T, Kinoshita JH (1983) Aldose reductase localization in human retinal mural cells. Invest Ophthalmol Vis Sci 24:1516–1519
2. Antonelli-Orlidge A, Saunders KB, Smith SR, D'Amore PA (1989) An activated form of transforming growth factor β is produced by cocultures of endothelial cells and pericytes. Proc Natl Acad Sci USA 86:4544–4548
3. Becker CG (1972) Demonstration of actomyosin in mesangial cells of the renal glomerulus. Am J Pathol 66:97–110
4. Bensley RR, Vimtrup BJ (1928) On the nature of the Rouget cells of capillaries. Anat Rec 39:37–55
5. Boya J (1976) An ultrastructural study of the relationship between pericytes and cerebral macrophages. Acta Anat (Basel) 95:598–608
6. Buzney SM, Frank RN, Robison WG (1975) Retinal capillaries: proliferation of mural cells *in vitro*. Science 190:985–986
7. Buzney SM, Massicotte SJ, Hetu N, Zetter BR (1983) Retinal vascular endothelial cells and pericytes. Differential growth characteristics *in vitro*. Invest Ophthalmol Vis Sci 24:470–480
8. Carlson EC (1989) Fenestrated subendothelial basement membranes in human retinal capillaries. Invest Ophthalmol Vis Sci 30:1923–1932
9. Castellot JJ Jr, Addonizio ML, Rosenberg R, Karnovsky MJ (1981) Cultured endothelial cells produce a heparinlike inhibitor of smooth muscle cell growth. J Cell Biol 90:372–379
10. Clark ER, Clark EL (1925) The relation of "Rouget" cells to capillary contractility. Am J Anat 35:265–282
11. Clark ER, Clark EL (1940) Microscopic observations on the extra-endothelial cells of living mammalian blood vessels. Am J Anat 66:39–49
12. Cohen MP, Frank RN, Khalifa AA (1980) Collagen production by cultured retinal pericytes. Invest Ophthalmol Vis Sci 19:90–94
13. Cotran RS (1965) The delayed and prolonged vascular leakage in inflammation. II. An electron microscopic study of the vascular response after thermal injury. Am J Pathol 46:589–620
14. Crocker DJ, Murad TM, Geer JC (1970) Role of the pericyte in wound healing. An ultrastructural study. Exp Mol Pathol 13:51–65
15. DeBault LE, Kahn LE, Frommes SP, Cancilla PA (1979) Cerebral microvessels and derived cells in tissue culture: isolation and preliminary characterization. In Vitro 15:473–487
16. DeNofrio D, Hoock TC, Herman IM (1989) Functional sorting of actin isoforms in microvascular pericytes. J Cell Biol 109:191–202
17. Feldman PS, Shneidman D, Kaplan C (1978) Ultrastructure of infantile hemangioendothelioma of the liver. Cancer 42:521–527
18. Folkman J, Haudenschild CC, Zetter BR (1979) Long-term culture of capillary endothelial cells. Proc Natl Acad Sci USA 76:5217–5221
19. Frank RN, Dutta S, Mancini MA (1987) Pericyte coverage is greater in the retinal than in the cerebral capillaries of the rat. Invest Ophthalmol Vis Sci 28:1086–1091
20. Fujimoto T, Singer SJ (1987) Immunocytochemical studies of desmin and vimentin in pericapillary cells of chicken. J Histochem Cytochem 35:1105–1115
21. Gitlin JD, D'Amore PA (1983) Culture of retinal capillary cells using selective growth media. Microvasc Res 26:74–80
22. Goldstein GW, Wolinsky JS, Csejtey JS, Diamond I (1975) Isolation of metabolically active capillaries from rat brain. J Neurochem 25:715–717
23. Herman IM, D'Amore PA (1985) Microvascular pericytes contain muscle and nonmuscle actin. J Cell Biol 101:43–52
24. Herman IM, Jacobson S (1988) *In situ* analysis of microvascular pericytes in hypertensive rat brains. Tissue Cell 20:1–12

25. Herman IM, Newcomb PM, Coughlin JE, Jacobson S (1987) Characterization of microvascular cell cultures from normotensive and hypertensive rat brains: pericyte-endothelial cell interactions *in vitro*. Tissue Cell 19:197–206
26. Hudes GR, Li W, Rockey JH, White P (1988) Prostacyclin is the major prostaglandin synthesized by bovine retinal capillary pericytes in culture. Invest Ophthalmol Vis Sci 29:1511–1516
27. Jotereau FV, LeDouarin NM (1978) The developmental relationship between osteocytes and osteoclasts: a study using the quail-chick nuclear marker in endochondral ossification. Dev Biol 63:253–265
28. Joyce NC, DeCamilli P, Boyles J (1984) Pericytes, like vascular smooth muscle cells, are immunocytochemically positive for cyclic GMP-dependent protein kinase. Microvasc Res 28:206–219
29. Joyce NC, Haire MF, Palade GE (1985) Contractile proteins in pericytes. I. Immunoperoxidase localization of tropomyosin. J Cell Biol 100:1379–1386
30. Joyce NC, Haire MF, Palade GE (1985) Contractile proteins in pericytes. II. Immunocytochemical evidence for the presence of two isomyosins in graded concentrations. J Cell Biol 100:1387–1395
31. Kelley C, D'Amore P, Hechtman HB, Shepro D (1987) Microvascular pericyte contractility *in vitro:* comparison with other cells of the vascular wall. J Cell Biol 104:483–490
32. Kelley C, D'Amore P, Hechtman HB, Shepro D (1988) Vasoactive hormones and cAMP affect pericyte contraction and stress fibers *in vitro*. J Muscle Res Cell Motil 9:184–194
33. Kennedy A, Frank RN, Varma SD (1983) Aldose reductase activity in retinal and cerebral microvessels and cultured vascular cells. Invest Ophthalmol Vis Sci 24:1250–1258
34. Khatami M, Li W, Rockey JH (1986) Kinetics of ascorbate transport by cultured retinal capillary pericytes. Invest Ophthalmol Vis Sci 27:1665–1671
35. Krause D, Vatter B, Dermietzel R (1988) Immunochemical and immunocytochemical characterization of a novel monoclonal antibody recognizing a 140 kD protein in cerebral pericytes of the rat. Cell Tissue Res 252:543–555
36. Krogh A (1929) The Anatomy and Physiology of Capillaries. Hafner, New York
37. Kuwabara T, Cogan DG (1963) Retinal vascular patterns. VI. Mural cells of the retinal capillaries. Arch Ophthalmol 69:492–502
38. Larson DM, Carson MP, Haudenschild CC (1987) Junctional transfer of small molecules in cultured bovine brain microvascular endothelial cells and pericytes. Microvasc Res 34:184–199
39. Le Beux YJ, Willemot J (1978) Actin- and myosin-like filaments in rat brain pericytes. Anat Rec 190:811–826
40. Majno G, Palade GE, Schoefl GI (1961) Studies on inflammation. II. The site of action of histamine and serotonin along the vascular tree: a topographic study. J Biophys Biochem Cytol 11:607–626
41. Mazanet R, Franzini-Armstrong C (1982) Scanning electron microscopy of pericytes in rat red muscle. Microvasc Res 23:361–369
42. Michael AF, Keane WF, Raj L, Vernier RL, Mauer SM (1980) The glomerular mesangium. Kidney Int 17:141–158
43. Nayak RC, Berman AB, George KL, Eisenbarth GS, King GL (1988) A monoclonal antibody (3G5)-defined ganglioside antigen is expressed on the cell surface of microvascular pericytes. J Exp Med 167:1003–1015
44. Orlidge A, D'Amore PA (1986) Cell specific effects of glycosaminoglycans on the attachment and proliferation of vascular wall components. Microvasc Res 31:41–53
45. Orlidge A, D'Amore PA (1987) Inhibition of capillary endothelial cell growth by pericytes and smooth muscle cells. J Cell Biol 105:1455–1462
46. Rhodin JAG (1968) Ultrastructure of mammalian venous capillaries, venules, and small collecting veins. J Ultrastruct Res 25:452–500
47. Rouget C (1873) Mémoire sur le developpement, la structure et les propriétes physiologiques des capillaires sanguins et lymphatiques. Arch Physiol Normale Pathol 5:603–661

48. Sato Y, Rifkin DB (1989) Inhibition of endothelial cell movement by pericytes and smooth muscle cells: activation of a latent transforming growth factor-β1-like molecular by plasmin during co-culture. J Cell Biol 109:309–315

49. Schor AM, Schor SL (1986) The isolation and culture of endothelial cells and pericytes from the bovine retinal microvasculature: a comparative study with large vessel vascular cells. Microvasc Res 32:21–38

50. Sedor JR, Abboud HE (1985) Histamine modulates contraction and cyclic nucleotides in cultured rat mesangial cells. J Clin Invest 75:1679–1689

51. Shabik M, de Oliveira F (1966) Studies on developing retinal vessels. X. Formation of the basement membrane and differentiation of intramural pericytes. Br J Ophthalmol 50:124–133

52. Sherer GK, Fitzharris TP, Faulk WP, LeRoy EC (1980) Cultivation of microvascular endothelial cells from human preputial skin. In Vitro 16:675–684

53. Sims DE (1986) The pericyte – a review. Tissue Cell 18:153–174

54. Skalli O, Pelte M-F, Peclet M-C, Gabbiani G, Gugliotta P, Bussolati G, Ravazzola M, Orci L (1989) Alpha-smooth muscle cell actin, a differentiation marker of smooth muscle cells, is present in microfilamentous bundle of pericytes. J Histochem Cytochem 37:315–321

55. Speiser P, Gittelsohn AM, Patz A (1968) Studies on diabetic retinopathy. III. Influence of diabetes on intramural pericytes. Arch Ophthalmol 80:332–337

56. Spitznas M, Reale E (1975) Fracture faces of fenestrations and junctions of endothelial cells in human choroidal vessels. Invest Ophthalmol Vis Sci 14:98–107

57. Stramm LE, Li W, Aguirre GD, Rockey JH (1987) Glycosaminoglycan synthesis and secretion by bovine retinal capillary pericytes in culture. Exp Eye Res 44:17–28

58. Sweet E, Abraham EH, D'Amore PA (1988) Functional evidence of gap junctions between capillary endothelial cells and pericytes *in vitro*. Invest Ophthalmol Vis Sci 29:109a

59. Tilton RG, Kilo C, Williamson JR, Murch DW (1979) Differences in pericyte contactile function in rat cardiac and skeletal muscle microvasculatures. Microvasc Res 18:336–352

60. Tilton RG, Miller EJ, Kilo C, Williamson JR (1985) Pericyte form and distribution in rat retinal and uveal capillaries. Invest Ophthalmol Vis Sci 26:68–73

61. Tilton RG, Faller AM, Hoffmann PL, Kilo C, Williamson JR (1987) Acellular capillaries and increased pericyte degeneration in the diabetic extremity. Front Diabetes 8:186–189

62. Voyta JC, Via DP, Butterfield CE, Zetter BR (1984) Identification and isolation of endothelial cells based on their increased uptake of acetylated low density lipoprotein. J Cell Biol 99:2034–2040

63. Wagner RC, Matthews ML (1975) The isolation of capillary endothelium from epididymal fat. Microvasc Res 10:286–297

64. Wallow IH, Burnside B (1980) Actin filaments in retinal pericytes and endothelial cells. Invest Ophthalmol Vis Sci 19:1433–1441

65. Weibel ER (1974) On pericytes, particularly their existence on lung capillaries. Microvasc Res 8:218–235

66. Zimmerman K (1923) Der feinere Bau der Blutcapillaren. Z Anat Entwicklungsgesch 68:29–109

Attachment Substrates for Smooth Muscle Cells*

J. Thyberg, U. Hedin, and B. A. Bottger

Introduction

The smooth muscle cell is the sole cell type present in the media of blood vessels and its morphological, biochemical, and functional properties have been described in considerable detail [8]. It originates from mesenchymal cells in the embryo and has initially a fibroblast-like structure with widespread cisternae of rough endoplasmic reticulum, a large Golgi complex, and a poorly developed myofilamentous system. Accordingly, its main functions during vasculogenesis are to proliferate and to produce extracellular matrix components. As the vessels approach their final size, these activities cease and the smooth muscle cell becomes a highly specialized contractile cell with a cytoplasm largely occupied by actin and myosin filaments [14, 87].

In the media of mature blood vessels, the smooth muscle cells are arranged in concentric layers [66]. Each cell is encircled by a basement membrane composed of collagen type IV, laminin, nidogen (entactin), and heparan sulfate proteoglycans [91, 92]. This proteinaceous network is likely to play an important role in the homeostasis of the cells. In part, it may do so by direct interaction with them, by regulating the transport of molecules to and from them, and by establishing a link with the surrounding extracellular matrix. The latter is primarily made up of elastic lamellae or fibers [68, 75], fibrils of collagen type I and III [4, 49] and chondroitin sulfate and dermatan sulfate proteoglycans [64, 99].

The terminal differentiation process depicted above is not irreversible. Thus, it has long been known that smooth muscle cells may return to a proliferative and secretory active state in the adult, for example, in association with atherosclerosis and hypertension [6, 54, 69, 79]. In a similar manner, it is well known that the proliferation of smooth muscle cells in culture is preceded by a shift from a highly differentiated, contractile phenotype to a less differentiated, synthetic phenotype [17, 27, 89, 90]. For this reason, the in vitro system has become a valuable experimental model for studies of smooth muscle cell biology as related to cardiovascular disease.

The first attempts to culture smooth muscle were made in the beginning of this century [55, 56]. However, it was not until the early 1970s that this technique became well established. From then on, it has been utilized in a large

* Supported by the Swedish Medical Research Council (06537), the Swedish Heart Lung Foundation, the King Gustaf V 80th Birthday Fund, and the funds of Karolinska Institutet

number of investigations covering a broad spectrum of biological questions [14, 18]. In particular, it has served a crucial role in the study of smooth muscle cell growth and the discovery of platelet-derived growth factor (PDGF) as a potent mitogen for cells of mesenchymal origin [70, 71]. Likewise, it has been of great value in exploring the participation of the smooth muscle cell in the synthesis and secretion of the extracellular matrix components of the vessel wall [13, 49, 99].

To understand the regulation of differentiated properties and growth of vascular smooth muscle cells in further detail, and the basis for their involvement in cardiovascular disease, the in vitro technique will undoubtedly continue to be a powerful tool. However, as in other cell systems, it will become increasingly important to work out defined, serum-free conditions that make it possible to design the experiments and interpret the results in a clearcut way [3]. So far this work is only in its beginning and many difficulties can be expected along the way. The ultimate goal is to culture the cells on a supportive layer of pure molecules and in a synthetic medium supplemented with a balanced mixture of pure nutrients, hormones, and growth factors. The substrate may be composed of adhesive glycoproteins normally present in the microenvironment of the cells or reaching them via the blood in connection with endothelial damage or malfunction. Another possibility is to use artificial or synthetically produced substrates that serve a similar role. Agents of physiological as well as pharmacological interest may be added to the medium.

The purpose of this chapter is to briefly review the application of attachment substrates in the culture of vascular smooth muscle cells. Special attention will be paid to fibronectin, an adhesive glycoprotein found in blood plasma and extracellular matrices [1, 37, 72], and laminin, a major component of the basement membrane [91, 92]. Other molecules that will be considered include vitronectin/serum spreading factor [74, 82], collagen [41, 52], thrombospondin [25, 44, 83], and tenascin/hexabrachion protein [24]. In addition, we will give a brief account of the principles for isolation and culture of smooth muscle cells, and mention some basic methods of study. The reader is also referred to more comprehensive reviews covering mammalian cell culture at large [3, 39, 63, 77, 88]. Because of the restricted space available, the list of references will be selective rather than complete. Likewise, the list of suppliers for special materials will mention but a few rather than all existing sources.

Description of Methods and Materials

Isolation and Culture of Smooth Muscle Cells

Most in vitro studies on vascular smooth muscle deal with cells isolated from large and medium-sized arteries [15, 18]. However, methods for preparation of cerebral microvascular [53] and lymphatic smooth muscle cells [40] have also been described. The main species used to date are rat, rabbit, pig, calf, monkey, and man. Depending on the age of the donor, the cells will be either in a synthetic (embryos and newborns) or a contractile (adults) phenotype at the start of culture.

There are two principal techniques to prepare arterial smooth muscle cells for culture, explantation and enzymatic isolation [15, 18]. In the first case, pieces of arterial media cleaned from intima and adventitia are placed in a Petri dish and submerged in a serum-supplemented nutrient medium. Within a few days, cells start to migrate out of the tissue fragments and within a few weeks a confluent cell layer is formed. For subculturing, the cells are detached with trypsin and reseeded in dishes or flasks. A disadvantage of this method is that a long time passes before a reasonable number of cells has accumulated, during which they will diverge from their in vivo state both morphologically and functionally. In the second procedure, pieces of arterial media are digested by enzymes to free the cells from surrounding extracellular matrix. The resulting single cell suspension is used to set up primary cultures. Collagenase is the main component of the enzyme solution and may be used either alone or in combination with elastase and hyaluronidase. If necessary, DNase may be added to remove nucleic acid released from degenerating cells. The advantage of this method is that a large number of cells can be produced in a relatively short time and in a state closely resembling that occurring in vivo. Moreover, if the cells need to be propagated before use, confluent cultures are formed in a much shorter time than with the explantation technique. However, the enzymes may have adverse effects on the cells and it is recommended that the toxicity of new enzyme batches is tested.

As a more detailed guideline for the reader, we will describe the procedure utilized in our laboratory for isolation of rat aortic smooth muscle cells. Medium F-12, originally composed for growth of fibroblasts at low serum concentration [31], is used throughout. It is supplemented with 10 mM each of the organic buffers HEPES and TES, pH 7.3 [23], 50 µg/ml of L-ascorbic acid [78], and 50 µg/ml of gentamycin sulfate (hereinafter referred to as medium F-12). Either 0.1% bovine serum albumin (BSA, globulin-free) or 10% newborn calf serum (NCS) is also added. As an alternative, medium MCDB 104 can be recommended [51].

The aorta is excised from 300- to 400-g Sprague-Dawley rats under carbon dioxide anesthesia and immersed in medium F-12/0.1% BSA. The adventitia is removed with fine forceps, the vessel cut open longitudinally with a pair of scissors, and the intima scraped off with a scalpel (avoid traction to minimize cell damage). The media is cut into smaller pieces (about 5 × 5 mm) and digested with 0.1% collagenase (Gibco BRL, Sigma type I or Ia) in medium F-12/ 0.1% BSA at 20 °C under continuous agitation (100–200 rpm), first for 1 h to remove possible remnants of adventitia and intima, and then for another 15–20 h with fresh enzyme solution to dissolve the tissue completely. The freed cells are passed through a nylon filter (mesh size 50 × 50 µm), rinsed twice with medium, and counted in a hemocytometer or an electronic cell counter (5–8 million cells are normally obtained from each aorta). Finally, the cells are seeded in primary cultures, either on specific substrates in medium F-12/0.1% BSA (30 000–50 000 cells/cm^2) or in plastic flasks in medium F-12/10% NCS (10 000–30 000 cells/cm^2). Cells grown to confluence in the presence of serum are detached by treatment with 0.1% trypsin (Difco 1 : 250) and 0.02% EDTA in Dulbecco's phosphate-buffered saline (PBS) without calcium and magnesium, rinsed twice with medium, and reseeded in secondary cultures.

Attachment Substrates

Nontransformed cells need to be attached to a solid substrate to survive and grow in vitro. Traditionally, this has been achieved by seeding them on a surface of glass or plastic (cell culture grade), both characterized by a negative charge. The glycoconjugates on the cell surface likewise have a negative charge and the binding of the cells to the substrate is mediated via a bridge of divalent cations or adhesive glycoproteins present in the serum added to the medium [58]. Coating of the culture vessels with polylysine or other positively charged polymers has been found to improve the attachment and growth of many cells at low serum concentration [50]. In this case, direct electrostatic interactions between the substrate and the cell surface are likely to be involved.

For the culture of cells under serum-free conditions, specific attachment glycoproteins are generally required. They may be derived from the blood, as in the case of fibronectin and vitronectin, or be components of the pericellular and extracellular matrix that surrounds the cells in vivo, as in the case of laminin and collagen [41, 42]. The use of these and related macromolecules in cell culture work has followed as a result of the accelerating development in the biochemistry of extracellular matrix components during recent years [22, 62], and the realization of their fundamental role in the regulation of growth and differentiated properties of cells [32, 95], including those of the vascular system [36]. An overview of the substrates dealt with here is given in Table 1.

Fibronectin and Fibronectin-Derived Peptides

Fibronectin is a 500-kDa glycoprotein found in blood plasma and extracellular matrices. It consists of two subunit chains linked by disulfide bonds close to their carboxy-terminal ends. Each subunit is divided into domains with varying binding specificities. Most of them interact with other plasma or extracellular proteins, such as fibrinogen/fibrin, heparin/heparan sulfate proteoglycans (HSPG), and collagen/gelatin. In the center of the molecule there is also one domain that interacts with cell-surface receptors belonging to the integrin superfamily of proteins. The amino acid sequence Arg-Gly-Asp (RGD) has been identified as the minimal cell-attachment site within this domain. Plasma fibronectin is produced by hepatocytes and tissue fibronectin by various epithelial and mesenchymal cells. A comparison of fibronectins from different sources reveals slight differences in molecular weight and chemical structure. These are mainly due to alternative splicing of the primary transcript and variations in the degree of glycosylation. Nevertheless, the main function of all fibronectins is the same, namely to act as a substrate for attachment and spreading of cells, either directly or by mediating a link to collagen matrices and fibrin clots [1, 37, 72].

Fibronectin can be isolated from blood plasma (about 300 µg/ml), cell culture media, and cell extracts with good yields. Most protocols are based on the affinity of the molecule for collagen/gelatin. Thus, it is possible to prepare electrophoretically pure fibronectin from human plasma by affinity chromatography on gelatin-agarose (Bio-Rad, Pharmacia LKB Biotechnology). A

Table 1. Overview of attachment substrates for smooth muscle cells

Molecule and basic chemical properties[a]	Major binding affinities	Sources for isolation
Fibronectin, dimeric 500 (250) kDa glycoprotein	Cell-surface receptors, collagen, fibrinogen/fibrin, heparin/HSPG	Blood plasma, cell cultures
Vitronectin, monomeric 75 kDa glycoprotein	Cell-surface receptors, heparin/HSPG, thrombin/antithrombin	Blood plasma or serum
Laminin, trimeric 900 (220, 220, 400) kDa glycoprotein	Cell-surface receptors, collagen heparin/HSPG, nidogen	EHS tumor, placenta
Collagen type I, triple helical 285–300 (95) kDa glycoprotein	Cell-surface receptors	Rat tail and other tissues, cell cultures
Collagen type IV, triple-helical 550–600 (185, 185, 175) kDa glycoprotein	Cell-surface receptors, heparin/HSPG, laminin, nidogen	EHS tumor, placenta
Thrombospondin, trimeric 450 (180) kDa glycoprotein	Collagen, fibrinogen/fibrin, fibronectin, heparin/HSPG, plasminogen, thrombin	Platelets, cell cultures
Tenascin, hexameric 1900 (320) kDa glycoprotein	Cell-surface receptors, chondroitin sulfate proteoglycans, fibronectin (?)	Chick brain or gizzard, cell cultures

[a] Approximate subunit molecular masses indicated in parentheses
HSPG, heparan sulfate proteoglycans

similar technique is utilized for cell culture media and cell layer extracts. If necessary, ion exchange chromatography on DEAE-cellulose can be used as an additional step to free the fibronectin from contaminating glycosaminoglycans. For detailed instructions the reader is referred to published procedures [73, 102]. Plasma fibronectin is also available commercially, for example, from Boehringer Mannheim, Calbiochem, Gibco BRL, Sigma, Telios, and Wako.

Functionally specific domains of fibronectin can be isolated by limited proteolysis with enzymes like plasmin, chymotrypsin, or trypsin, and subsequent purification by adsorptive chromatography, ion exchange chromatography, or gel filtration [72]. A large-scale procedure based on digestion with thermolysin and hydroxyapatite chromatography has also been described [9]. The fragments used in our laboratory include a 105-kDa cell-binding fragment [101], a 70-kDa collagen-binding fragment [60], and a 31-kDa heparin-binding fragment [101]. A set of proteolytic fragments of human plasma fibronectin covering all major domains in the molecule is available commercially from Telios.

To prepare cell culture substrates, fibronectin and the proteolytic fragments are dissolved in PBS (pH 7.3) at 10–20 µg/ml, added onto a surface (0.2 ml/cm^2) of plastic (culture dishes) or glass (coverslips), and allowed to adsorb for 1–2 h (or longer) at 20 °C. The surface is then rinsed twice with PBS and left in medium F-12/0.1% BSA for at least 15 min. The last step is included to block nonspecific binding sites before the seeding of cells [33, 34].

RGD-containing synthetic peptides can be produced in an automated peptide synthesizer and then further purified by reverse-phase HPLC. Several such peptides are also available commercially, for example, from Peninsula, Sigma, and Telios. Substrates are prepared by linkage of the peptide to a surface of polystyrene, agarose, or polyacrylamide [11, 61]. Here, we describe the procedure used in our laboratory for smooth muscle cells [35]. Culture dishes are incubated with 1 mg/ml of BSA in Dulbecco's PBS (pH 7.3) for 2 h at 20 °C and rinsed once with PBS. The resulting albumin coat is derivatized with 0.6 μmol/ml of N-succinimidyl 3-(2-pyridyldithio) propionate (SPDP, Pharmacia LKB Biotechnology) in PBS for 2 h, rinsed twice with PBS, and reduced with 50 mM dithiothreitol in 0.1 M sodium acetate buffer (pH 4.5) for 1 h. In parallel, the peptide is incubated with SPDP in PBS at equimolar concentration (0.6 μmol/ml) for 1 h. The latter solution is then added to the BSA-coated dishes and left for 10–15 h at 20 °C. This leads to the formation of an intermolecular conjugate by linking SPDP molecules bound to amino groups in BSA and the peptide by a disulfide bond [16, 35]. If the peptide contains cysteine (a sulfhydryl group), it can be added directly to the SPDP-treated albumin coat. In the end, the dishes are rinsed three times with PBS before the seeding of cells.

Vitronectin

Vitronectin (also termed serum spreading factor, S-protein, or epibolin) is a 75-kDa glycoprotein found in blood plasma and tissues. It promotes attachment and spreading of cells in vitro and modulates blood coagulation and complement-mediated cytolysis [38, 74]. It contains an RGD sequence but binds to a cell-surface receptor distinct from the fibronectin receptor [65]. Vitronectin can be isolated from plasma or serum by glass bead chromatography followed by concanavalin A-agarose, DEAE-agarose, and heparin-agarose chromatography [82] or glass bead chromatography followed by FPLC on a Mono Q column [74]. Recently, a simpler and more rapid method based on heparin affinity chromatography was developed [103]. Commercial sources for vitronectin include Calbiochem, Telios, and Wako.

For cell culture work vitronectin is handled in a similar way as fibronectin (see above). However, it should be noted that vitronectin in most cases is more potent than fibronectin on a weight basis and thus can be used at a lower concentration in the preparation of attachment substrates (1–5 μg/ml).

Laminin

Laminin is the main noncollagenous constituent of basement membranes. It is a 900-kDa glycoprotein composed of three chains (A, B1 and B2) linked by disulfide bonds and shaped as a cross with three short arms and one long arm. The molecule is divided into functional domains and interacts with other basement membrane components as well as cells. The 150-kDa glycoprotein nidogen (entactin) binds in a stable manner close to the center of the cross and has been proposed to form a link between laminin and collagen type IV. A major binding site for heparin and heparan sulfate proteoglycans (HSPG) is present at the end of the long arm. The basis for the interaction of laminin with cells is still incompletely known. At least two cell-binding sites have been described. One of them is located close to the center of the cross and includes the sequence Tyr-Ile-Gly-Ser-Arg (YIGSR). The other is located at the end of the long arm and includes the sequence Arg-Gly-Asp (RGD), also found in the cell-binding domain of fibronectin [91, 92].

The Engelbreth-Holm-Swarm (EHS) sarcoma and other transplantable rodent tumors are the most frequent sources for isolation of laminin. The large-scale procedures used in these cases are based on precipitation/extraction with sodium chloride at varying concentrations, followed by DEAE-cellulose chromatography and agarose gel filtration. Intact laminin-nidogen complexes are isolated by extraction with EDTA in a physiologic buffer and subsequent agarose gel filtration. Purification of smaller amounts of laminin from cell culture media can be accomplished by sequential chromatography on gelatin-agarose (to remove fibronectin) followed by heparin-agarose (Bio-Rad, Pharmacia LKB Biotechnology). Cell-binding and heparin-binding fragments of laminin are prepared by digestion with elastase or pepsin followed by agarose gel filtration or heparin-agarose chromatography [46, 93]. Laminin is available commercially from Gibco BRL, Sigma, and Telios, and YIGSR-containing laminin peptides from Peninsula.

For cell culture work laminin and laminin peptides are treated in a similar way as fibronectin (see above). We normally use laminin at about two times higher concentration than fibronectin in the preparation of attachment substrates (20–40 µg/ml). With laminin-nidogen complexes, it is advisable to use calcium- and magnesium-free PBS to avoid self-aggregation [59]. If problems appear in adsorbing laminin to glass, this may be overcome by covering the glass surface with silane and attaching the protein covalently to this layer with glutaraldehyde. The same technique can be applied also for other attachment factors [2].

Collagen and Collagen Gels

Collagen is the most abundant protein in mammals and a major component of all extracellular matrices. It is made up of three alpha chains (60–240 kDa) wound around each other in a triple-helix. To date, at least 20 types of genetically distinct alpha chains have been recognized, forming at least eleven types of collagen with variable tissue distribution and supramolecular structure [52]. Of these, type I and III are the predominant in blood vessels. They are assembled into cross-banded fibrils and provide the tissue with tensile strength. Type IV is present in the basement membrane beneath the endothelium and around the smooth muscle cells, where it forms a complex three-dimensional network. Small amounts of collagen type V, VI, and VIII have also been detected in the vessel wall, but their macromolecular organization remains poorly understood [4, 49].

Collagen improves the attachment, growth, and differentiation of epithelial as well as mesenchymal cells and has found wide use in cell culture [41, 42]. The methodology for isolation and characterization of the different collagens has recently been reviewed [28]. The main procedures are based on extraction with salt or dilute acid, limited pepsin digestion, and selective salt precipitation. Further purification is achieved by affinity chromatography, ion exchange chromatography, or gel filtration. The collagens most frequently used in cell culture, type I and IV, are available commercially (Boehringer Mannheim, Gibco BRL, Sigma, and Telios). A film of collagen type I can be made by dissolving the protein in dilute acetic acid (0.5 M), add it to the dishes $(2-10 \ \mu g/cm^2)$, and air-dry (overnight). Alternatively, the protein is allowed to adsorb to the plastic surface from a more concentrated solution $(1-3 \ mg/ml)$ and excess fluid withdrawn after a few hours. In both cases, the dishes are rinsed two to three times with medium before the seeding of cells. To prepare a collagen gel, the protein is dissolved in dilute acetic acid $(3-5 \ mg/ml)$ and added to the dishes. A gel is formed when the pH is brought to neutrality. This is accomplished by placing the dishes in a closed container with a small amount of ammonium hydroxide, or by adding tenfold concentrated medium containing an appropriate amount of 10 M sodium hydroxide. Another technique is to dialyze the acid collagen solution against PBS (pH 7.4) at 4 °C and then warm it to 37 °C to gel. For further practical details the reader is referred to the reviews of Kleinman et al. [41, 42]. In our studies, collagen type IV is dissolved either in PBS (pH 7.3) or 0.5 M carbonate buffer (pH 9.4) and then treated in a similar way as fibronectin (see above).

An interesting recent finding is that purified collagen type IV, laminin, and heparan sulfate proteoglycans (HSPG) are able to reassemble into basement membrane-like structures when incubated together in vitro [30].

Thrombospondin

Thrombospondin is a 450-kDa glycoprotein stored in platelet alpha granules and released during platelet degranulation. It is composed of three disulfide-linked monomers with binding affinities for fibrinogen/fibrin, plasminogen, thrombin, fibronectin, collagen type IV, laminin, and heparin/heparan sulfate proteoglycans. It has been demonstrated in many tissues, including blood vessels, and is synthesized by a variety of normal and transformed cells in culture, including endothelial and smooth muscle cells. These properties make it a highly interesting molecule in relation to cardiovascular disease [25, 44, 83].

Platelet concentrates constitute the most convenient source for isolation of thrombospondin, but cell culture media are also possible to use. The platelets are stimulated to degranulate by treatment with thrombin or the calcium ionophore A23187. After centrifugation, the supernatant is subjected to sequential affinity chromatography on columns of gelatin-agarose and heparin-agarose. Thrombospondin eluted from the second column is further purified by gel filtration [76]. Alternatively, the supernatant from the degranulated platelets is passed over columns of heparin-agarose and gelatin-agarose and purified by FPLC on columns of Superose 12 and Mono Q [21]. Attachment substrates are prepared by dissolving thrombospondin (50 µg/ml) in a physiologic buffer with $1-2$ mM calcium chloride, allowing the protein to adsorb to a surface of glass or plastic, and finally rinsing with a BSA-containing buffer before the seeding of cells [45, 67, 96].

Tenascin

Tenascin (also called brachionectin, cytotactin, hexabrachion protein, myotendinous antigen, and various other names) is a large oligomeric glycoprotein of extracellular matrices. It is a hexameric, multidomain molecule made up of 320-kDa subunits (large species variations) linked together by disulfide bonds to form a characteristic six-armed structure. It is synthesized in a tissue- and time-specific manner during embryonic development, rare in most mature organs, but strongly expressed in a variety of tumors. Dense connective tissues like tendons and ligaments represent the most stable locations. The arterial wall is among the other sites where it has been demonstrated also during adult life. It has been suggested to function in cell adhesion and stimulation of cell growth, but few details are known so far [20, 24, 26].

Tenascin can be isolated from media of cultured cells (glioma cell lines, fibroblasts) or tissue homogenates (chick brain or gizzard) by immunoaffinity chromatography on a column of antibodies coupled to agarose beads. It can also be purified from culture media by ammonium sulfate precipitation and glycerol gradient centrifugation [24]. To prepare attachment substrates tenascin is dissolved in PBS (pH 7.3) at 100 µg/ml and allowed to adsorb to plastic dishes for 30 min or longer. To remove unbound material and block unspecific binding sites, the dishes are rinsed with PBS/1% BSA before the seeding of cells [26].

Synthetic, Nonbiological Materials

Synthetic materials such as nylon, silicone, and elastin-like polymers can be expected to find increasing use as cell culture substrates in the future. They may be applied both in basic research and to produce living grafts for replacement of smaller blood vessels. Earlier work in this field includes growth of smooth muscle cells on silicone rubber exposed to repeated stretching/recoiling [12], studies of attachment site formation in smooth muscle cells seeded on collagen-hydroxyethylmethacrylate hydrogels [94], and in vitro construction of a blood vessel model from isolated vascular cells, collagen gels, and a nylon mesh to provide mechanical support [98].

Basic Methods of Study

Light and Electron Microscopy

Routine examination of the cultures is done in an inverted microscope with phase contrast optics. During primary culture under serum-free conditions, enzymatically isolated smooth muscle cells initially elongate and subsequently spread out on the substrate (Fig. 1). In the presence of serum, the cells rapidly grow into multiple overlapping layers and a hill and valley pattern is established. Transmission electron microscopy indicates that cells in a contractile phenotype have a cytoplasm largely occupied by myofilaments coalescing in dense bodies, mitochondria, and numerous surface caveolae. On the other hand, cells in a synthetic phenotype are dominated by widespread cisternae of rough endoplasmic reticulum and a large Golgi complex. Within 1–2 weeks of culture, a basement membrane and smaller amounts of extracellular matrix with collagen fibrils and elastin-like aggregates are often detected around the cells [33, 34, 89, 90].

Immunocytochemistry

Immunocytochemistry serves as a valuable aid in the identification and characterization of smooth muscle cells (Fig. 2). Antibodies against smooth muscle actin [29, 86], smooth muscle myosin [7], and the intermediate filament proteins desmin and vimentin [43] are primary tools in this context. Thus, during in vitro cultivation the cytoskeleton of smooth muscle cells develops features similar to those observed in the fetus and in atherosclerotic lesions [85]. Antibodies against basement membrane components like laminin and type IV collagen may also be helpful to distinguish smooth muscle cells from fibroblasts [91, 92]. Several of the antibodies mentioned above are available commercially (see addresses of suppliers for special materials).

Assay of DNA Synthesis and Cellular Proliferation

To follow the initiation of DNA synthesis, the cultures are exposed to tritiated thymidine (1–2 µCi/ml) and then rinsed with medium. For liquid scintillation counting we precipitate macromolecular material with cold 5% trichloroacetic acid (TCA), rinse the dishes three times with cold 5% TCA, and dissolve the cells in 0.1 M potassium hydroxide. Aliquots of the lysates are mixed with scintillation fluid and radioactivity is determined in a liquid scintillation spectrometer. For autoradiography we fix the cells (grown on glass coverslips) in 2% buffered glutaraldehyde, dehydrate in ethanol (70%–100%), and air-dry. The coverslips are mounted on glass slides, dipped in Kodak NTB2 emulsion, left in the dark for 1–2 days at 4°C, and the film developed in Kodak D-19. The cells are stained with 1% methylene blue and the fraction labeled nuclei is determined by counting 300–500 cells on each coverslip. In selecting the time of exposure to the radioactive precursor, it should be noted that quiescent

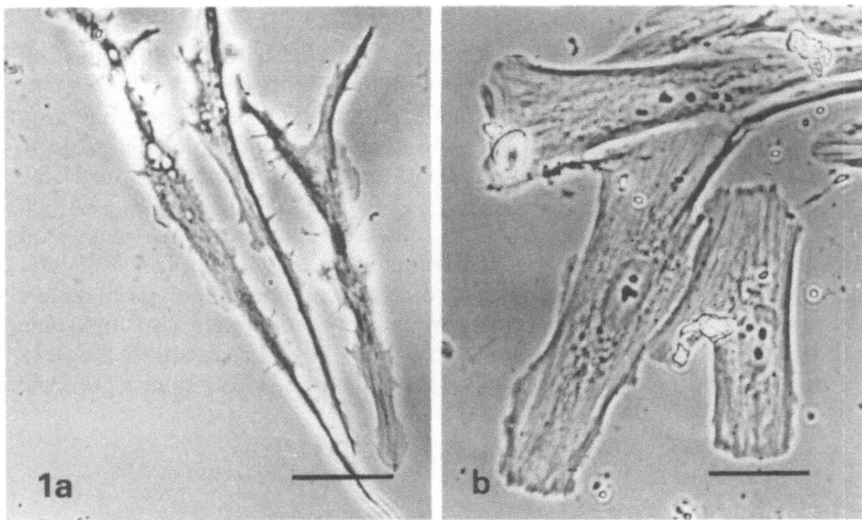

Fig. 1 a, b. Smooth muscle cells enzymatically isolated from adult rat aorta and grown on a substrate of human plasma fibronectin in serum-free medium for **a** 2 or **b** 4 days. Phase-contrast micrographs. *Bars,* 20 μm

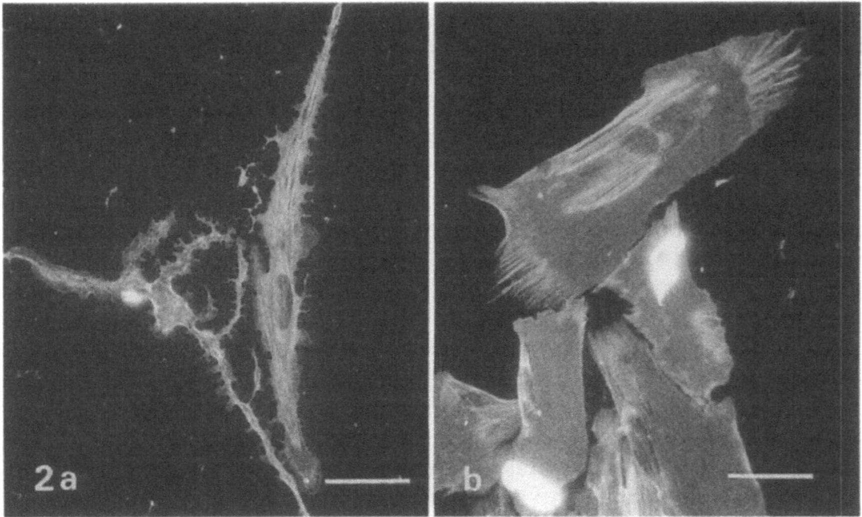

Fig. 2 a, b. Smooth muscle cells enzymatically isolated from adult rat aorta and grown on a substrate of human plasma fibronectin in serum-free medium for **a** 1 or **b** 4 days. Indirect immunofluorescence microscopy with primary antibodies against smooth muscle α-actin and fluorescein isothiocyanate-labeled secondary antibodies. Early in culture α-actin-containing filaments fill out large parts of the cytoplasm (**a**). However, already after a few days there is a dramatic reorganization of the cytoskeleton with a more restricted distribution of α-actin (**b**). *Bars,* 20 μm

smooth muscle cells (serum-free primary cultures or serum-starved subcultures) show a lag phase of 10–15 h when stimulated with PDGF, and even longer with other mitogens. Although more time-consuming, the autoradiographic technique has the advantage of allowing direct visual inspection of the cultures. In addition, the results are not influenced by possible effects of the experimental treatments on the uptake and intracellular concentration of the labeled nucleoside in the cells.

To follow cellular proliferation, the cells are detached by treatment with 0.1% trypsin and 0.02% EDTA in calcium- and magnesium-free PBS and counted in a hemocytometer or an electronic cell counter at appropriate intervals. In serum-free cultures the cells show an increased vulnerability and some of them may disintegrate in response to trypsin. As an alternative, they can then be counted directly on the dishes in a microscope (with a square graticule in the ocular).

Discussion

Effects of Attachment Proteins on Smooth Muscle Phenotype

Atherosclerosis and hypertension are the two main diseases of the cardiovascular system and both include proliferation of smooth muscle cells as a central element [6, 54, 69, 79]. The regulation of smooth muscle replication is therefore a major topic in cardiovascular research, and the cell culture technique has been established as an important experimental tool in this work [15, 18]. Studies in several laboratories have indicated that the smooth muscle cell has to go through an overall change in its differentiated properties before it is able to synthesize DNA and divide [17, 27, 89, 90]. This process represents a return to a state similar to that existing in the embryo and the young growing organism, and is referred to as a modulation from a contractile to a synthetic phenotype. It was found to take place in media containing either plasma-derived serum or whole blood serum and to be independent of the presence of PDGF or other growth factors [19, 89]. Later on, fibronectin was identified as one of the principal plasma constituents responsible for the shift in smooth muscle phenotype [33]. It was shown to exert its effect in substrate form and via the RGD sequence in the cell-binding domain of the molecule [34, 35], evidently by interaction with a cell-surface receptor belonging to the integrin superfamily of proteins [10]. In contrast, substrates of the basement membrane components laminin and collagen type IV retained the cells in a contractile phenotype, although endogenous production of fibronectin eventually led to a shift into a synthetic phenotype [34]. Preliminary observations suggest that basement membrane components may also stimulate the cells to return from a synthetic to a contractile phenotype (Fig. 3).

The phenotypic modulation does not itself initiate cell growth but makes the smooth muscle cells able to replicate their DNA and divide in response to peptide mitogens, among which PDGF appears to be the most important: PDGF > fibroblast growth factor, (FGF) > epidermal growth factor (EGF) > insulin-like growth factor-I (IGF-I; unpublished observations).

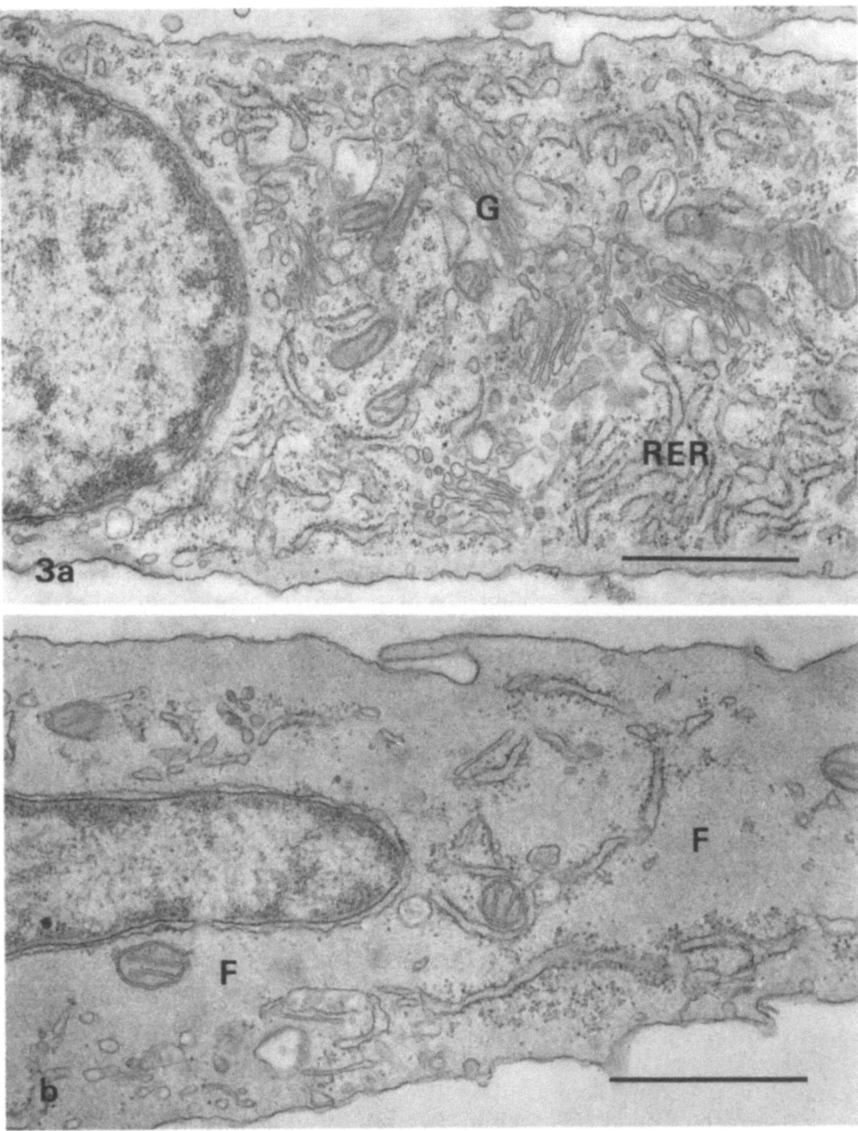

Fig. 3a, b. Smooth muscle cells enzymatically isolated from adult rat aorta, growth to confluence in serum-containing medium, trypsinized, and grown in serum-free medium on a substrate of fibronectin for **a** 2 days or on a mixed substrate of laminin and collagen type IV for **b** 1 day. The cell in **a** is in a synthetic phenotype and has an extensive rough endoplasmic reticulum (*RER*) and a large Golgi complex (*G*). The cell in **b** has at least partly returned to a contractile phenotype and shows poorly developed synthetic organelles but a more elaborate myofilamentous system (*F*), here appearing as diffuse areas in the cytoplasm. Electron micrographs. *Bars,* 1 µm

Moreover, when stimulated with an exogenous mitogen synthetic state cells start to produce a PDGF-like mitogen and to promote their growth in an autocrine manner [57]. Recent studies have indicated that this is coupled to expression of mRNA for the A chain of PDGF and expression of PDGF α- and β-receptors (unpublished observations) [81, 84]. In analogy with these results, smooth muscle cells of newborn animals [48, 80] and of atherosclerotic lesions [5, 47, 97, 100] have been found to express PDGF genes and produce PDGF-like proteins.

Future Directions

Serum-free culture of vascular smooth muscle on specific attachment substrates is up to now only in its very beginning. In the future, this technique will be of great potential value to sort out the mechanisms involved in the control of the differentiated properties of smooth muscle cells. Here, a few of the many interesting questions to be analyzed are listed: Are there blood components other than fibronectin which induce a transition of the cells from a contractile to a synthetic phenotype? What is the signal transduction machinery that brings about the change in cellular phenotype? How do basement membrane components act to withhold the contractile phenotype? Are cells which have gone through the transition into a synthetic phenotype able to return to a contractile phenotype? What is the link between cellular phenotype, the ability to respond to peptide growth factors, and autocrine production of such molecules? How do smooth muscle cells interact with other cells found in the vessel wall, such as endothelial cells, macrophages, and lymphocytes? Exploration of these and related questions will deepen our knowledge of the biology of the smooth muscle cell and form a basis for further examination of its role in cardiovascular disease.

Concluding Remarks

Smooth muscle cells build up the media of blood vessels and take active part in the development of atherosclerosis and hypertension. In vitro cultivation represents a powerful technique to study the control of differentiated properties and growth of these cells. Preferably, they should be seeded in a synthetic medium supplemented with pure nutrients, hormones, and growth factors. For this sake, attachment substrates prepared from plasma proteins like fibronectin and vitronectin or extracellular matrix components like laminin, collagen, thrombospondin, and tenascin are used. This approach offers great advantages in experimental design and makes it possible to analyze cellular behavior under defined conditions. It is anticipated that this methodology will help to widen our understanding of the role of the smooth muscle cell in cardiovascular disease. In a longer perspective, it may also help to work out better methods for the prevention and treatment of this major group of human health disorders.

References

1. Akiyama SK, Yamada KM (1987) Fibronectin. Adv Enzymol Relat Areas Mol Biol 59:1–57
2. Aplin JD, Hughes RC (1981) Protein-derivatised glass coverslips for the study of cell-to-substratum adhesion. Anal Biochem 113:144–148
3. Barnes DW, Sirbasku DA, Sato GH (eds) (1984) Methods for preparation of media, supplements, and substrata for serum-free animal cell culture, vol 1. Liss, New York
4. Barnes MJ (1985) Collagens in atherosclerosis. Collagen Relat Res 5:65–97
5. Barrett TB, Benditt EP (1988) Platelet-derived growth factor gene expression in human atherosclerotic plaques and normal artery wall. Proc Natl Acad Sci USA 85:2810–2814
6. Benditt EP (1977) The origin of atherosclerosis. Sci Am 236:74–85
7. Benzonana G, Skalli O, Gabbiani G (1988) Correlation between the distribution of smooth or non muscle myosins and α-smooth muscle actin in normal and pathological soft tissues. Cell Motil Cytoskeleton 11:260–274
8. Bohr DF, Somlyo AP, Sparks HV Jr (eds) (1980) The cardiovascular system, vascular smooth muscle II. American Physiological Society, Bethesda (Handbook of physiology)
9. Borsi L, Castellani P, Balza E, Siri A, Pellecchia C, de Scalzi F, Zardi L (1986) Large-scale procedure for the purification of fibronectin domains. Anal Biochem 155:335–345
10. Bottger BA, Hedin U, Johansson S, Thyberg J (1989) Integrin-type fibronectin receptors of rat arterial smooth muscle cells: isolation, partial characterization and role in cytoskeletal organization and control of differentiated properties. Differentiation 41:158–167
11. Brandley BK, Schnaar RL (1988) Covalent attachment of an Arg-Gly-Asp sequence peptide to derivatizable polyacrylamide surfaces: support of fibroblast adhesion and long-term growth. Anal Biochem 172:270–278
12. Buck RC (1983) Behaviour of vascular smooth muscle cells during repeated stretching of the substratum in vitro. Atherosclerosis 46:217–223
13. Burke JM, Ross R (1979) Synthesis of connective tissue macromolecules by smooth muscle. Int Rev Connect Tissue Res 8:119–157
14. Burnstock G (1981) Development of smooth muscle and its innervation. In: Bülbring E, Brading AF, Jones AW, Tomita T (eds) Smooth muscle: an assessment of current knowledge. Arnold, London, pp 431–457
15. Campbell JH, Campbell GR (eds) (1987) Vascular smooth muscle in culture. CRC, Boca Raton
16. Carlsson J, Drevin H, Axén R (1978) Protein thiolation and reversible protein-protein conjugation. N-succinimidyl-3-(2-pyridyldithio) propionate, a new heterobifunctional reagent. Biochem J 173:723–737
17. Chamley JH, Campbell GR, McConnell JD, Gröschel-Stewart U (1977) Comparison of vascular smooth muscle cells from adult human, monkey and rabbit in primary culture and in subculture. Cell Tissue Res 177:503–522
18. Chamley-Campbell J, Campbell GR, Ross R (1979) The smooth muscle cell in culture. Physiol Rev 59:1–61
19. Chamley-Campbell JH, Campbell GR, Ross R (1981) Phenotype-dependent response of cultured aortic smooth muscle to serum mitogens. J Cell Biol 89:379–383
20. Chiquet-Ehrismann R, Mackie EJ, Pearson CA, Sakakura T (1986) Tenascin: an extracellular matrix protein involved in tissue interactions during fetal development and oncogenesis. Cell 47:131–139
21. Clezardin P, Hunter NR, Lawler JW, Pratt DA, McGregor JL, Pepper DS, Dawes J (1986) Structural and immunological comparison of human thrombospondins isolated from platelets and from culture supernatants of endothelial cells and fibroblasts. Evidence for a thrombospondin polymorphism. Eur J Biochem 159:569–579
22. Cunningham W (ed) (1987) Structural and contractile proteins, extracellular matrix. Methods Enzymol 144/145

23. Eagle H (1971) Buffer combinations for mammalian cell culture. Science 174:500–503
24. Erickson HP, Lightner VA (1988) Hexabrachion protein (tenascin, cytotactin, brachionectin) in connective tissues, embryonic brain, and tumors. Adv Cell Biol 2:55–90
25. Frazier WA (1987) Thrombospondin: a modular adhesive glycoprotein of platelets and nucleated cells. J Cell Biol 105:625–632
26. Friedlander DR, Hoffman S, Edelman GM (1988) Functional mapping of cytotactin: proteolytic fragments active in cell-substrate adhesion. J Cell Biol 107:2329–2340
27. Fritz KE, Jarmolych J, Daoud AS (1970) Association of DNA synthesis and apparent dedifferentiation of aortic smooth muscle cells in vitro. Exp Mol Pathol 12:354–362
28. Furuto DK, Miller EJ (1987) Isolation and characterization of collagens and procollagens. Methods Enzymol 144:41–61
29. Gown AM, Vogel AM, Gordon D, Lu PL (1985) A smooth muscle-specific monoclonal antibody recognizes smooth muscle actin isozymes. J Cell Biol 100:807–813
30. Grant DS, Leblond CP, Kleinman HK, Inoue S, Hassell JR (1989) The incubation of laminin, collagen IV, and heparan sulfate proteoglycan at 35 °C yields basement membrane-like structures. J Cell Biol 108:1567–1574
31. Ham RG (1965) Clonal growth of mammalian cells in a chemically defined, synthetic medium. Proc Natl Acad Sci USA 53:288–293
32. Heaysman JEM, Middleton CA, Watt FM (eds) (1987) Cell behaviour: shape, adhesion and motility. J Cell Sci [Suppl] 8
33. Hedin U, Thyberg J (1987) Plasma fibronectin promotes modulation of arterial smooth-muscle cells from contractile to synthetic phenotype. Differentiation 33:239–246
34. Hedin U, Bottger BA, Forsberg E, Johansson S, Thyberg J (1988) Diverse effects of fibronectin and laminin on phenotypic properties of cultured arterial smooth muscle cells. J Cell Biol 107:307–319
35. Hedin U, Bottger BA, Luthman J, Johansson S, Thyberg J (1989) A substrate of the cell-attachment sequence of fibronectin (Arg-Gly-Asp-Ser) is sufficient to promote transition of arterial smooth muscle cells from a contractile to a synthetic phenotype. Dev Biol 133:489–501
36. Herman IM (1987) Extracellular matrix-cytoskeletal interactions in vascular cells. Tissue Cell 19:1–19
37. Hynes RO (1986) Fibronectins. Sci Am 254:42–51
38. Izumi M, Yamada KM, Hayashi M (1989) Vitronectin exists in two structurally and functionally distinct forms in human plasma. Biochim Biophys Acta 990:101–108
39. Jakoby WB, Pastan IH (eds) Cell culture. Methods Enzymol 58
40. Johnston MG, Walker MA (1984) Lymphatic endothelial and smooth-muscle cells in tissue culture. In Vitro 20:566–572
41. Kleinman HK, Klebe RJ, Martin GR (1981) Role of collagenous matrices in the adhesion and growth of cells. J Cell Biol 88:473–485
42. Kleinman HK, Luckenbill-Edds L, Cannon FW, Sephel GC (1987) Use of extracellular matrix components for cell culture. Anal Biochem 166:1–13
43. Kocher O, Skalli O, Bloom WS, Gabbiani G (1984) Cytoskeleton of rat aortic smooth muscle cells. Normal conditions and experimental intimal thickening. Lab Invest 50:645–652
44. Lawler J (1986) The structural and functional properties of thrombospondin. Blood 67:1197–1209
45. Lawler J, Weinstein R, Hynes RO (1988) Cell attachment to thrombospondin: the role of Arg-Gly-Asp, calcium, and integrin receptors. J Cell Biol 107:2351–2361
46. Ledbetter SR, Kleinman HK, Hassell JR, Martin GR (1984) Isolation of laminin. In: Barnes DW, Sirbasku DA, Sato GH (eds) Methods for preparation of media, supplements, and substrata for serum-free animal cell culture, vol 1. Liss, New York, pp 231–238
47. Libby P, Warner SJC, Salomon RN, Birinyi LK (1988) Production of platelet-derived growth factor-like mitogen by smooth-muscle cells from human atheroma. N Engl J Med 318:1493–1498

48. Majesky MW, Benditt EP, Schwartz SM (1988) Expression and developmental control of platelet-derived growth factor A-chain and B-chain/sis genes in rat aortic smooth muscle cells. Proc Natl Acad Sci USA 85:1524–1528
49. Mayne R (1986) Collagenous proteins of blood vessels. Arteriosclerosis 6:585–593
50. McKeehan WL (1984) Use of basic polymers as synthetic substrata for cell culture. In: Barnes DW, Sirbasku DA, Sato GH (eds) Methods for preparation of media, supplements, and substrata for serum-free animal cell culture, vol 1. Liss, New York, pp 209–213
51. McKeehan WL, McKeehan KA, Hammond SL, Ham RG (1977) Improved medium for clonal growth of human diploid fibroblasts at low concentrations of serum protein. In Vitro 13:399–416
52. Miller EJ, Gay S (1987) The collagens: an overview and update. Methods Enzymol 144:3–41
53. Moore SA, Strauch AR, Yoder EJ, Rubenstein PA, Hart MN (1984) Cerebral microvascular smooth muscle in tissue culture. In Vitro 20:512–520
54. Munro JM, Cotran RS (1988) The pathogenesis of atherosclerosis: atherogenesis and inflammation. Lab Invest 58:249–261
55. Murray MR (1965) Muscle. In: Willmer EN (ed) Cells and tissues in culture. Methods, biology and physiology, vol 2. Academic, London, pp 311–372
56. Murray MR, Kopech G (1953) A bibliography of the research in tissue culture 1894/1950, 2 vols. Academic, New York
57. Nilsson J, Sjölund M, Palmberg L, Thyberg J, Heldin C-H (1985) Arterial smooth muscle cells in primary culture produce a platelet-derived growth factor-like protein. Proc Natl Acad Sci USA 82:4418–4422
58. Panina GF (1985) Monolayer growth systems: multiple processes. In: Spier RE, Griffiths JB (eds) Animal cell biotechnology, vol 1. Academic, London, pp 211–242
59. Paulsson M (1988) The role of Ca^{2+} binding in the self-aggregation of laminin-nidogen complexes. J Biol Chem 263:5425–5430
60. Perris R, Johansson S (1987) Amphibian neural crest cell migration on purified extracellular matrix components: a chondroitin sulfate proteoglycan inhibits locomotion on fibronectin substrates. J Cell Biol 105:2511–2521
61. Pierschbacher M, Hayman EG, Ruoslahti E (1983) Synthetic peptides with cell attachment activity of fibronectin. Proc Natl Acad Sci USA 80:1224–1227
62. Piez KA, Reddi AH (eds) (1984) Extracellular matrix biochemistry. Elsevier, New York
63. Pollack R (ed) (1981) Readings in mammalian cell culture. Cold Spring Harbor Laboratory, Cold Spring Harbor
64. Poole AR (1986) Proteoglycans in health and disease: structures and functions. Biochem J 236:1–14
65. Pytela R, Pierschbacher MD, Ruoslahti E (1985) A 125/115-kDa cell surface receptor specific for vitronectin interacts with the arginine-glycine-aspartic acid adhesion sequence derived from fibronectin. Proc Natl Acad Sci USA 82:5766–5770
66. Rhodin JAG (1980) Architecture of the vessel wall. In: Bohr DF, Somlyo AP, Sparks HV Jr (eds) The cardiovascular system, vascular smooth muscle II. American Physiological Society, Bethesda, pp 1–31 (Handbook of physiology)
67. Roberts DD, Sherwood JA, Ginsburg V (1987) Platelet thrombospondin mediates attachment and spreading of human melanoma cells. J Cell Biol 104:131–139
68. Rosenbloom J (1987) Elastin: an overview. Methods Enzymol 144:172–196
69. Ross R (1986) The pathogenesis of atherosclerosis – an update. N Engl J Med 314:488–500
70. Ross R, Vogel A (1978) The platelet-derived growth factor. Cell 14:203–210
71. Ross R, Raines EW, Bowen-Pope DF (1986) The biology of platelet-derived growth factor. Cell 46:155–169
72. Ruoslahti E (1988) Fibronectin and its receptor. Annu Rev Biochem 57:375–413
73. Ruoslahti E, Hayman EG, Pierschbacher M, Engvall E (1982) Fibronectin: purification, immunochemical properties, and biological activities. Methods Enzymol 82:803–831

74. Ruoslahti E, Suzuki S, Hayman EG, Ill CR, Pierschbacher MD (1987) Purification and characterization of vitronectin. Methods Enzymol 144:430–437
75. Sandberg LB, Soskel NT, Leslie JG (1981) Elastin structure, biosynthesis, and relation to disease states. N Engl J Med 304:566–579
76. Santoro SA, Frazier WA (1987) Isolation and characterization of thrombospondin. Methods Enzymol 144:438–446
77. Sato GH, Ross R (eds) (1979) Hormones and cell culture. Cold Spring Harbor Conf Cell Proliferation 6
78. Schwartz E, Bienkowski RS, Coltoff-Schiller B, Goldfischer S, Blumenfeld OO (1982) Changes in the components of extracellular matrix and in growth properties of cultured aortic smooth muscle cells upon ascorbate feeding. J Cell Biol 92:462–470
79. Schwartz SM, Campbell GR, Campbell JH (1986) Replication of smooth muscle cells in vascular disease. Circ Res 58:427–444
80. Seifert RA, Schwartz SM, Bowen-Pope DF (1984) Developmentally regulated production of platelet-derived growth factor-like molecules. Nature 311:669–671
81. Sejersen T, Betsholtz C, Sjölund M, Heldin C-H, Westermark B, Thyberg J (1986) Rat skeletal myoblasts and arterial smooth muscle cells express the gene for the A chain but not the gene for the B chain (c-sis) of platelet-derived growth factor (PDGF) and produce a PDGF-like protein. Proc Natl Acad Sci USA 83:6844–6848
82. Silnutzer J, Barnes DW (1984) Human serum spreading factor (SF): assay, preparation, and use in serum-free cell culture. In: Barnes DW, Sirbasku DA, Sato GH (eds) Methods for preparation of media, supplements, and substrata for serum-free animal cell culture, vol 1. Liss, New York, pp 245–268
83. Silverstein RL, Leung LLK, Nachman RL (1986) Thrombospondin: a versatile multifunctional glycoprotein. Arteriosclerosis 6:245–253
84. Sjölund M, Hedin U, Sejersen T, Heldin C-H, Thyberg J (1988) Arterial smooth muscle cells express platelet-derived growth factor (PDGF) A chain mRNA, secrete a PDGF-like mitogen, and bind exogenous PDGF in a phenotype- and growth state-dependent manner. J Cell Biol 106:403–413
85. Skalli O, Bloom WS, Ropraz P, Azzarone B, Gabbiani G (1986) Cytoskeletal remodeling of rat aortic smooth muscle cells in vitro: relationships to culture conditions and analogies to in vivo situations. J Submicrosc Cytol 18:481–493
86. Skalli O, Ropraz P, Trzeciak A, Benzonana G, Gillessen D, Gabbiani G (1986) A monoclonal antibody against α-smooth muscle actin: a new probe for smooth muscle differentiation. J Cell Biol 103:2787–2796
87. Somlyo AV (1980) Ultrastructure of vascular smooth muscle. In: Bohr DF, Somlyo AP, Sparks HV Jr (eds) The cardiovascular system, vascular smooth muscle II. American Physiological Society, Bethesda, pp 33–67 (Handbook of physiology)
88. Spier RE, Griffiths JB (eds) (1985) Animal cell biotechnology, vols 1, 2. Academic, London
89. Thyberg J, Palmberg L, Nilsson J, Ksiazek T, Sjölund M (1983) Phenotype modulation in primary cultures of arterial smooth muscle cells. On the role of platelet-derived growth factor. Differentiation 25:156–167
90. Thyberg J, Nilsson J, Palmberg L, Sjölund M (1985) Adult human arterial smooth muscle cells in primary culture. Modulation from contractile to synthetic phenotype. Cell Tissue Res 239:69–74
91. Timpl R (1989) Structure and biological activity of basement membrane proteins. Eur J Biochem 180:487–502
92. Timpl R, Dziadek M (1986) Structure, development, and molecular pathology of basement membranes. Int Rev Exp Pathol 29:1–112
93. Timpl R, Paulsson M, Dziadek M, Fujiwara S (1987) Basement membranes. Methods Enzymol 145:363–391
94. Toselli, P, Faris B, Oliver P, Franzblau C (1984) Ultrastructural studies of attachment site formation in aortic smooth muscle cells cultured on collagen-hydroxyethyl-methacrylate hydrogels. J Ultrastruct Res 86:252–261
95. Trelstad RL (ed) (1984) The role of extracellular matrix in development. Liss, New York

96. Tuszynski GP, Rothman V, Murphy A, Siegler K, Smith L, Smith S, Karczewski J, Knudsen KA (1987) Thrombospondin promotes cell-substratum adhesion. Science 236:1570–1573
97. Walker LN, Bowen-Pope DF, Ross R, Reidy MA (1986) Production of platelet-derived growth factor-like molecules by cultured arterial smooth muscle cells accompanies proliferation after arterial injury. Proc Natl Acad Sci USA 83:7311–7315
98. Weinberg CB, Bell E (1986) A blood vessel model constructed from collagen and cultured vascular cells. Science 231:397–400
99. Wight TN (1989) Cell biology of arterial proteoglycans. Arteriosclerosis 9:1–20
100. Wilcox JN, Smith KM, Williams LT, Schwartz SM, Gordon D (1988) Platelet-derived growth factor mRNA detection in human atherosclerotic plaques by in situ hybridization. J Clin Invest 82:1134–1143
101. Woods A, Couchman JR, Johansson S, Höök M (1986) Adhesion and cytoskeletal organisation of fibroblasts in response to fibronectin fragments. EMBO J 5:665–670
102. Yamada KM, Akiyama SK (1984) Preparation of cellular fibronectin. In: Barnes DW, Sirbasku DA, Sato GH (eds) Methods for preparation of media, supplements, and substrata for serum-free animal cell culture, vol 1. Liss, New York, pp 215–230
103. Yatohgo T, Izumi M, Kashiwagi H, Hayashi M (1988) Novel purification of vitronectin from human plasma by heparin affinity chromatography. Cell Struct Funct 13:281–292

Methods of Culturing Vascular Smooth Muscle Cells on Microcarriers *

P.F. Davies and S.E. O'Connor

Introduction

Vascular smooth muscle cells, like other anchorage-dependent cells, will die if maintained in suspension without attachment to a substratum. They are therefore usually cultured on tissue culture plastic (in flasks and Petri dishes) from which they can be released by the action of trypsin : EDTA. In contrast, the use of microcarrier beads provides a *mobile* substratum for cell growth and manipulation. Furthermore, microcarrier techniques provide a large surface area for endothelial culture in a small volume of beads, thereby allowing perfusion columns of microcarriers containing large numbers of smooth muscle cells to be constructed and, if scale-up of tissue culture is desired, batches of cells may be grown in suspension culture in the research laboratory on an economical scale.

Commercially Available Microcarriers

The earliest microcarriers were derived from DEAE Sephadex [1]. Great strides have been made over the past 10 years to improve the physical characteristics of the surface of microcarriers and to develop a number of alternative materials for their manufacture. Currently, there are a variety of commercial microcarriers available. These include: Biocarriers (BioRad; porous polyacrylamide), Biosilon (Nunc, Denmark; solid polystyrene), Cytodex 1, 2, and 3 (Pharmacia; porous dextran), Cytospheres (Lux; solid polystyrene), Gelibeads (KC Biologicals; porous cross-linked gelatin), Superbeads (Flow Laboratories; porous dextran), Biospheres (Whatman; solid or porous materials) and Rapid Cell (ICN; solid or porous materials). The surfaces of these microcarriers exhibit either a low positive charge or low negative charge, both of which promote optimal cell attachment and spreading [2]. Prior to use, several of the porous kind require hydration which involves multiple washings to prepare the microcarriers for cell attachment. When prepared swollen beads are stored, agents added to prevent bacterial contamination must be removed by dialysis before seeding with cultured cells. These minor inconveniences are offset in some cases by excellent cell attachment properties conferred by collagen-coating, e.g., Cytodex 3.

* This study was funded by grants to Dr. Davies from the National Heart, Lung, and Blood Institutes of the National Institutes of Health (HL 36049 and HL 36028)

In contrast to porous microcarriers, nonporous solid microcarriers are conveniently supplied in a sterilized, dry form and can be used immediately. This laboratory has used primarily nonporous Biosilon microcarriers and to a lesser extent, porous collagen-coated Cytodex 3. The techniques outlined below therefore apply primarily to our experience with these microcarriers. Standard procedures for the use of Cytodex 3 are well documented in the handbook [6] available from Pharmacia, and work well for vascular cells. Our choice of Biosilon is based upon the similarity in composition to tissue culture plastic, a nonporous composition (thereby eliminating absorption of biochemically interesting molecules within the bead), their transparency, resistance to shrinkage upon air drying, and extractability into organic solvents allowing fixation, staining, and subsequent processing for electron microscopy [see 3–5].

Stirrer Systems

To promote efficient nutrient, gaseous, and metabolite diffusion between cells and medium, microcarriers are usually maintained in suspension culture or in roller bottles. In our experience and that of others [6], cell yields are poorer in roller bottle culture than in spinner and rod-stirred cultures.

Some consideration should be given to the type of stirrer system employed. A number of configurations available for suspension of transformed cells work poorly when used for microcarrier-borne cells such as vascular smooth muscle. We have found that a very gentle stirring system manufactured by Techne (Techne MCS-104 system) works well for the microcarrier culture of vascular cells, monocytes, peritoneal macrophages, 3T3 cells, and skin fibroblasts. The action of the stirring system, shown in Fig. 1, is effected by a rod suspended in a flask with an indented base. The bulb at the end of the rod contains a magnet that is drawn around by another magnet placed in the MCS unit. The flask configuration forces medium to circulate in a manner which uniformly suspends the microcarriers. Horizontal and vertical circulations create a constantly moving solution without a stationary zone beneath the stirrer, a common fault with flat-bottomed stirrer flasks. The whole unit fits conveniently in a standard tissue culture incubator and accommodates four 1-l flasks. Stirring systems have been reviewed by Hirtenstein et al. [7]. Several are composed of simple paddle stirrers which generate extremely uneven shear forces as a function of radius of the paddle. Although we have had limited success with this paddle configuration using solid microcarriers, cells on Cytodex 3 (collagen-coated) can be grown reasonably successfully in such systems.

It is important to maintain a significant volume of gas in the stirrer flask to allow for efficient exchange of oxygen and CO_2. Typically, a volume of 500 ml of medium is used in a 1-l stirrer flask. Prior to use, the inner surface should be treated with a silicon solution (Prosil-28, SCM; 1% solution in distilled water) followed by a thorough rinsing of the flasks with distilled water. The silicon coating minimizes cell attachment to the flask surface. The coating procedure should be repeated approximately every month. Flasks must be loosely assembled and autoclaved before use.

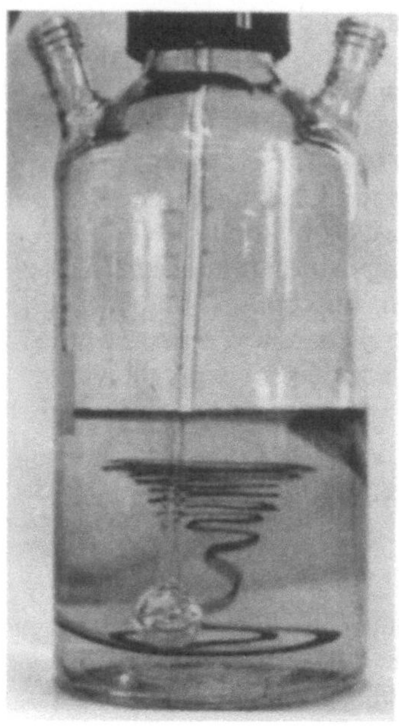

Fig. 1. Stirring system for the gentle suspension culture of anchorage-dependent cells upon microcarriers (Techne MCS-104). The helical path followed by an ink drop to mark the flow patterns is shown. [From 14]

Unstirred Microcarriers: Short-Term Culture

In some situations, only a small number of smooth muscle cells on microcarriers may be required for a particular experiment. It is often convenient to initiate small microcarrier cultures in static rather than stirred conditions. In such a situation, microcarriers can be conveniently grown in tissue culture dishes providing that the dish surface has been treated to prevent cell attachment. This may be accomplished by the use of bacteriological plastic or by coating standard tissue culture dishes with the inert polymer, hydroxyethyl methacrylate (Hydron NCC, Interferon Sciences) as a 1.2% solution in 95% ethanol/5% water. A stock solution is prepared by adding Hydron crystals to the ethanol/water and stirring at room temperature for several hours. The solution is simply layered onto the dish surface, poured off, and the dish allowed to dry in the laminar flow hood under UV light (usually overnight). Approximately 0.2 g of Biosilon microcarriers are placed in 1-ml complete culture medium in a Hydron-coated 35-mm diameter culture dish for 10 min. The dish is gently rocked to sink the beads. A minimum of 1.2×10^6 smooth muscle cells (obtained by any standard isolation procedure) is added in a volume of approximately 1 ml of medium to the culture dish which is rocked gently back and forth in order to mix the cells and microcarriers. This concentration is equivalent to an inoculation density of approximately 20 cells per bead. Because the system is not stirred, extensive bridging of smooth muscle cells between microcarriers rapidly occurs, and it is recommended that the medium be changed every 24 h and if possible, the cells be used within 48 h.

Prolonged stationary culture will result in the development of large clumps of cell/microcarriers restricting diffusion of gases and metabolites and resulting in cell death. This is more of a problem with smooth muscle cells and fibroblasts than for epithelioid cells which resist forming multilayers.

Procedure for the Culture of Vascular Smooth Muscle Cells on Microcarriers in a Stirred System

Smooth muscle cells obtained by either primary culture techniques [8], or by nonenzymatic explant culture [9], or following trypsinization of subcultures, are suitable for use on microcarriers. All procedures are carried out in a laminar flow hood. The technique is described for solid polystyrene microcarriers (e.g., Biosilon) which are supplied dry and sterile.

1. Transfer 250 ml of culture medium to a 1-l stirrer flask precoated with silicon as outlined above. Typically the medium will contain 10% serum, although this should be increased to 20% for primary cells or for other cells with low plating efficiency.
2. Under sterile conditions, weigh out 6 g of microcarriers and transfer them to the sterile flask. Agitate gently to submerge the beads. Allow to stand for 30 min in the incubator at 37 °C. The purpose of this presoaking in a growth medium is to enhance cell attachment by coating the beads with cold insoluble globulin present in the serum. Such amphoteric bridging proteins link the negatively charged bead surface to the net negatively charged cell surface, thereby considerably improving plating efficiency.
3. Inoculate the stirrer flask with between 2×10^6 and 7×10^6 smooth muscle cells per gram of beads; this initial plating density corresponds to between 5 and 20 cells per bead and will yield a confluent culture of about 150 cells per bead within 7–10 days. An example of a growth curve for aortic smooth muscle cells obtained by the explant technique is shown in Fig. 2. The cell number varies with the size distribution of the beads, the age of the culture, and the extent of bead clumping (see below); 1 g of dry Biosilon microcarriers contains approximately 3.7×10^5 beads to provide a total surface area of 500 cm^2.

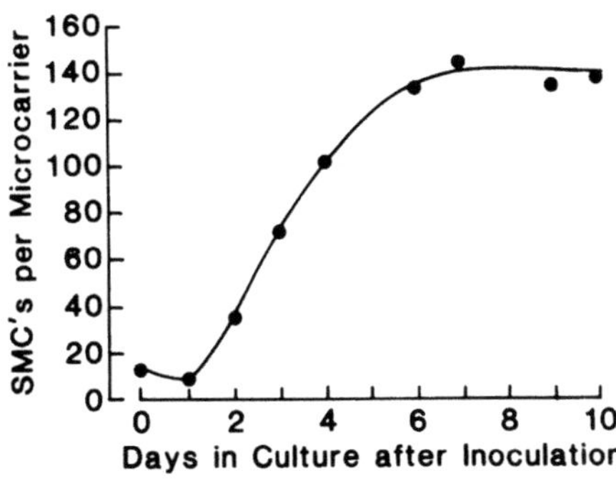

Fig. 2. Growth curve expressed as cells per microcarrier (see text) of bovine aortic smooth muscle cells and Biosilon microcarriers. Initial plating density was 10 cells/bead. The curve plateau is at approx. 140 cells/bead representing a confluent monolayer. [From 5]

4. Immediately after inoculation, the stirring unit is started for 1 min and then switched off. After 1 h, the stirrer may be intermittently activated (if automation is available, a cycle of 3 min on, 2 min off, is recommended) for up to 24 h. While this procedure distributes the cells more evenly throughout the microcarriers, it is not essential and perfectly adequate cultures can be obtained by one or two agitations in the 6-h period following initial inoculation. In practice, it is convenient to seed the cells in the morning, intermittently activating the stirrer throughout the day, and turning on the stirrer continuously at the end of the day. Immediately before activating the stirrer continuously, fill the flask with tissue culture medium to a final volume of 500 ml.

Observing the Cells

A stirrer speed of 40–50 rpm in the Techne stirrer system is optimal to maintain the microcarriers in suspension while minimizing shear effects. For Cytodex 3 microcarriers, a stirrer speed of 18 rpm has been reported optimal for endothelial cells [10] and is also suitable for smooth muscle cells. The observation of microcarriers by phase microscopy shows only cells at the periphery of the bead (Fig. 3); those over the bulk of the surface of the bead are not visible. Nevertheless, an approximation of whether the seeding has been successful or not can be obtained, since the image at the periphery of the bead represents a random orientation and is representative of the distribution on that particular microcarrier. To visualize the cells in more detail, however, it is necessary to fix and stain them:

1. Remove an aliquot of microcarriers from the stirrer flask using a wide bore pipette to prevent damage to the cells. Transfer to a 15-ml conical plastic tube or similar container. The microcarriers (density approximately 1.04 g/ml) will settle quickly to the bottom of the tube.
2. Remove the supernatant as close to the beads as is practical and discard. Rinse the microcarrier "pellet" several times by the addition of a buffered salt solution, e.g., Hank's balanced salt solution, removing the supernatant after each rinse. As the solutions are added down the side of the tube, the microcarriers will be disturbed from the pellet.
3. Add approximately 2 ml 95% ethanol to fix the cells on the microcarriers. Allow to fix for 5 min at room temperature.
4. Remove ethanol and replace with hematoxylin solution for 4 min. If a commercially available hematoxylin staining solution is not used, a working solution can be prepared using the following protocol. With the aid of heat, dissolve 50 g of hematoxylin crystals in 50 ml of 100% alcohol and 100 g of alum in 1 l distilled water. Remove from heat and mix the two solutions together. Bring the solution to a rapid boil (less than 1 min with stirring) and remove from heat. Add 2.5 g mercuric oxide (red) slowly and reheat to a simmer until it becomes a dark purple. Immediately plunge the beaker into ice water. The stain is ready to use when cool. Addition of 3 ml glacial acetic acid per 100 ml of solution increases the precision of the nuclear stain. NB: Exercise caution in handling these volatile solutions in the presence of heat.

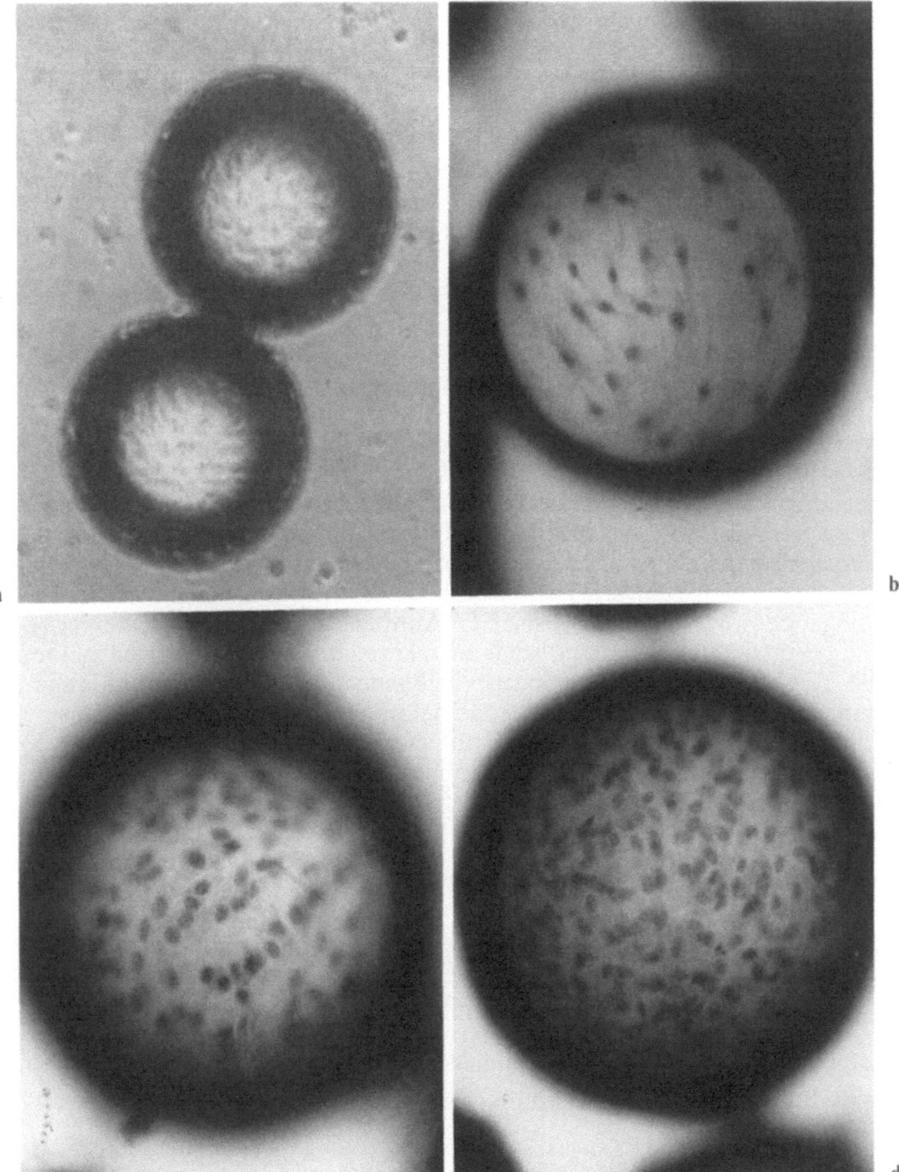

Fig. 3. a Appearance of vascular smooth muscle cells at subconfluent density (90 cells/bead) by phase-contrast microscopy. Only cells at the periphery are visible. Unstained, $\times 175$. **b–d** Vascular smooth muscle cells stained with hematoxylin at increasing cell densities. Average cell/bead ratios in each culture were 60 (**b**), 110 (**c**), and 155 (**d**) respectively; $\times 280$. [From 5]

5. Remove the hematoxylin supernatant from the microcarriers and fill the tube with deionized water. After the beads have settled, remove as much of the deionized water as possible and replace with fresh deionized water. The beads should appear dark purple when the cells are confluent; they will be a lighter color at lesser cell densities.
6. Place an aliquot of stained beads onto a microscope slide and observe under bright field optics. Manipulation of the microscope condenser and diaphragm is usually necessary to obtain good images, especially at moderate to high magnifications.

Measurement of Cell Numbers

A useful estimate is the number of cells per bead. Since the weight of the beads is known and the average number of beads per gram is also known, the cell to bead ratio can be used to calculate the total number of cells in the stirrer flask. The measurement of cell numbers is similar to that used with standard tissue culture flasks because the cells are readily trypsinized from the polystyrene microcarriers. A sample of microcarriers is placed in a conical tube as for the staining procedure (above). The cells are removed from the microcarriers by adding a known volume of trypsin : EDTA using standard trypsinization protocols, gently vortexed, and the beads are allowed to settle to the bottom of the tube. The released cells remain in suspension much longer than the beads and can therefore be removed with the supernatant. The cell number in the supernatant is determined using a hemocytometer or a Coulter counter (Coulter Electronics, Hialeah, Fla.). In order to measure the number of beads in the same sample, the supernatant is completely removed and a known volume of water is added to the beads. The mixture is vortexed vigorously and an aliquot removed before the beads have time to settle. The aliquot is transferred to a slide or small culture dish and the number of beads counted. The total number of beads in the original aliquot can then be calculated knowing the volume of the aliquot and the volume of the suspension. The cell to bead ratio can then be calculated (see Fig. 2).

If total cell protein is required, the microcarrier cells are dissolved in either 0.1% sodium dodecyl sulfate (30 min at room temperature) or 1 N NaOH with occasionally vortexing. The insoluble beads settle, and the supernatant is assayed using the Lowry procedure or similar cell protein assay.

Morphology of Smooth Muscle Cells on Microcarriers

At low plating densities, individual smooth muscle cells assume a fusiform morphology with random orientation; as the beads become more heavily colonized, however, the fusiform shape becomes less apparent (Fig. 3). In conventional culture, vascular smooth muscle cells at confluence typically form a "hill and valley" conformation [9]. In microcarrier cultures, however, this morphology is rarely noted, although multilayers will form on occasion between beads if there is extensive bridging of cells. The shear forces generated in a stirred suspension may prevent the formation of multilayers in the microcarrier cultures. Once a confluent density is reached, it is common for two or more microcarriers to become bridged by smooth muscle cells. Two and three bead clumps are found in almost all cultures regardless of their age, but they do not represent a significant problem because there is adequate diffusion of nutrients and waste products in and out of the clump. Large multibead clumps, however, may develop if the culture is left static for long periods of time or is stirred at too slow a speed.

Preparation for Electron Microscopy

Both scanning (SEM) and transmission electron microscopy (TEM) can be performed on cells attached to microcarriers. The use of solid beads minimizes shrinkage, whereas porous beads are less suitable because they shrink during dehydration.

Preparation for Scanning Electron Microscopy

1. Fix an aliquot of microcarriers in 2% glutaraldehyde in buffered salt solution, pH 7.4, for 1 h at 4°C.
2. Wash the microcarriers twice with deionized water.
3. Gradually exchange cell water with alcohol by replacement with increasing concentrations of ethanol. Start with 10% ethanol and increase at 10% increments up to 100% ethanol. Each solution should be added for 10 min at room temperature.
4. The microcarriers are then dried using the critical-point technique from ethanol/CO_2 or can be air dried directly from 100% ethanol.
5. The dried microcarriers are transferred to sticky copper tape mounted on an SEM stub and coated with a thin layer (approximately 200 Å) of gold-palladium (Hummer IV Sputter Coater, Technics Instruments, Alexandria, Va.), and viewed in a standard scanning electron microscope.

Preparation for Transmission Electron Microscopy

The procedure is a modification of that published by Sargent et al. [11]. Following fixation and washing, as described above for SEM, the microcarriers in buffered salt solution are warmed to 45°C in a water bath; 2% agar in phosphate buffered saline is prepared and also warmed to 45°C. A few drops of 1% eosin are added to make the agar visible during processing; 1 to 2 ml of the agar solution is added to the sample. The sample in agar solution is then gently mixed using a wooden stick.

The sample is centrifuged at 1000 rpm for 2 min. After the agar block is formed, the sample is removed by cutting the centrifuge tube and flooding with phosphate buffered saline. The excess agar is cut off the agar block. The block of agar containing the sample is diced into 1-mm cubes which are then transferred to a glass vial. Following postfixation for 1 h in 0.2 M cacodylate buffer containing 1% osmium tetroxide, the cubes are washed in three changes (10 min each) of 0.2 M cacodylate buffer. The samples are then washed three times (5 min each) with distilled water and gently shaken in 0.2 M cacodylate buffer containing 1% (w/v) uranyl acetate. After 1 h of staining, cubes are washed in three changes of distilled water and processed through graded ethanol solutions (50%, 10 min; 70%, 10 min; 95%, 10 min; 100%, 10 min; 100%, 15 min). The samples are then processed through three changes of 1,2 epoxy propane and finally embedded in resin [12]. The changeover from dehydration fluid to embedding fluid is made in three stages: 1/1 dehydrating/embedding fluid for 30 min; 1/3 dehydrating/embedding fluid for 30 min; 100% embedding fluid for 1 h; 100% embedding fluid overnight.

Samples are cured for at least 10 h in a 70 °C oven. Semithin sections are cut and stained with 0.1% toluidine blue in 1% sodium borate and examined with the light microscope. Ultrathin sections are prepared from selected areas, picked up on copper grids, and stained with 7.5% uranyl acetate in 50% ethanol followed by lead citrate stain [13]. The sections are examined in the transmission electron microscope, usually at 60 KV.

References

1. Van Wezel AL (1967) Growth of cell strains and primary cells on microcarriers in homogeneous culture. Nature 216:64
2. Gebb C, Hirtenstein MD, Clark JM, Lindgren G, Uretblad P (1982) Alternative surface for microcarrier culture of animal cells. Dev Biol Stand 50:93
3. Davies PF (1981) Microcarrier culture of vascular endothelium. Exp Cell Res 134:365
4. Davies PF (1982) Microcarrier cultures in vascular endothelial research. Dev Biol Stand 50:125
5. O'Connor SE, Davies PF (1987) Microcarrier culture of vascular smooth muscle cells. In: Campbell JH, Campbell GR (eds) Vascular smooth muscle in culture, vol 1. CRC, Boca Raton, p 23
6. Pharmacia Fine Chemicals (1982) Microcarrier cell culture: principle and methods. Pharmacia Fine Chemicals, Uppsala, Sweden
7. Hirtenstein MD, Clark JM, Gebb C (1982) A comparison of various laboratory scale culture configurations for microcarrier culture of animal cells. Dev Biol Stand 50:73
8. Chamley-Campbell JH, Campbell GR, Ross R (1981) Phenotype-dependent response of cultured aortic smooth muscle to serum mitogens. J Cell Biol 89:378
9. Ross R (1971) The smooth muscle cell. II. Growth of smooth muscle in culture and formation of elastin fibers. J Cell Biol 50:172
10. Cocks JM, Angus JA, Campbell JM, Campbell GR (1985) Release and properties of EDRF from endothelial cells in culture. J Cell Physiol 123:310
11. Sargent GF, Sims TA, McNeish AS (1980) The use of polystyrene microcarriers to prepare cell monolayers for transmission electron microscopy. J Microsc 122:204
12. Spurr AR (1969) A low viscosity epoxy embedding medium for electron microscopy. J Ultrastruct Res 26:31
13. Sato J (1968) Lead citrate stain in electron microscopy. Electron Microsc 17:158
14. De Bruyne NA, Morgan BJ (1981) Am Lab: June issue

Growth Factors for Smooth Muscle Cells

U. Janßen-Timmen, P. Salbach, R. Gronwald,
and A.J.R. Habenicht

Introduction

The growing interest in smooth muscle cell (SMC) biology originates in morphological studies demonstrating that SMCs are prominent constituents of the "proliferative early lesion" of atherosclerosis. This disease still accounts for more than 50% of all deaths in highly industrialized countries [22, 23, 27, 36, 37]. In addition, abnormal SMC proliferation has been observed in several other clinically important diseases such as subacute and chronic forms of arthritis and in graft versus host disease in transplanted kidneys [30; reviewed by 9]. While the proliferation of SMCs during embryonal angiogenesis is confined to the media of the arterial wall [32], the irregular growth of SMCs in atherosclerosis is characterized by the presence of multiple layers of proliferating SMCs in the intima [23, 27]. The mechanisms underlying the accumulation of SMCs in the intima are not yet known but are assumed to be the result of the interaction between the SMCs, monocytes/macrophages, lymphocytes, and endothelial cells [23, 27]. These morphological characteristics of the early lesions of atherosclerosis resemble in many respects a chronic inflammatory reaction [8, 30].

One approach to the study of the growth factors that regulate SMC growth has been the use of tissue culture [reviewed by 24]. Many findings concerning the biology of SMCs have initially been obtained by studying cultured fibroblasts which have served as an important model system for SMCs [23]. However, not all characteristics of SMCs in vivo are maintained in vitro [5] and not all characteristics of SMCs in vitro are expressed in fibroblasts. Thus, SMCs, to a certain extent, undergo a dedifferentiation process when put into a culture dish. Therefore, cautious interpretation of the results obtained by in vitro studies of SMCs is required. Tissue culture techniques have allowed investigators to gain insight into the growth factor requirements of SMCs [reviewed by 23, 27], to elucidate the autocrine mechanisms of growth of both normal and malignant cells including SMCs [reviewed by 10], to determine the structure and function of growth factor receptors [reviewed by 38], to study the growth factor-dependent signaling pathways (reviewed by [28], to identify "immediate early genes" that are activated in response to growth factors (reviewed by [4]), and to identify factors responsible for growth factor receptor expression [8, 22].

This brief review focuses on growth factors for SMCs and fibroblasts and in particular on their principal growth factor, the platelet-derived growth factor (PDGF). We will also discuss the effects of transforming growth factor

β (TGF-β), interleukin I (IL-1), insulin-like growth factors (IGFs), bombesin, and thrombospondin on the growth of SMCs.

Tissue culture techniques for SMCs have been described in detail elsewhere and therefore will not be covered here [24]. Instead, at the end of this chapter we have included a protocol for the preparation of plasma-derived serum (PDS) that will allow investigators to arrest the SMCs in the G_0 stage of the cell cycle. PDS does not contain significant amounts of PDGF and other growth factors that are released from platelets during platelet aggregation. The use of PDS is of major importance for SMC research because cell cycle arrest of SMCs is a prerequisite to the study of the effects of individual growth factors in the cell cycle. In addition to the information provided in this brief review, readers are referred to a recently published volume that covers further aspects of atherogenesis, SMCs, and fibroblast biology [8a].

PDGF is the Principal Growth Factor for SMCs

One of the most significant discoveries in the field of SMC biology was made in 1974 when Ross et al. [25] found that SMCs that had been explanted from monkey aortas were unable to divide in culture medium supplemented with homologous PDS. SMCs that had been maintained in PDS containing media, instead of monkey blood serum, acquired a flat morphology, and flow cytometry analyses showed that the cells were arrested in the $G_{0/1}$ phase of the cell cycle. All of the growth-promoting activity of monkey blood serum could be restored by adding platelet extracts to the quiescent growth arrested SMCs [26]. Thus, when platelet extracts were added, the cells acquired a more spindlelike morphology, typical of rapidly growing SMCs. Flow cytometry analyses demonstrated that the addition of the platelet factors led to recruitment of up to 90% of the cell population from the G_0 into the G_1 phase [26]. Subsequent biochemical studies led to the discovery of PDGF as the principal mitogen for connective tissue-derived cells [reviewed by 9]. Although SMCs can also be arrested in the G_0 stage of the cell cycle by simply lowering the concentration of serum in the culture medium to 0.5%, the use of PDS as a means for cell cycle arrest is thought to represent a more physiological condition than "serum starvation" [26]. Thus, the use of PDS and of growth factors such as PDGF and TGF-β in cultures of SMCs represent an important research tool for the study of the role of individual growth factors in SMC biology (see below).

Autocrine Mechanisms of Growth of SMCs

SMCs derived from the developing rat aorta [32] as well as from the injured rat carotid artery [3] have been shown to secrete large amounts of PDGF into the culture medium. When assayed for the expression of PDGF receptors, the same cells bound much less ^{125}I-PDGF compared with control SMCs [32]. Furthermore, in situ hybridization [36] and immunocytochemical techniques [30] demonstrated the presence of β-PDGF receptors in intimal SMCs of human atherosclerotic plaques.

These results indicated that PDGF expression is developmentally regulated. They also suggest that PDGF functions as an autocrine growth factor for SMCs during embryogenesis, atherogenesis, and wound healing. Further support for an autocrine function of PDGF comes from experiments with human fibroblasts. Thus, when treated with either PDGF or EGF, these cells respond with a transient increase in the expression of PDGF A-chain mRNA and protein [19]. Since fibroblasts also express PDGF receptors, it is possible that the PDGF/EGF-induced expression of the A-chain gene leads to autocrine growth regulation. If PDGF can act as an autocrine growth factor [reviewed by 10], then agents other than PDGF that are capable of inducing expression of PDGF should also lead to a mitogenic response. Recent evidence for such an indirect type of growth regulation has been obtained for IL-1 in aortic SMCs [21]. In this study, IL-1 stimulated the transient accumulation of PDGFα-chain mRNA and the number of PDGF receptors were concomitantly downregulated. Moreover, the mitogenic response of the cells to IL-1 could be blocked by anti-PDGF-AA antibodies. This indicates that IL-1, similarly to TGF-β (see below) and tumor necrosis factor might belong to a group of peptide factors that stimulate growth of SMCs by affecting the rate of PDGF gene transcription.

PDGF Acts in Concert with Other Growth Factors

While PDGF is a potent growth factor for SMCs [25] and other cells of mesenchymal origin [11], it does not exert its maximal mitogenic activity in the absence of plasma proteins [35]. The requirement of plasma proteins for maximal PDGF-induced mitogenesis was shown by the ability of platelet poor plasma to strongly support the mitogenic effect of PDGF [20, 35]. These studies led to a concept of cell cycle regulation where PDGF induces a state of "competence" which is followed by a phase of "progression" in the cell cycle that requires the presence of "progression factors." The concept that peptide factors can act as "competence" or "progression" factors is based on the hypothesis that a sequence of distinct events mediated by the sequential action of several peptide factors is required for maximal mitogenesis [reviewed in 31]. The underlying molecular mechanisms of these events remain to be fully identified, but there is reason to believe that insulin-like growth factors (IGF-1, IGF-2) are among the factors that support PDGF mitogenic activity. For example, in porcine aortic SMCs, secretion of IGF-1 is stimulated by PDGF [6], and the mitogenic response of SMCs towards PDGF could be significantly reduced by anti-IGF-1 antibodies.

These results suggest that IGFs secreted in response to PDGF can stimulate cell proliferation in an autocrine or paracrine fashion and that they play an important role in cell cycle progression. Similar effects have been described for EGF. In Swiss 3T3 cells, the apparent affinity and number of EGF receptors for EGF binding is reduced by PDGF pretreatment [28], a process that has been termed "receptor transmodulation". The functional significance of EGF receptor transmodulation has not been clearly delineated, but it is noteworthy that an increase in the number of EGF receptors and a concomitant decrease of the affinity for EGF has been observed in fetal mouse tissues during gestation [1]. This suggested that receptor affinity correlates with the state of cellular differentiation.

Bombesin

Bombesin [reviewed in 28] is a tetradecapeptide first isolated from frog skin. Bombesin belongs to a family of peptide factors including gastrin-releasing peptide and the neuromedins. Bombesin has been shown to be a potent mitogen for Swiss 3T3 cells [29]. Similarly to PDGF, the initiation of DNA synthesis by bombesin can be observed in the absence of other proteins, but it is synergistically enhanced in the presence of insulin and PDGF. A general similarity in the mechanisms of action of PDGF and bombesin is indicated by the ability of both growth factors to induce a variety of identical intracellular events [reviewed in 28].

However, several observations distinguish the signal transduction pathway induced by bombesin from that induced by PDGF. Thus, it has been demonstrated that phospholipase Cγ is phosphorylated at tyrosine residues in response to PDGF and EGF but not in response to bombesin [18]. Furthermore, the bombesin-dependent induction of the c-myc protooncogene and subsequent mitogenesis could be blocked by pertussis toxin indicating that a pertussis toxin sensitive guanine nucleotide binding protein was involved in the mechanism of action of bombesin. In contrast both c-myc activation and mitogenesis in response to PDGF were pertussis toxin insensitive [12]. Taken together, these results clearly show that while both bombesin and PDGF are potent mitogens the bombesin-induced signalling pathway can be distinguished from that induced by PDGF.

Growth Regulation of Fibroblasts and SMCs by TGF-β

TGF-β has been shown to be a growth-promoting as well as a growth-inhibiting factor, depending on the cell type and the experimental conditions [34]. For example, sparse SMCs are growth inhibited whereas dense cultures are growth stimulated by TGF-β [14]. The underlying mechanisms for these opposing TGF-β effects remain to be defined. However, indirect effects of TGF-β on the expression of both EGF and PDGF receptors have been demonstrated. For the EGF receptor it has been shown that TGF-β induces a decrease in EGF receptor affinity [17] and concomitantly an increase in EGF receptor number [2]. These combined effects lead to a diminished mitogenic response towards EGF. Another report showed that, in 3T3 cells, TGF-β differentially regulates the expression of the two PDGF receptor subunits, α and β [8]. Binding experiments using radiolabeled AA or BB homodimers showed that the α-subunit was undetectable 12 h after TGF-β treatment whereas the expression of the β-subunit was stimulated. This apparent differential regulation of distinct PDGF receptor subunits resulted in a loss of mitogenic responsiveness towards PDGF-AA, but an unaltered mitogenic response towards PDGF-AB or

PDGF-BB. Taken together these results indicate that TGF-β might affect the growth of its target cells by interfering with the activity of several known growth factors at the level of transmodulation of growth factor receptors.

Thrombospondin

Proteins of the extracellular matrix appear to profoundly affect SMC growth. For example, the homotrimeric glycoprotein thrombospondin (TS) has been shown to be transiently integrated into the extracellular matrix of cultured rat aortic SMCs. TS is specifically synthesized and secreted by SMCs in response to submaximal PDGF concentrations. TS gene transcription is induced by PDGF alone, and this effect of PDGF is greatly enhanced by the translation inhibitor, cycloheximide [15]. This PDGF-dependent "superinduction" of TS gene transcription in the presence of cycloheximide indicates that TS belongs to the family of "immediate early genes" whose transcription rate is stimulated as a direct result of the PDGF-dependent signaling pathway [reviewed by 4]. While TS by itself does not stimulate cellular growth, it acquires potent mitogenic activity when EGF is present in the culture medium [13, 16]. Heparin, an inhibitor of TS integration into the SMC extracellular matrix, prevents by 80% the mitogenic activity induced by a combination of TS and EGF. Furthermore, a monoclonal antibody directed against TS inhibited by 50% the proliferative response of SMCs towards 5% serum. The results indicate that the mitogenic activity of TS and, possibly, other extracellular matrix proteins, may be mediated by their participation in the remodeling of extracellular matrix during cell proliferation.

Addendum

Availability of Growth Factors for SMCs

All growth factors discussed in this chapter are easily obtainable from a variety of companies in highly purified form and/or as recombinant peptides isolated from bacterial, yeast, or mammalian expression systems. In some cases, radio-labeled growth factors can also be obtained for in vitro binding studies. In addition antibodies against different growth factors are available. Finally, complementary DNA sequences (cDNAs) encoding selected growth factors and/or their receptors can be purchased from the American Type Culture Collection of human DNA probes and libraries.

It should be noted, however, that the information provided by the various companies concerning the concentration dependence and solutions in which the growth factors are to be dissolved are based on standard assay systems using model cell systems. The concentrations used in these systems are not necessarily applicable to SMCs in culture. Therefore, it is necessary to test each individual growth factor in SMC cultures in order to obtain optimal results. Furthermore, the concentrations needed to obtain an optimal response may not only depend on the cell species used but also on the presence of growth factor binding proteins in the respective culture medium.

Preparation of Plasma-Derived Serum (PDS)

Materials and Preparations

Requirements: precooled centrifuge with appropriate rotor (30 000 g), precooled 50-cc syringes containing 1/10 volume of 3.8% Na citrate, 1 M $CaCl_2$, 50-cc centrifuge tubes, 37°C waterbath, 19-gauge butterfly sets, and ten donors who have had a limited breakfast (no butter, eggs, or cream).

Precautions

In order to obtain optimal results it is important to conform to the following rules:

1. Remove platelets and white blood cells *immediately* after blood drawing.
2. After spinning down platelets and white blood cells leave a 1-cm layer of plasma above the buffy coat to avoid any contamination of PDS with platelets or other white blood cells.
3. If there is any sign of blood clotting in the syringe or the centrifuge tube, discard it rather than pool its contents with other tubes.
4. Use only tissue culture washed glassware to avoid toxicity.

Blood Drawing

1. Prepare 50-cc syringes containing 1/10 of the final volume of 3.8% Na citrate on ice.
2. Precool centrifuge and centrifuge tubes.
3. Draw 45 cc blood/syringe, invert syringe several times immediately, and put citrated blood on ice.

Centrifugation

1. Divide citrated blood into precooled centrifuge tubes and balance.
2. Centrifuge avoiding any unnecessary delay at 4°C 30 000 $\times g$ $\times$ 20 min.
3. Remove plasma carefully using a pipette and pool in glass beaker or appropriate tubes. Avoid aspirating contents of buffy coat.

Preparation of Plasma

1. Add back 1/50 of plasma volume of 1 mM $CaCl_2$, mix well, aliquot in centrifuge tubes and balance for subsequent spin, then incubate 2 h 37°C.

2. Inactivate complement by incubating plasma for 30 min at 56 °C.
3. Remove clotted contents from the sides of the centrifuge tubes and spin at 30000 g for 30 min at 4 °C.
4. Decant, filter through a 0.22-μm filtration unit, aliquot, and store at –20 °C.

References

1. Adamson ED, Meek J (1984) The ontogeny of epidermal growth factor receptors during mouse development. Dev Biol 103:62–70
2. Assoian RK (1985) Biphasic effects of type β transforming growth factor on epidermal growth factor receptors in NRK fibroblasts. J Biol Chem 260:9613–9617
3. Barrett TB, Benditt EP (1988) Platelet-derived growth factor gene expression in human atherosclerotic plaques and normal artery wall. Proc Natl Acad Sci USA 85:2810–2814
4. Bravo R (1989) Growth factor-inducible genes in mouse fibroblasts. In: Habenicht AJR (ed) Growth factors, differentiation factors, and cytokines. Springer, Berlin Heidelberg New York
5. Chamley-Campbell J, Campbell GR, Ross R (1979) The smooth muscle cell in culture. Physiol Rev 59:1–61
6. Clemmons DR, van Wyk JJ (1985) Evidence for a functional role of endogenously produced somatomedin-like peptides in the regulation of DNA synthesis in cultured human fibroblasts and porcine smooth muscle cells. J Clin Invest 75:1914–1918
7. Gronwald RGK, Grant FJ, Haldeman BA, Hart CE, O'Hara PJ, Hagen FS, Ross R, Bowen-Pope DF, Murray MJ (1988) Cloning and expression of a cDNA coding for the human platelet-derived growth factor receptor: evidence for more than one receptor class. Proc Natl Acad Sci USA 85:3435–3439
8. Gronwald RGK, Seifert RA, Bowen-Pope DF (1989) Differential regulation of expression of two platelet-derived growth factor receptor subunits by transforming growth factor-β. J Biol Chem 264:8129–8125
8a. Habenicht AJR (ed) (1989) Growth factors, differentiation factors, and cytokines. Springer, Berlin Heidelberg New York
9. Habenicht AJR, Salbach P, Blattner C, Janßen-Timmen U (1989) Platelet-derived growth factor – formation and biological activities. In: Habenicht AJR (ed) Growth factors, differentiation factors, and cytokines. Springer, Berlin Heidelberg New York
10. Heldin C-H, Westermark B (1989) Autocrine stimulation of growth of normal and transformed cells. In Habenicht AJR (ed) Growth factors, differentiation factors, and cytokines. Springer, Berlin Heidelberg New York
11. Kohler N, Lipton A (1974) Platelets as a source of fibroblast growth-promoting activity. Exp Cell Res 87:297–301
12. Letterio JL, Coughlin SR, Williams LT (1986) Pertussis toxin-sensitive pathway in the stimulation of c-myc expression and DNA synthesis by bombesin. Science 234:1117–1119
13. Majack RA, Cook SC, Bornstein P (1986) Control of smooth muscle cell growth by components of the extracellular matrix: autocrine role for thrombospondin. Proc Natl Acad Sci USA 83:9050–9054
14. Majack RA (1987) Beta-type transforming growth factor specifies organizational behaviour in vascular smooth muscle cell cultures. J Cell Biol 105:465–471
15. Majack RA, Mildbrandt J, Dixit VM (1987) Induction of thrombospondin messenger RNA levels occurs as an immediate primary response to platelet-derived growth factor. J Biol Chem 262:821–825
16. Majack RA, Goodman LA, Dixit VM (1988) Cell surface thrombospondin is functionally essential for vascular smooth muscle cell proliferation. J Cell Biol 106:415–422
17. Massague J (1985) Transforming growth factor β modulates the high affinity receptors for epidermal growth factor and transforming growth factor a. J Cell Biol 100:1508–1514

18. Meisenhelder J, Suh P-G, Rhee SG, Hunter T (1989 Phospholipase C$_{-\gamma}$ is a substrate for the PDGF and EGF receptor. Cell 57:1109–1122
19. Paulsson Y, Hammacher A, Heldin C-H, Westermark B (1987) Possible positive autocrine feedback in the prereplicative phase of human fibroblasts. Nature 328:715–717
20. Pledger WJ, Stiles CD, Antoniades HN, Scher CD (1977) Induction of DNA synthesis in Balb/c 3T3 cells by serum components: reevaluation of the commitment process. Proc Natl Acad Sci USA 74:4481–4485
21. Raines EW, Dower SK, Ross R (1989) Interleukin-1 mitogenic activity for fibroblasts and smooth muscle cells is due to PDGF-AA. Science 243:393–396
22. Raines EW, Bowen-Pope DF, Ross R (1989) Platelet-derived growth factor. In: Sporn MB, Roberts AB (eds) Peptide growth factors and their receptors. Springer, Berlin Heidelberg New York (Handbook of experimental pharmacology)
23. Ross R (1986) The pathogenesis of atherosclerosis – an update. N Engl J Med 314:488–500
24. Ross R, Kariya B (1980) Morphogenesis of vascular smooth muscle in atherosclerosis and cell culture. In: Bohr D (ed) Circulation, vascular smooth muscle. American Physiological Society, Bethesda (Handbook of physiology)
25. Ross R, Glomset J, Kariya B, Harker L (1974) A platelet-dependent serum factor that stimulates the proliferation of arterial smooth muscle cells in vitro. Proc Natl Acad Sci USA 71:1207–1210
26. Ross R, Nist C, Kariya B, Rivest MJ, Raines E, Callis J (1978) Physiological quiescence in plasma-derived serum: influence of platelet-derived growth factor on cell growth in culture. J Cell Physiol 97:497–508
27. Ross R, Raines EW, Bowen-Pope DF (1986) The biology of platelet-derived growth factor. Cell 46:155–169
28. Rozengurt E (1986) Early signals in the mitogenic response. Science 234:161–166
29. Rozengurt E, Sinnett-Smith J (1983) Bombesin stimulation of DNA synthesis and cell division in cultures of Swiss 3T3 cells. Proc Natl Acad Sci USA 80:2936–2940
30. Rubin K, Hansson GK, Rönnstrand L, Claesson-Welsh L, Fellström B, Tingström A, Larsson E, Klaresgok L, Heldin C-H, Terracio L (1988) Induction of B-type PDGF receptors for platelet-derived growth factor in vascular inflammation: possible implications for development of vascular proliferative lesions. Lancet 1:1353–1356
31. Scher CD, Shepard RC, Antoniades HN, Stiles CD (1979) Platelet-derived growth factor and the regulation of the mammalian fibroblast cell cycle. Biochim Biophys Acta 560:217–241
32. Seifert RA, Schwartz SM, Bowen-Pope DF (1984) Developmentally regulated production of platelet-derived growth factor-like molecules. Nature 311:669–671
33. Sejersen T, Betsholtz C, Sjölund M, Heldin C-H, Westermark B, Thyberg J (1986) Rat skeletal myoblasts and arterial smooth muscle cells express the gene for the A-chain but not for the B-chain of platelet derived growth factor (PDGF) and produce a PDGF-like protein. Proc Natl Acad Sci USA 83:6844–6848
34. Sporn MB, Roberts AB (1988) Peptide growth factors are multifunctional. Nature 332:217–219
35. Vogel A, Raines E, Kariya B, Rivest MJ, Ross R (1978) Coordinate control of 3T3 cell proliferation by platelet-derived growth factor and plasma components. Proc Natl Acad Sci USA 75:2810–2814
36. Wilcox JN, Smith KM, Williams LT, Schwarz SM, Gordon D (1988) Platelet-derived growth factor mRNA detection in human atherosclerotic plaques by in situ hybridization. J Clin Invest 82:1134–1143
37. Wilcox JN, Schwartz S, Gordon D (1989) Local production of platelet-derived growth factors in the human atherosclerotic plaque. In: Habenicht AJR (ed) Growth factors, differentiation factors, and cytokines. Springer, Berlin Heidelberg New York
38. Williams LT (1989) Signal transduction by the platelet-derived growth factor receptor. Science 243:1564–1570

List of Suppliers

Abbot Hospitals Inc.
North Chicago, IL 60064,USA

Accurate Chemical and Scientific Corp.
300 Shames Dr., Westbury, NY 11590, USA

ADAPS Inc.
Dedham, MA, USA

American Type Tissue Culture Collection 12301 Parklawn Drive,
Rockville, MD 20852, USA (or Bethesda?)

A/S Nunc Tissue Culture Supplies
Kamstrup, DK-4000, Roskilde, Denmark

Atlanta Chemie und Handelsgesellschaft mbH, Carl-Benz-Str. 7,
D-6900 Heidelberg, FRG

Bayer AG
Bayerwerk, D-5090 Leverkusen, FRG

Becton Dickinson (Falcon)
2 Bridgewater Lane, Lincoln Park, NJ 07035, USA

Becton Dickinson GmbH
Postfach 10 16 29, D-6900 Heidelberg, FRG

Becton Dickinson Labware
1950 William Drive, Oxnard, CA 93030, USA

Behring Diagnostics
P.O. Box 12087, San Diego, CA 92114-4180, USA

Bellco Biotechnology
340 Erudo Road, Vineland, NJ 08360, USA

Bender und Hobein GmbH
Buchbrunnenweg 26, D-7900 Ulm, FRG

Bethesda Research Labs Inc.
P.O. Box 6009, Gaithersburg, MD 20877, USA

Bio Cell Consulting
Terrassenweg 8, CH-4203 Grellingen, Switzerland

BioGenex Laboratories
6549 Sierra Lane, Dublin, CA, USA

Bio-Rad Laboratories
1414 Harbor Way, Richmond, CA 94804, USA

Biochrom KG (Distributor for Worthington)
Leonorenstr. 2–6,
D-1000 Berlin 46

Biological Industries
Beit Ha'Emek, Israel

Biomed Labordiagnostik
Bruckmannring 28, D-8042 Oberschleißheim, FRG

Biomedical Engineering
University of Iowa Hospital, Iowa City, IA 52242, USA

Biomedical Technologies Inc.
378 Page Street, Stoughton MA 20072, USA

BioPolymers Inc.
309 Farmington Avenue, Farmington, CT 06032, USA

Blood Transfusion Service Central Laboratory, Amsterdam, The Netherlands

Boehringer Mannheim Biochemicals
7941 Castleway Dr., Indianapolis, IN 46250, USA

Boehringer Mannheim GmbH Biochemica
Postfach 31 01 20, D-6800 Mannheim 31, FRG

Brinkman Canada Ltd.
50 Galaxy Blvd, Rexdale, Ontario M9W 4Y5, Canada

C. R. Bard
Calbiochem, Box 12087, San Diego, CA 92112, USA

Camon
Bahnstr. 9a, D-6200 Wiesbaden, FRG

Cell-Tak (see BioPolymers Inc.)

Central Laboratory of the Blood Transfusion Service
Plesmanlaan 125, NL-1066 CX Amsterdam, The Netherlands

Clonetics Corporation
9620 Chesapeake Drive, San Diego, CA 92123, USA

Cole Parmer Chicago
7425 North Oak Park Avenue, Chicago, IL 60648, USA

Collaborative Research Inc.
Two Oak Park, Bedford, MA 01730, USA

Collagen Corporation
2500 Faber Place, Palo Alto, CA 94303, USA

Connaught Medical Research Laboratories
Willowdale, Ontario, Canada

Corning Glass Works
MP-21-5-8, Corning, NY 14831, USA

Costar Corporation
One Alewife Center, Cambridge, MA 02139, USA

Coulter Electronics Inc.
6001 West 20th, Hialeah, FL 33010, USA

Dako GmbH
Postfach 70 04 07, D-2000 Hamburg 70, FRG

Dakopatts a/s
Produktionsvej 42, P.O. Box. 1359, DK-2600 Glostrup, Denmark

Deutsche L'air Liquide
Burgstr. 6, D-6200 Wiesbaden, FRG

Dow-Corning Corporation
Midland, MI 48686-0994, USA

Duke Scientific Corp.
1135 Dora San Antonio Rd., P.O. Box 5005, Palo Alto, CA 94303, USA

ENZO Biochem Inc.
325 Hudson Street, New York, NY 10013, USA

Ethicon GmbH
Robert-Koch-Str. 1, D-2000 Norderstedt, FRG

E.Y. Laboratories Inc.
P.O. Box 1787, San Mateo, CA 94401, USA

Falcon (see Becton Dickinson)

Fisher Scientific Co.
711 Forbes Avenue, Pittsburgh, PA 15219, USA

Flow Laboratories Ltd.
P.O. Box 17, Irvine, Ayrshire KA12 8NB, Scotland, UK

Flow Laboratories Inc.
7655 Old Springhouse Road, McLean, VA 22102, USA

Flow Laboratories GmbH
Postfach 12 49, D-5309 Meckenheim, FRG

Gelman Sciences Ltd.
10 Arrowden Road, Brackmills, Northampton NN4 OEZ, England, UK

Gibco Inc.
3175 Staley Road, Grand Island, NY 14072, USA

Gibco Ltd.
P.O. Box 35, Paisley PA3 4EF, Scotland, UK

Gibco/BRL GmbH
Postfach 12 12, D-7514 Eggenstein, FRG

Graticules Ltd.
Tonbridge, Kent, UK

Greiner und Söhne GmbH
Postfach 13 20, D-7440 Nürtingen, FRG

Heraeus GmbH
Postfach 15 53, D-6450 Hanau, FRG

Hoechst AG
Postfach 80 03 20, D-6230 Frankfurt 80, FRG

Hyclone Laboratories, Inc.
1725 South State Hwy. 89-9, Logan, UT 84321, USA

ICN Immuno Biologicals
P.O. Box 1200, Lisle, IL 60532, USA

ICN Biochemicals Inc.
P.O. Box 28050, Cleveland, OH 44128, USA

Immunotech
Luminy case 915, F-13288 Marseille, Cedex 9, France

Interferon Sciences
New Brunswick, NJ, USA

Jim's Instrument Manufacturing Inc.
Iowa City, IA 52242, USA

KC-Biologicals Inc.
P.O. Box 14848, Lenexa, KS 66215, USA

Kontes
Spruce St., P.O. Box 729, Vineland, NJ 08360, USA

Langenbrinck
Mozartstraße, D-7830 Emmendingen, FRG

Leo Pharmaceuticals
Copenhagen, Denmark

Leo Pharmaceutical Products
P.O. Box 51, NL-1380 AB Weesp, The Netherlands

Lutz KG
Postfach 14 55, D-6890 Wertheim, FRG

Lux (see Nunc)

Lux,
distributed by Miles Scientific, Naperville, Illinois USA

Merck
Postfach 41 19, D-6100 Darmstadt, FRG

Merck-Schuchardt
Eduard-Buchner-Straße 13, D-8011 Hohenbrunn, FRG

Micro-Optik Nosch
Heigelinstr. 5, D-7000 Stuttgart, FRG

Miles Laboratories Inc.
30W 475 North Aurora Road, Naperville, IL 60566, USA

Millipore Corporation
8 Ashby Road, Bedford, MA 01730, USA

Molecular Probes
4849 Pitchford Avenue, Eugene, OR 97402, USA

Nitex (supplied by Tetko)

Nucleopore GmbH
Falkenweg 47, D-7400 Tübingen, FRG

Nucleopore Corporation
7035 Commerce Circle, Pleasanton, CA 94566-3249, USA

Nunc Inc.
2000 Aurora Road, Naperville, IL 60566, USA

Nunc GmbH
Hagenauer Str. 21a, D-6200 Wiesbaden-Biebrich, FRG

Nunc A/S
Postbox 280, Kamstrupvej 90, Kamstrup, DK-4000 Roskilde, Denmark

Oxoid Deutschland
Am Lippelacis 6, D-4230 Wesel, FRG

Paesel GmbH
(Distributor for Biomedical Technologies, Stoughton, MA, USA;
and Ventrex Laboratories Inc, Portland, ME, USA)
Borsigallee 6, D-6000 Frankfurt/Main 63, FRG

PCR Inc.
P.O. Box 1466, Gainesville, FL 32602, USA

Peninsula Laboratories
611 Taylor Way, Belmont, CA 94002, USA

Percell Bioloytica AB
S-22370 Lund, Sweden

Pharmacia LKB Biotechnology Inc.
800 Centennial Ave., Piscataway, NJ 08854, USA

Pharmacia LKB Biotechnology AB
Björkgatan 30, S-75182 Uppsala, Sweden

Pharmacia LKB Biotechnologie
Munzinger Str. 9, D-7800 Freiburg i. Br., FRG

Polysciences Inc.
400 Valley Rd., Warrington, PA 18976, USA

Progen Biotechnik GmbH
Im Neuenheimer Feld 519, D-6900 Heidelberg, FRG

Renner GmbH
Riedstr. 6, D-6701 Dannstadt-Schauernheim 1, FRG

Riedel-de Haen AG
Wunstorfer Str. 40, D-3016 Seelze, FRG

Sanbio b.v.
P.O. Box 540, NL-5400 AM Uden, The Netherlands

Sasco
D-8011 Putzbrunn, FRG

Scherfe System
Arlstr. 63, D-7402 Kirchentellinsfurt, FRG

SCM Corporation
Gainesville, FL, USA

Sebak
Hollerbach 20, D-8359 Aidenbach, FRG

Serva Fine Biochemicals Inc.
200 Shames Drive, Westbury, NY 11590, USA

Serva Feinbiochemica GmbH
Postfach 10 52 62, D-6900 Heidelberg 1, FRG

Sigma GmbH
Grünwalder Weg 30, D-8024 Deisenhofen, FRG

Sigma Israel
Petah Tiqua, Israel

Sigma Chemical Co.
P.O. Box. 14508, St Louis, MO 63178, USA

Spectrum Medical Industries Inc.
60916 Terminal Annex, Los Angeles, CA 90054, USA

Sterilin Ltd.
Sterilin House, Clockhouse Lane, Feltham TW14 8QS, UK

Techne Inc.
3700 Brunswick Pike, Princeton, NJ 08540-6192, USA

Techne (Cambridge) Ltd.
Duxford, Cambridge CB2 4PZ, UK

Technics Instruments
Alexandria, Virginia, USA

Tecnomara Deutschland
Ruhberg 4, D-6301 Fernwald, FRG

Telios Pharmaceuticals
2090 Science Park Road, San Diego, CA 92121, USA

Tetko
420 Sawmill River Road, Elmsford, NY 10523, USA

Thomae
Postfach 17 55, D-7950 Biberach/Riss, FRG

Vangard International Inc.
111-a Green Grove Road, Neptune, NJ 07754, USA

Ventrex Laboratories Inc.
217 Read Street, Portland, ME 04103, USA

Viobin Corporation
Montecello, IL, USA

Wako Pure Chemical Industries
10 Doshomachi 3-Chome, Higashi-Ku, Osaka 541, Japan

Whatman Biosystems
Hillsboro, Oregon, USA

Wheaton Scientific
1000 North 10th Street, Milville, NJ 08332, USA

Worthington Biochemical Corporation
Halls Mill Road, Freehold, NJ 07728, USA

This is a list of suppliers mentioned in this book. It is not intended to be a
complete list of suppliers for cell culture techniques.

Subject Index

If you have any concerns about our products,
you can contact us on
ProductSafety@springernature.com

In case Publisher is established outside the EU,
the EU authorized representative is:
**Springer Nature Customer Service Center GmbH
Europaplatz 3, 69115 Heidelberg, Germany**

Printed by Libri Plureos GmbH
in Hamburg, Germany